Power Systems

More information about this series at http://www.springer.com/series/4622

Kyo-Beum Lee · June-Seok Lee

Reliability Improvement Technology for Power Converters

Springer

Kyo-Beum Lee
Department of Electrical and Computer Engineering
Ajou University
Suwon
Korea (Republic of)

June-Seok Lee
Korea Railroad Research Institute (KRRI)
Uiwang-si
Korea (Republic of)

ISSN 1612-1287 ISSN 1860-4676 (electronic)
Power Systems
ISBN 978-981-10-4991-0 ISBN 978-981-10-4992-7 (eBook)
DOI 10.1007/978-981-10-4992-7

Library of Congress Control Number: 2017943172

Printed on acid-free paper

This Springer imprint is published by Springer Nature
The registered company is Springer Nature Singapore Pte Ltd.
The registered company address is: 152 Beach Road, #21-01/04 Gateway East, Singapore 189721, Singapore

Preface

Power electronics technology, which was used only for motor drives in the past, is currently being used in numerous industries including renewable energy in household appliances, automobiles, and energy storage devices. This change is due in part to the increased interest in improving the efficiency of electronic and electric devices. The main reason is that research on power electronics has been carried out in many universities and research institutes, and researchers equipped with skills in the field of power electronics have been exported across the industry. Based on this, it can be said that the power electronics technology related to low- and medium-capacity systems has matured considerably. The existing power electronics technology has evolved into a highly reliable technology based on the development experiences and know-how of researchers. However, due to recent increases in energy demand within the industry, high-pressure/high-capacity systems are required. As a result, new power electronics topologies are being developed and studied. There are two highly representative research areas in the field of multilevel converters. Multilevel converters, mainly in Europe and China, have been applied to high-voltage/high-capacity wind power generation systems, and related research is being actively conducted. In addition, the development of HVDC (high-voltage DC) transmission systems is also a representative field in terms of the application of multilevel converters. This book includes the basic introduction of the most widely studied and used multilevel converter topologies, and it introduces the most reliable three-level converter technology, which is the most similar to a two-level converter. In addition, this book introduces reliable technology for cascaded H-bridge inverters in the last chapter. Three-level converters and cascaded H-bridge inverters have been applied to high-voltage systems mainly due to the advantages of multilevel converters. The cascaded H-bridge converter is expanding its application range to high-power and high-voltage systems. However, three-level converters have recently been used in low voltage/medium capacity systems and have been expanding their applications as well. Two-level converters, which have been studied in the field of motor drives in the past, are typical topologies with high reliability based on accumulated technology. On the other hand, although three-level converters and cascaded H-bridge inverter have been applied in the past.

However, since the field is narrow and there are not many cases, their reliability-related technology is lacking. This book introduces the concept of multilevel converters and basic knowledge of three-level converters and cascaded H-bridge converter to students that want to study multilevel converters as well as power electronics engineers who want to apply three-level converters and cascaded H-bridge inverter to their field. They will also be introduced to techniques for dealing with a variety of reliability-related issues.

Suwon, Korea (Republic of) Kyo-Beum Lee
Uiwang-si, Korea (Republic of) June-Seok Lee

Contents

Chapter 1
Three-Level Converter Overview

1.1 Three-Level Converters

The development of power electronics coincides with the development of power semiconductor devices. Since the use of AC motors has increased since the 1980s, power conversion devices composed of power semiconductor devices have been steadily developed. In the early 2000s, there was a large increase in renewable energy industries such as wind power generation and photovoltaic power generation, which require connections with AC systems. This has widened the application fields of the power conversion apparatuses used only in existing AC motors.

In accordance with the demands of industry, power semiconductor devices have been continuously increased in terms of pressure and capacity. As a result, various power conversion topologies have been proposed. The most basic method among the proposed power conversion topologies is to connect a switching device in series and operate the converter at a high voltage as shown in Fig. 1.1a. Another method is to implement a new topology of the multi-level converter type such as Fig. 1.1b. When a switching device is connected in series to increase the voltage and capacity of a converter, the operation of the converter is the same as that of a two-level converter, and synchronization of the on-off operation of the switching elements connected in series is required. The multi-level converters produce output voltages with multiple levels by properly operating each of the switching elements. The output voltage with multiple levels has a smaller total harmonic distortion (THD) than a converter implemented by series-connected switching devices. This is advantageous in reducing filter size in grid-connected voltage converters and in reducing noise in converters driving an AC motor.

Multilevel converters are divided into Cascaded H-bridge converters, Flying Capacitor converters, and Neutral-Point-Clamped (NPC) converters as shown in Figs. 1.1 and 1.2 [1–3]. Of these, the NPC converters were first developed by Nabae in 1981. The NPC three-level converter is one of the most actively studied type of

K.-B. Lee and J.-S. Lee, *Reliability Improvement Technology for Power Converters*, Power Systems, DOI 10.1007/978-981-10-4992-7_1

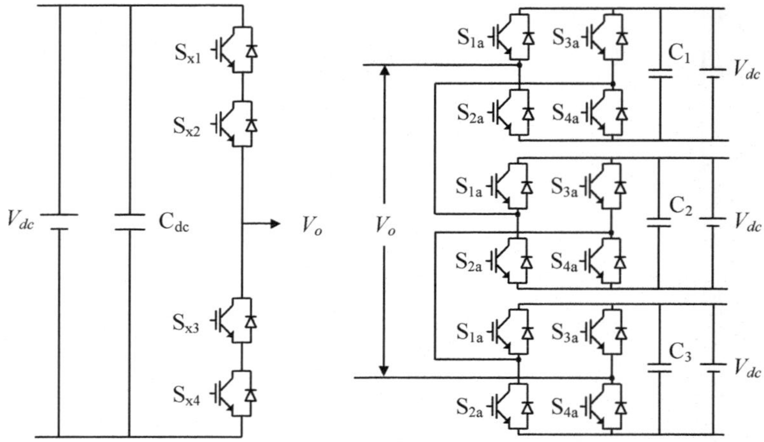

Fig. 1.1 High voltage and high capacity system using, a series connection method of switches (L), cascaded H-bridge topology (R)

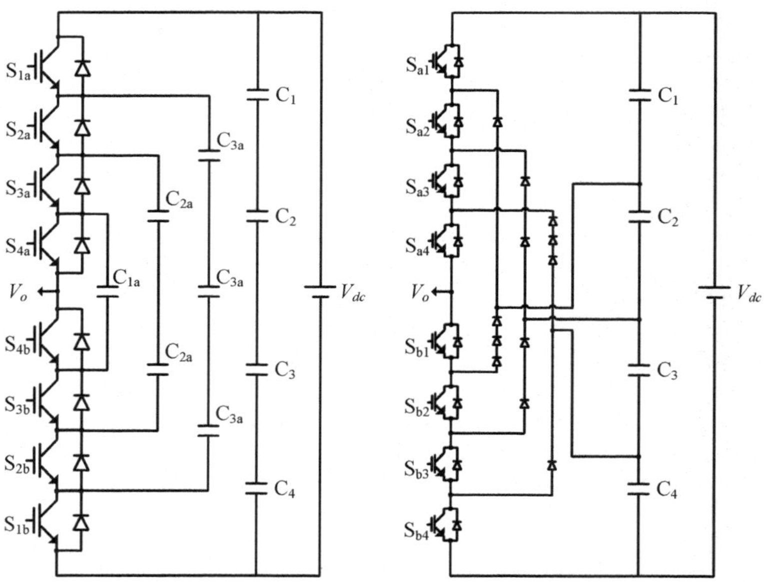

Fig. 1.2 Flying capacitor converter (L), NPC converter (R)

multi-level converter, and NPC three-level converter-based power converters have been applied to many products. Meanwhile, NPC four-level or higher level converters are not practical.

Fig. 1.3 ABB 4 kV/2 MW NPC Three-Level Converter (ACS 1000)

ABB's ACS1000 which is shown in Fig. 1.3 is a NPC three-level converter that can drive up to 4 kV/2 MW and uses an Integrated Gate Commutated Thyristor (IGCT) switching device. In 2012, Samsung Heavy Industries (Korea) developed a power conversion system based on a NPC three-level converter for 3.3 kV/7 MW offshore wind power generation as a national support project. As can be seen from the capacity of these commercial products, multi-level converters have been suitable for high-voltage and large-capacity power converters. However, recently they are also applied to small-capacity products with high performance requirements such as low total voltage harmonic distortion (THD) and high efficiency [4–7]. A typical product is the SUNNY TRIPOWER 12000TL 12 kW power converter shown in Fig. 1.4 for solar power generation system which was launched by SMA.

Fig. 1.4 Power converter for photovoltaic power generation system using a SMA's NPC three-level converter (SUNNY TRIPOWER 12000TL)

Since the application of three-level converters has expanded, Infineon, Semikron, and Vincotech, which are known worldwide as switching device and module manufacturers, have released and sold specific modules for three-level converter as shown in Fig. 1.5. The recently introduced modules for three-level converter are mainly composed of switching devices with rated voltage of 600 V, and the rated capacity is steadily decreasing. In recent years, three-level converters have been adopted based on the advantages mentioned above in various industrial fields where two-level converters were mainly used. As a result, many control technologies related to three-level converters have been developed, which are generalized for immediate application in the industry.

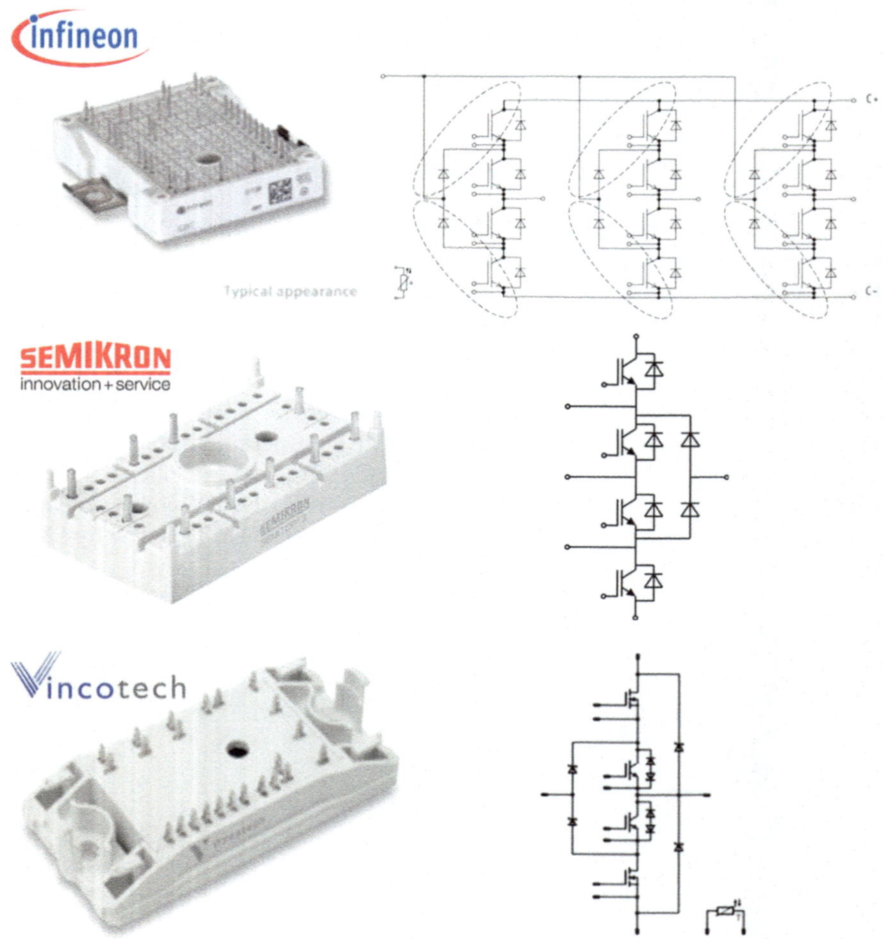

Fig. 1.5 Productized three-level converter module (as of 2015): Infineon Inc. FS3L50R07W2H3F (650V/50A), Semikron Inc. SK20MLI066 (600V/50A), Vincotech Inc. 10-FZ06NRA099FS (600V/18A)

It is clear that three-level converters are complicated in terms of structure and control compared to a two-level converter because the three-level converters consist of switches more than the two-level converter. Therefore, it is necessary to take measures to improve the reliability of the power conversion systems, which are increasingly complicated.

1.2 Three-Level Converter Types

Three-level converters include the NPC three-level converter and the T-type three-level converter mentioned above. Both of these topologies differ in terms of their structure, several performances, and design procedure. The following comparison between these topologies is based on a three-phase system.

1.2.1 NPC (Neutral-Point Clamped) Converter

The NPC three-level converter is structurally composed of 12 switches and 6 clamping diodes as shown in Fig. 1.6. The dc capacitor is divided into two parts, and the midpoint between the upper and lower parts is called the neutral-point. The six diodes are clamping diodes that connect the output of each leg of the converter to the neutral-point.

The NPC three-level converter is based on a series connection of four switches in each leg. Compared with a two-level converter, the rated-voltage of the used switch can be reduced by half, making it suitable for a high voltage and large capacity system.

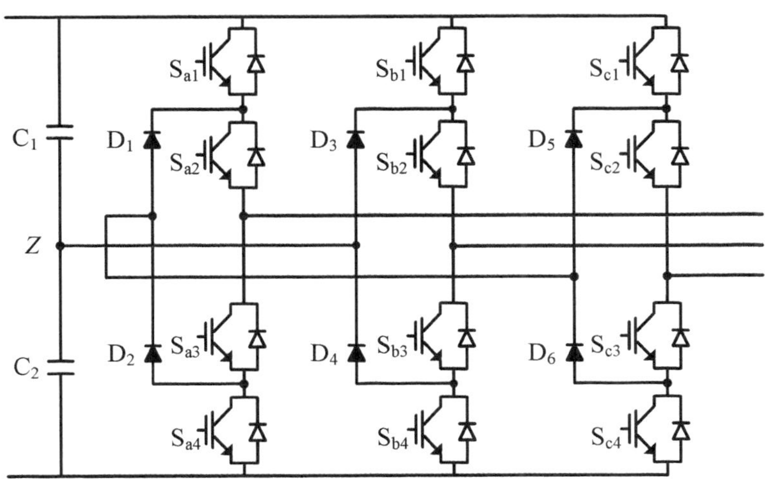

Fig. 1.6 NPC three-level converter

1.2.2 T-Type Three-Level Converter

T-type three-level converter is also called ANPC (Advanced NPC) three-level converter. Like the NPC three-level converter, the T-type three-level converter has two capacitors divided into upper and lower parts, and its consists of a total of 12 switches as shown in Fig. 1.7. In a T-type three-level converter, a bidirectional switch connected to the neutral-point or two switches connected in series is used to connect the neutral-point of the converter to the output of each leg.

T-type three-level converter is structurally similar to the conventional two-level converter. The rated-voltage of the switches S_{x1} and S_{x4} is designed to be equal to the rate-voltage of the switches used in two-level converter. On the other hand, the switches S_{x2} and S_{x3} which are connected to neutral-point have the rated-voltage which is half of the switches S_{x1} and S_{x4}. The T-type three-level converter has a disadvantage that it is difficult to apply T-type three-level converter to high-voltage, high-capacity power conversion sytem since the rated-voltage of the external switches S_{x1} and S_{x4} is not low and is the same as the two-level converter, unlike that of the NPC three-level converter. However, in a field that is relatively sensitive to the addition of switching elements, T-type three-level converter that does not use a clamping diode is advantageous.

In addition, the T-type three-level converter achieves a higher power conversion efficiency than the NPC three-level converter by using SiC (Silicon Carbide) switches and diodes as shown in Fig. 1.8. The efficiency is improved because the switching loss of a SiC switch is smaller than that of a Si (silicon) switch, and the on-voltage of a SiC diode is smaller than that of an IGBT (Insulated Gate Bipolar Transistor) as shown in Fig. 1.9 [8].

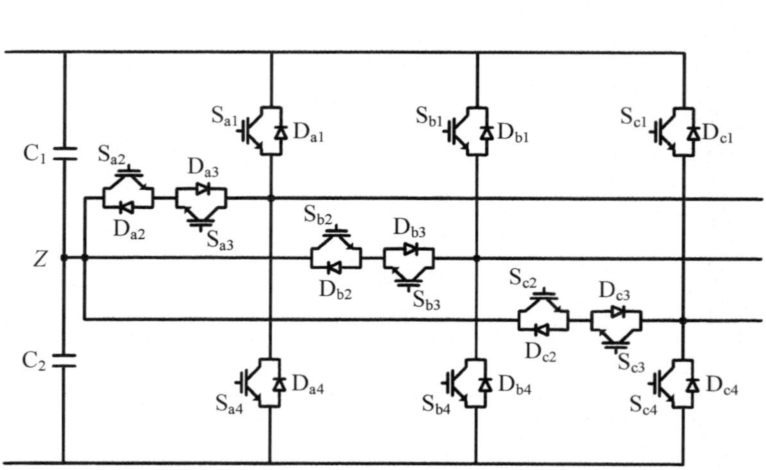

Fig. 1.7 T-type three-level converter

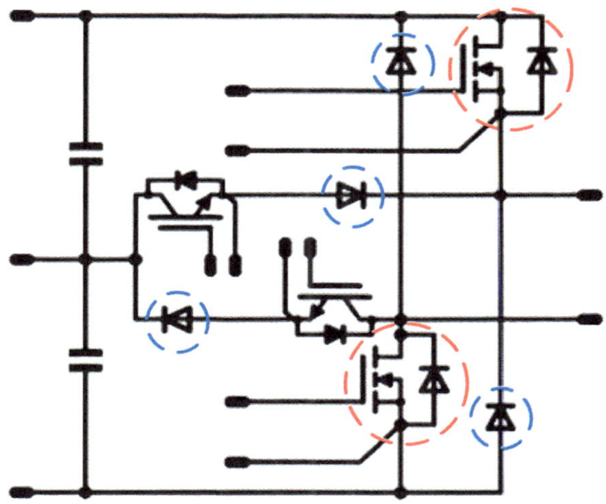

Fig. 1.8 SiC T-type three-level converter module with a diodes to switch

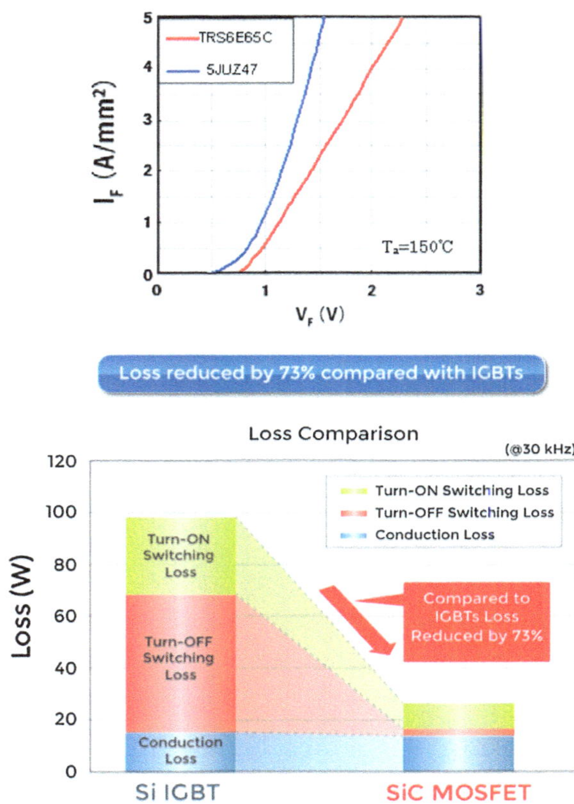

Fig. 1.9 SiC device characteristics

1.2.3 *Comparison Between Two-Level Converter and NPC/T-Type Three-Level Converters*

In order to analyze the efficiency of two-level converter and NPC/T-type three-level converters, Infineon switch characteristics are used.

- Two-level converter: Infineon IKW25T120 (1200 V)
- NPC three-level converter: Infineon IKW30N60T (600 V)
- T-type three-level converter: External switch Infineon IKW25T120 (1200 V).

Internal switch Infineon Corporation IKW30N60T (600 V). Converter operating conditions: DC stage (700 V), three-phase output voltage (325 V), and output current (20.5 A).

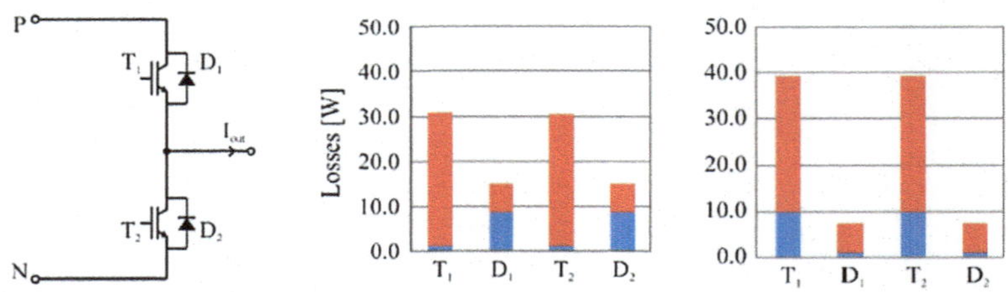

(a) Switching frequency 32 kHz: inverter mode loss of a two-level converter (L), rectifier mode loss (R)

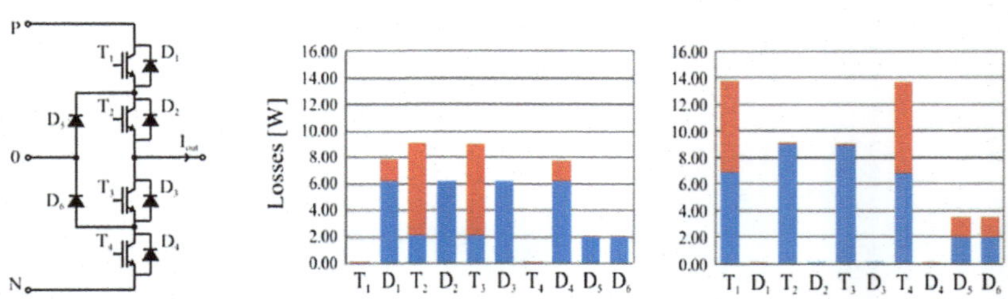

(b) Switching frequency 48 kHz: inverter mode loss of a NPC three-level converter (L), rectifier mode loss (R)

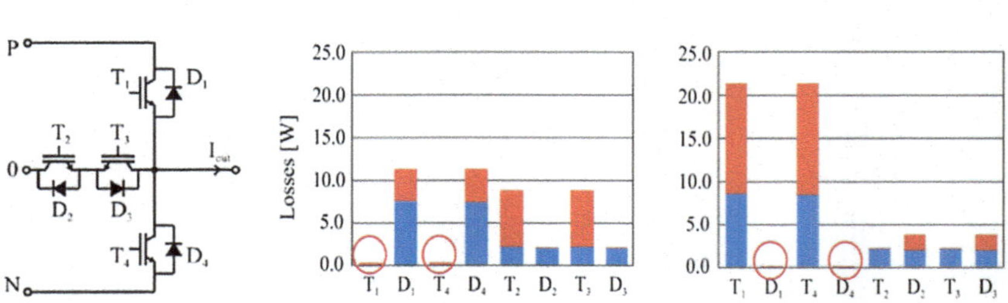

(c) Switching frequency 32 kHz: inverter mode loss of a T-type three-level converter (L), rectifier mode loss (R)

Fig. 1.10 Loss comparison between two-level and three-level converters (reprinted from [9], Fig. 5)

In order to compare the efficiency, the switching loss and conduction loss for each of the switches are considered and a loss analysis is performed in the inverter mode and rectifier mode according to the operation of the converter and the analysis results are shown in Fig. 1.10. When operating in the inverter mode in a two-level converter, the loss generated by the switch is larger than that of the diode; and when operating in the rectifier mode, the loss occurring in the diode is larger than that of the switch. The losses in the inverter mode in the NPC three-level converter and T-type three-level converter are also greater than those generated by the diode in the switch. In addition, the reverse loss in the rectifier mode is large. T-type three-level converter has no associated losses due to the absence of a clamping diode. However, since the external switches T_1 and T_4 of the T-type three-level converter have twice the rated voltage compared with the rated voltages of the external switches T_1 and T_4 of the NPC three-level converter, the efficiency improvement caused by absence of clamping diodes is mitigated. The efficiency values of the NPC three-level converter and the T-type three-level converter are different depending on the switching frequency. The efficiency of the T-type three-level converter is better when the switching frequency is operated at 20 kHz or more.

1.3 Three-Level Converter Operation Principle

The operation principles of the NPC three-level converter and the T-type three-level converter are the same. When the gate signal input to the switch of the NPC three-level converter can be used as the gate signal of the switch of the T-type three-level converter. In this chapter, the operation principle of three-level converters is explained based on the NPC three-level converter.

1.3.1 I/O Characteristics

The input-output characteristics of a three-level converter can be found by comparing the input-output characteristics of a two-level converter. In Fig. 1.11, the input voltage of the two-level converter refers to the DC voltage (V_{dc}), and the output voltage (V_{xz}) has a value of $V_{dc}/2$ or $-V_{dc}/2$ depending on the switching state as shown in Fig. 1.11a. The input voltage of the three-level converter refers to the same DC voltage (V_{dc}) as the two-level converter. Since the three-level converter has a switching state connected to the neutral point, an output voltage of $V_{dc}/2$, 0 or $-V_{dc}/2$ can be generated as shown in Fig. 1.11b. Two-level converter and three-level converter depend on the number of output voltages. In the NPC three-level converter, the output voltage of the three levels is determined by the on-off operation of each switch. When the first switch S_{x1} and the second switch S_{x2} are turned on and the third switch S_{x3} and the fourth switch S_{x4} are turned off, the output is connected to the top of the DC stage, and the output voltage is $V_{dc}/2$.

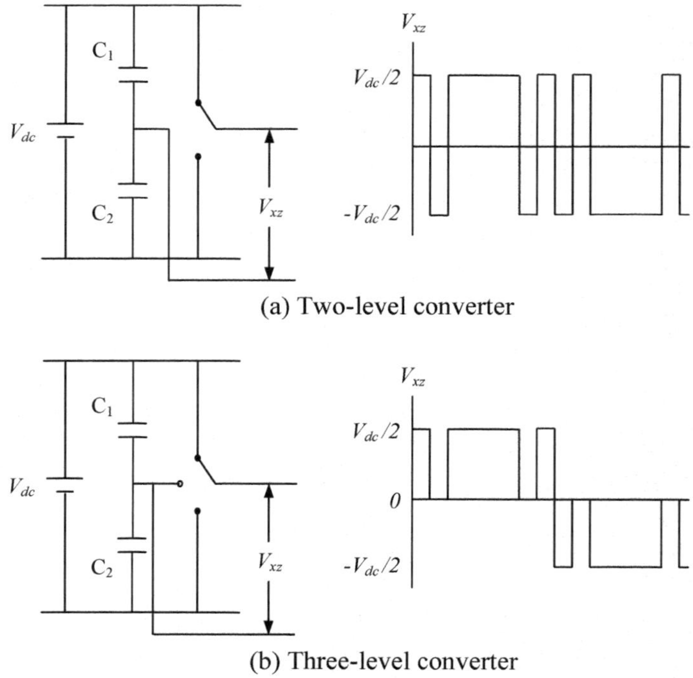

Fig. 1.11 Comparison of two-level and three-level converters

Conversely, when the first switch S_{x1} and the second switch S_{x2} are turned off and the third switch S_{x3} and the fourth switch S_{x4} are turned on, the output voltage has a value of $-V_{dc}/2$. Finally, when the first switch S_{x1} and the fourth switch S_{x4} are turned off and the second switch S_{x2} and the third switch S_{x3} are turned on, the output voltage has a value of zero. It is summarized in Fig. 1.12 and the output voltage of the third level is defined as P, N or O depending on the switch operation state, and can be expressed as shown in Table 1.1.

Table 1.1 Operation characteristics in accordance to the ON/OFF state of each switch

Switching state	State of switching device				Output voltage V_{xz} (x = a, b, c)
	S_{x1}	S_{x2}	S_{x3}	S_{x4}	
P	ON	ON	OFF	OFF	$+V_{dc}/2$
O	OFF	ON	ON	OFF	0
N	OFF	OFF	ON	ON	$-V_{dc}/2$

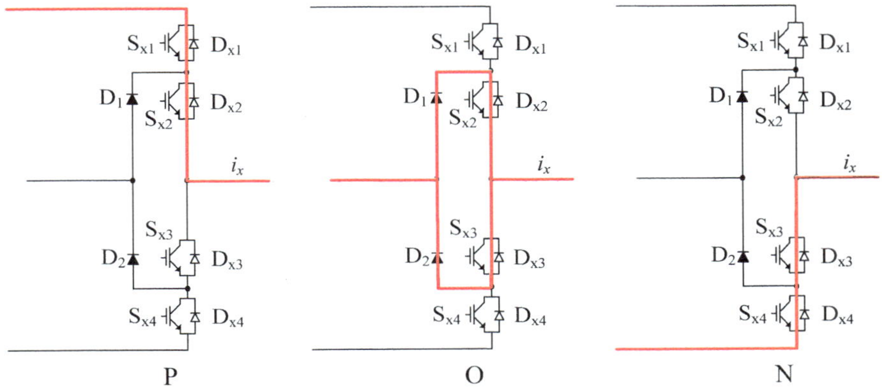

Fig. 1.12 Circuit connection diagram in accordance with the ON/OFF state of a switch

1.3.2 Output Voltage of Three-Leg Three-Level Inverters

In a three-leg NPC three-level converter, the switching state of each leg can be represented as P, N or O. Therefore, it is possible to represent the three-leg output voltage using the switching states of the three legs. There are 27 switching states that can be represented by the three-leg output voltage, and these 27 switching states can be expressed as vectors as shown in Fig. 1.13 and Table 1.2.

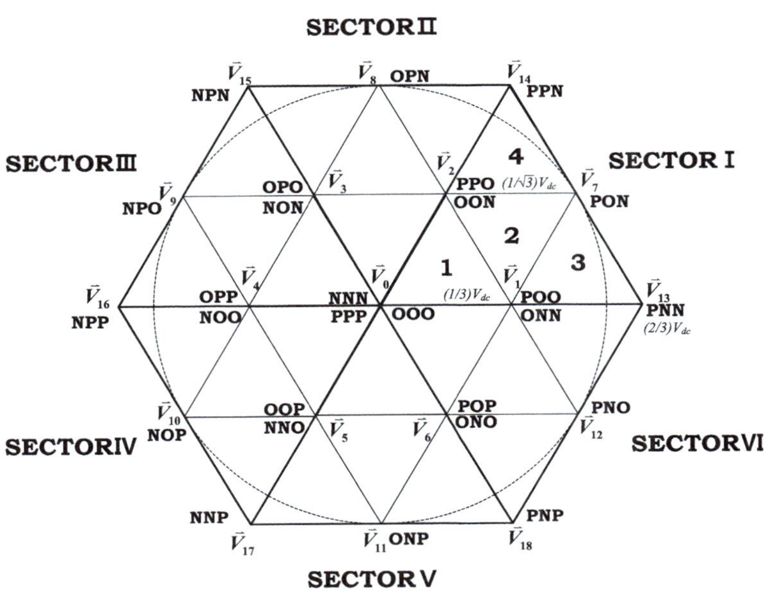

Fig. 1.13 Voltage vectors of a three-level converter (reprinted from [10])

Table 1.2 Switching states and voltage vectors

Space vector		Switching state		Types of vector	Magnitude
V_0		[POO][OOO][NNN]		Zero vector	0
		P-type	N-type	Small vector	$\frac{1}{3}V_{dc}$
V_1	V_{1P}	[POO]	–		
	V_{1N}	–	[ONN]		
V_2	V_{2P}	[PPO]	–		
	V_{2N}	–	[OON]		
V_3	V_{3P}	[OPO]	–		
	V_{3N}	–	[NON]		
V_4	V_{4P}	[OPP]	–		
	V_{4N}	–	[NOO]		
V_5	V_{5P}	[OOP]	–		
	V_{5N}	–	[NNO]		
V_6	V_{6P}	[POP]	–		
	V_{6N}	–	[ONO]		
V_7		[PON]		Medium vector	$\frac{\sqrt{3}}{3}V_{dc}$
V_8		[OPN]			
V_9		[NPO]			
V_{10}		[NOP]			
V_{11}		[ONP]			
V_{12}		[PNO]			
V_{13}		[PNN]		Large vector	$\frac{2}{3}V_{dc}$
V_{14}		[PPN]			
V_{15}		[NPN]			
V_{16}		[NPP]			
V_{17}		[NNP]			
V_{18}		[PNP]			

A vector diagram of the three-level converter is composed of a large vector, a medium vector, a small vector, and a zero vector. Small vectors have two types, P-type and N-type.

1.3.3 Space Vector PWM (SVPWM)

The voltage vector diagram of a three-level converter is divided into six sectors. The six sectors are expressed as Sectors I–VI. Each of these sectors can be divided into four parts, which are expressed as Regions 1–4. In the space vector PWM method, the command voltage vector is determined by the output voltage of the three-level converter by the following four steps.

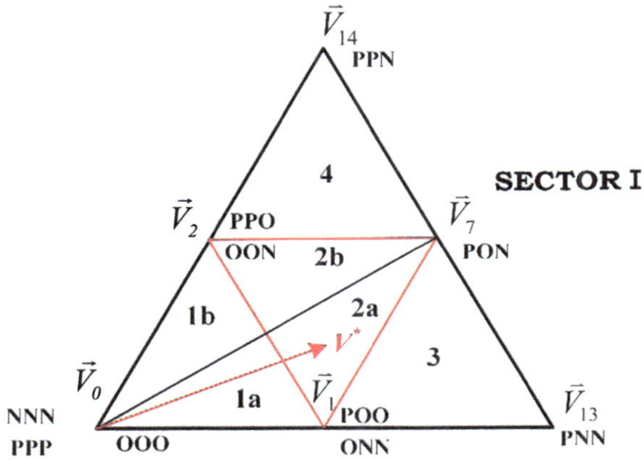

Fig. 1.14 Command voltage vector in Sector I

(a) Determine the Sector and Region of the command voltage vector.

In Fig. 1.14, the Sector and Region to which the command voltage vector (V^*) belongs are determined to be Sector I and Region 2.

(b) Determine the effective vector for generating the command voltage vector.

If Sector I-Region 2 is determined in accordance with the command voltage vector (V^*) is determined to be Sector I-Region 2, the voltage vectors V_1, V_2, and V_7 that are closest to the command voltage vector (V^*) are selected as valid vectors that are the output voltage vectors. The command voltage vector (V^*) can be selected with different criteria for various purposes. The method of selecting adjacent vectors shown here exhibits excellent performance in terms of output voltage ripple.

(c) Calculate the effective time of the selected effective vectors

Prior to the calculation of the effective time of the selected effective vectors, the command voltage vector can be expressed using the magnitude (V_{ref}) and phase (θ) as shown in the following equation.

$$\vec{V}_{ref} = V_{ref} e^{j\theta} \tag{1.1}$$

The following equation is arrived at following equation when the magnitude and phase of the selected voltage vectors V_1, V_2, and V_7 are used.

$$\vec{V}_1 = \frac{1}{3} V_{dc}, \quad \vec{V}_2 = \frac{1}{3} V_{dc} e^{j\pi/3}, \quad \vec{V}_7 = \frac{\sqrt{3}}{3} V_{dc} e^{j\pi/6} \tag{1.2}$$

The effective time of the selected valid vectors is calculated so that the average output voltage during a switching period (T_s) becomes equal to the command voltage vector (V^*). Therefore, the following equations must be satisfied.

$$\vec{V}_{ref}T_s = \vec{V}_1T_1 + \vec{V}_2T_2 + \vec{V}_7T_7 \tag{1.3}$$

$$T_s = T_1 + T_2 + T_7 \tag{1.4}$$

where T_1, T_2, and T_7 represent the effective switching time of the selected voltage vector V_1, V_2, and V_7.

The following equation can be obtained by substituting Eqs. (1.1) and (1.2) into Eq. (1.3).

$$V_{ref}e^{j\theta}T_s = \frac{1}{3}V_{dc}T_1 + \frac{\sqrt{3}}{3}V_{dc}e^{j\pi/6}T_2 + \frac{1}{3}V_{dc}e^{j\pi/3}T_7 \tag{1.5}$$

Equation (1.5) is divided into the real part (Re) and the imaginary part (Im) as shown below.

$$\begin{aligned} &V_{ref}(\cos\theta + j\sin\theta)T_s \\ &= \frac{1}{3}V_{dc}T_1 + \frac{\sqrt{3}}{3}V_{dc}(\cos\frac{\pi}{6} + j\sin\frac{\pi}{6})T_2 + \frac{1}{3}V_{dc}(\cos\frac{\pi}{3} + j\sin\frac{\pi}{3})T_7 \end{aligned} \tag{1.6}$$

$$\begin{aligned} \text{Re} &: T_1 + \frac{3}{2}T_2 + \frac{1}{2}T_7 = 3\frac{V_{ref}}{V_{dc}}(\cos\theta)T_s \\ \text{Im} &: \frac{3}{2}T_2 + \frac{\sqrt{3}}{2}T_7 = 3\frac{V_{ref}}{V_{dc}}(\sin\theta)T_s \end{aligned} \tag{1.7}$$

When Eq. (1.7) is substituted into Eq. (1.4), the effective switching time of the V_1, V_2, and V_7 voltage vectors in Sector I-Region 2 can be calculated as follows.

$$\begin{aligned} T_1 &= T_s[1 - \frac{2V_{ref}}{V_{dc}}\sin\theta] \\ T_2 &= T_s[\frac{2V_{ref}}{V_{dc}}\sin(\frac{\pi}{3} + \theta) - 1] \\ T_7 &= T_s[1 - \frac{2V_{ref}}{V_{dc}}\sin(\frac{\pi}{3} - \theta)] \end{aligned} \tag{1.8}$$

The selected valid vectors selected in Sector I-Region 2 are generated for the calculated effective switching time. In Sector I, the effective vectors and effective switching time depending on the Region are calculated as shown in Table 1.3.

Table 1.3 The selected valid vector and effective switching time in Sector I

Region	T_x		T_y		T_z	
1	V_1	$T_s[\frac{2V_{ref}}{V_{dc}}\sin(\frac{\pi}{3}-\theta)]$	V_0	$T_s[1-\frac{2V_{ref}}{V_{dc}}\sin(\frac{\pi}{3}+\theta)]$	V_2	$T_s[\frac{2V_{ref}}{V_{dc}}\sin\theta]$
2	V_1	$T_s[1-\frac{2V_{ref}}{V_{dc}}\sin\theta]$	V_7	$T_s[\frac{2V_{ref}}{V_{dc}}\sin(\frac{\pi}{3}+\theta)-1]$	V_2	$T_s[1-\frac{2V_{ref}}{V_{dc}}\sin(\frac{\pi}{3}-\theta)]$
3	V_1	$T_s[2-\frac{2V_{ref}}{V_{dc}}\sin(\frac{\pi}{3}+\theta)]$	V_7	$T_s[\frac{2V_{ref}}{V_{dc}}\sin\theta]$	V_{13}	$T_s[\frac{2V_{ref}}{V_{dc}}\sin(\frac{\pi}{3}-\theta)-1]$
4	V_{14}	$T_s[\frac{2V_{ref}}{V_{dc}}\sin\theta-1]$	V_7	$T_s[\frac{2V_{ref}}{V_{dc}}\sin(\frac{\pi}{3}-\theta)]$	V_2	$T_s[2-\frac{2V_{ref}}{V_{dc}}\sin(\frac{\pi}{3}+\theta)]$

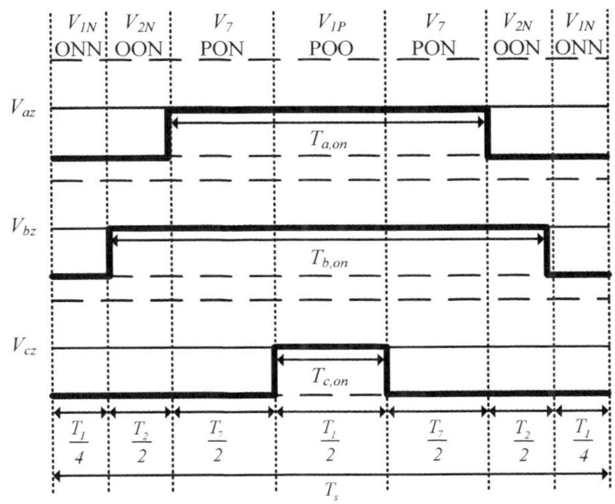

Fig. 1.15 Switching sequence in sector I-region 2a

(d) Determine the output order of the selected valid vectors

There are two considerations in the output order of the selected valid vectors. The first is that one switching should be done within a switching period T_s to have a constant switching frequency. The second consideration is that since N-type and P-type can be selected as small vectors, the N-ype and P-type small vectors should be arranged relatively evenly. Sector I-Region 2 outputs valid vectors according to Regions 2a and 2b in different orders. Figure 1.15 shows the valid vector and its effective switching time in Sector I-Region 2a. The output order of the valid vectors is V_{1N}, V_{2N}, V_7, V_{1P}, V_7, V_{2N} and V_{1N}, and the small vector is divided into P-type and N-type. The output state of each leg changes with the output order of the effective vectors according to the time, and each of the four phases switches on and off according to the output state. Since the effective switching time is calculated so that the average output voltage during a switching period (T_s) is equal to the command voltage vector (V^*), the output order of the valid vectors does not need to be considered.

However, having a constant switching frequency results in ease of implementation and clarifies the frequency selection of the attenuation target in the filter design. In addition, placing the P-type and N-type small vectors evenly helps maintain the neutral-point voltage at equilibrium.

1.3.4 Carrier-Based PWM

The space vector PWM method can achieve various control purposes by changing the voltage vector selection method. DTC (Direct Torque Control) [11], which is used in motor control and outputs one switching state during a switching cycle, can also be said to be a kind of space voltage PWM method. However, the space vector PWM method requires complicated equations and processes before the output voltage is generated. One possible solution to this drawback is to use the carrier-based PWM method. The switching state of each switch is determined simply by comparing the triangular carrier and the command voltage as shown in Fig. 1.16, and various control purposes can be achieved depending on how the command voltage is changed.

The command voltage for the carrier-based PWM method in the three-level converter can be generated by applying Sinusoidal PWM (SPWM), PWM methods using the offset voltage, and Third Harmonic Injection PWM (TH-PWM).

In this chapter, the PWM method using the offset voltage shown in Fig. 1.17 is introduced.

The three-leg command voltages ($V_{a,ref}$, $V_{b,ref}$, and $V_{c,ref}$) are represented as follows.

$$\begin{aligned} V_{a,ref} &= V_{mag}\cos(2\pi f_f t) \\ V_{b,ref} &= V_{mag}\cos(2\pi f_f t - 2\pi/3) \\ V_{c,ref} &= V_{mag}\cos(2\pi f_f t + 2\pi/3) \end{aligned} \quad (1.9)$$

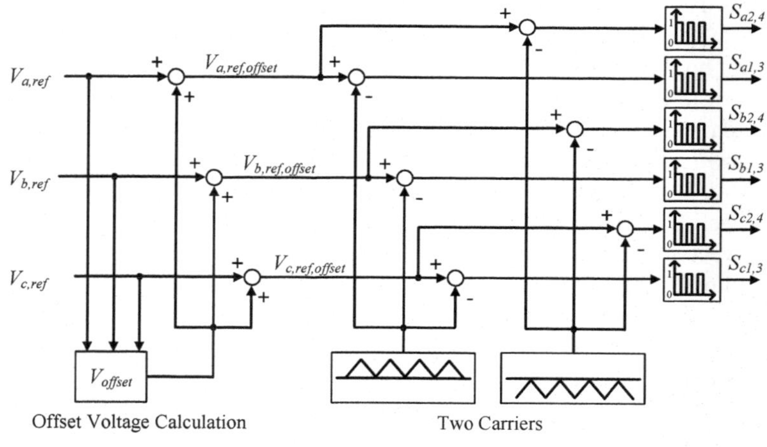

Fig. 1.16 Block diagram of the carrier-based PWM method using the offset voltage

Here, V_{mag} represents the magnitude of the command voltage, and f_f represents the fundamental wave frequency. The offset voltage (V_{offset}) is used to increase the voltage modulation range and can be expressed as

$$V_{offset} = -\frac{V_{ref,max} + V_{ref,min}}{2} \tag{1.10}$$

Here, $V_{ref,max}$ represents the largest value among the three-leg command voltages, and $V_{ref,min}$ represent the smallest voltage among the three-leg command voltages. The offset voltage (V_{offset}) calculated in Eq. (1.10) is added to the three-leg command voltage in Eq. (1.9).

$$\begin{aligned} V_{a,ref,offset} &= V_{a,ref} + V_{offset} \\ V_{b,ref,offset} &= V_{b,ref} + V_{offset} \\ V_{c,ref,offset} &= V_{c,ref} + V_{offset} \end{aligned} \tag{1.11}$$

The method of generating the command voltage in the PWM method using the offset voltage is the same as in the two-level converter. However, unlike the two-level converter in which the command voltage is compared to one carrier, in a three-level converter, two carriers are compared with the command voltage to produce an output voltage.

The on-off operation of the four switches in one leg is determined by comparing the command voltage with two carriers placed in parallel, as shown in Fig. 1.18a and is defined as follows.

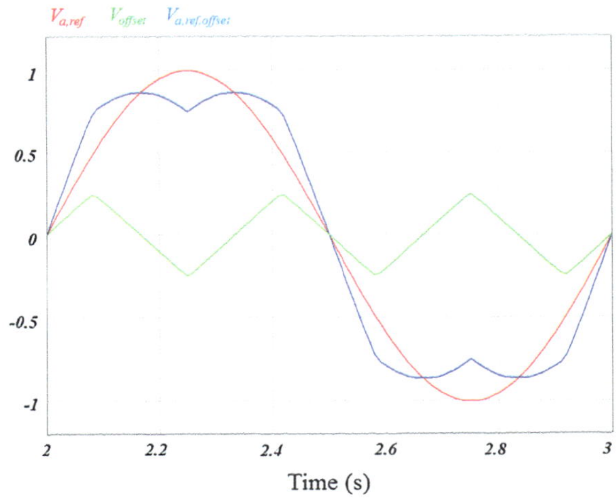

Fig. 1.17 Comparison of the command voltages for SPWM and PWM method using offset voltage

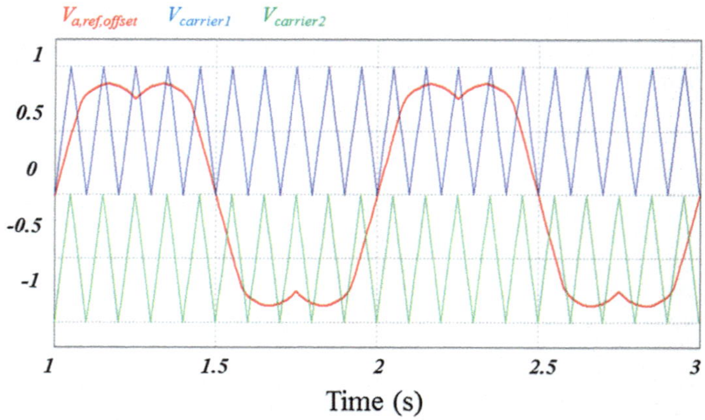

(a) Two carriers and one command voltage

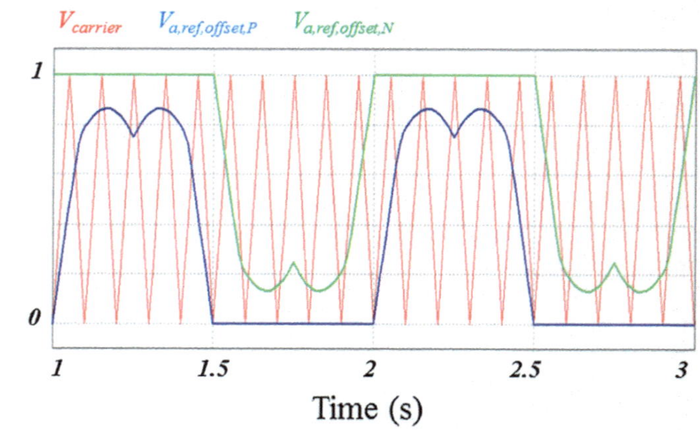

(b) One carrier and two command voltages

Fig. 1.18 PWM method using carrier waves in a three-level converter

(a) The above carrier waves determine the complementary switch states of S_{x1} and S_{x3}

- Carrier < command voltage: S_{x1} (ON), S_{x3} (OFF)
- Carrier > command voltage: S_{x1} (OFF), S_{x3} (ON)

(b) The following carrier waves determine the complementary switch states of S_{x2} and S_{x4}

- Carrier < command voltage: S_{x2} (ON), S_{x4} (OFF)
- Carrier > command voltage: S_{x2} (OFF), S_{x4} (ON)

When realizing an offset voltage PWM method in a real system, a carrier function provided by an MCU (Micro Controller Unit) can be used. However, this carrier function cannot be implemented in parallel with a single triangular function. To solve this problem, two command voltages are generated, as shown in Fig. 1.18b, so that only one carrier can be used. One of the two command voltages is compared to the carrier for the complementary on-off operations of S_{x1} and S_{x3} and the other command voltage is compared with the carrier for the complementary on-off operations of S_{x2} and S_{x4}.

1.3.5 Neutral-Point Voltage Fluctuation

The three-level converter has two DC capacitors. The input and output characteristics discussed above are based on the assumption that half of the total DC capacitor voltage is applied to each of the DC capacitors (neutral-point voltage equilibrium state).

Neutral-point voltage refers to the difference between two DC capacitor voltages. The PWM methods of a three-level converter need to consider the change in the neutral-point voltage, and each switching state causes the change in the neutral-point voltage to be different as shown in Fig. 1.19. The voltage vector directly affecting the neutral-point voltage is a small vector. When the three-level converter operates as an inverter, the P-type small vector reduces the upper capacitor voltage and decreases the lower capacitor voltage. On the other hand, the N-type small vector increases both the upper capacitor voltage and the lower capacitor voltage. The medium vector is not directly related to the change in the neutral-point voltage. However it indirectly affects the change in the neutral-point voltage depending on the state of the output current. In addition, the large vector does not affect the change of the neutral-point voltage by generating the same voltage change in the two DC capacitor voltages. Furthermore, the zero vector does not change the neutral-point voltage. In a three-level rectifier, the change in the two DC capacitor voltages due to the small and medium vectors appears opposite the three-level inverter.

Even if neutral-point voltage equilibrium is considered in the PWM method of the three-level converter, the imbalance of the neutral voltage can be caused by various reasons. The differences in the capacitance of DC short capacitors generated in the manufacturing process, the characteristics of each switching device, and mismatches of switching on-off points all contribute to neutral-point voltage imbalance.

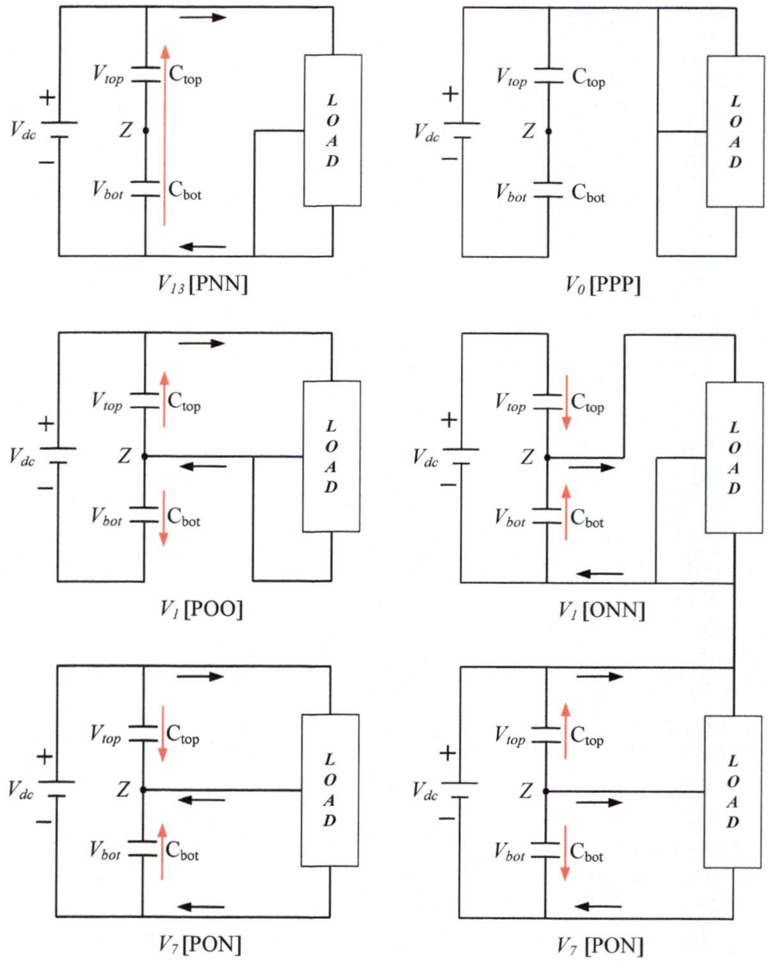

Fig. 1.19 Neutral point voltage change depending on the voltage vector type of a three-level converter

1.4 Three-Level Converter Reliability Research Field

Three research areas have been considered for reliability improvement of three-level converters. The first area is the switch fault diagnosis and tolerance control of three-level converters, the second one is the neutral-point voltage ripple reduction of three-level converters, and the third is related to the leakage current reduction of three-level converters.

1.4.1 Fault Diagnosis and Tolerance Control of Three-Level Converters

Compared to a two-level converter, the number of switches in a three-level converter is twice as large. Generally, when the number of switches in a power conversion apparatus increases, the probability of a fault in this power conversion apparatus also increases. This results in inoperability of the power conversion apparatus, which can result in a disruption of the plant and an interruption of the power supply. In fact, switch faults of power converters are found to make up about 21% of faults after capacitor and PCB faults as shown in Fig. 1.20. Research has been actively conducted in two areas to solve the problems caused by switch faults. The first research area is the fault diagnosis of switches. Switch fault diagnosis must accurately determine whether or not the switch of a power converter has failed and where it is located. This information is then used for the tolerant control or maintenance of the switch in which the fault occurred. It is a more required technology in a power conversion apparatus composed of many switches. The second research area is the use of tolerant control. Tolerant control is a technique for minimizing the loss caused by the faults of power converters. The operation of a power converter using tolerant control cannot satisfy all requirements in the steady state. However, it enables basic operations to be performed even in the event of a fault. Tolerant control can be utilized more easily in power converters consisting of many switches, such as three-level converters. An open-loop switch fault diagnosis method for the open-circuit switch fault of three-level converter is presented in Chap. 2 of this book, while tolerant control methods for open-circuit switch failures are explained in Chap. 3. The open-circuit switch fault diagnosis method and tolerant control methods presented in Chaps. 2 and 3 do not require additional elements or devices since they only use the surviving switches in the three-level converter.

1.4.2 Neutral-Point Voltage Ripple Reduction

In three-level converters, the neutral-point voltage refers to the voltage difference between the upper and lower capacitors of the DC stage. The imbalance of the neutral-point voltage caused by those mentioned in Sect. 1.3.5 results in a voltage imbalance of the DC component. The voltage imbalance of the DC component at the neutral-point results in switch breakdown and a THD increase of the output current. Therefore, various neutral-point voltage balance control methods are proposed for solving this problem. In order to use a three-level converter, the neutral-point voltage balance control must be applied.

There is also an imbalance of the AC component at the neutral-point voltage fluctuation. This means a ripple of the neutral-point voltage. Neutral-point voltage ripple is not caused by the problems mentioned above. However, it is determined by the PWM method and operating condition of the three-level converter. Increasing

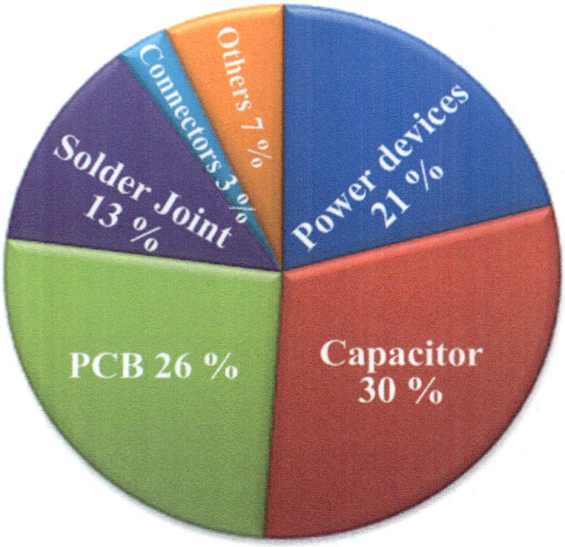

Fig. 1.20 Power converter fault type distribution (reprinted from [13], Fig. 2)

Table 1.4 Allowable ripple current depending on the capacitor capacity (Samyoung, RFA Series)

Rated voltage (V)	Capacitance (μF)	Magnitude (Diameter × height)	Permissible current ripple (Arms, 120 Hz)
350	*3300*	63.5 × 120	11.1
	3900	63.5 × 130	12.8
	4700	63.5 × 160	15.2
	7400	76.5 × 120	14.7
	5600	63.5 × 170	17.3
	5600	76.5 × 130	16.9
450	*2200*	63.5 × 120	9.1
	2700	63.5 × 130	10.6
	2700	76.5 × 120	11.2
	3300	63.6 × 160	12.7
	3300	76.5 × 130	13.0
	3900	63.5 × 170	14.4
	4700	76.5 × 160	16.7
	5600	76.5 × 170	21.1
	5600	89 × 160	19.9

the neutral-point voltage ripple results in a performance penalty, such as increases in the THD of the output current like the neutral-point voltage imbalance of the DC component. In addition, a large neutral-point voltage ripple means that the voltage

Table 1.5 Calculation of capacitor allowable ripple current with frequency change factor (Samyoung, RFA Series)

Frequency (Hz)	60	120	300	1 k	3 k
Factor	0.8	*1.0*	1.1	1.3	1.4

ripple and the injected current ripple of the two DC capacitors are large. A capacitor has the amount of allowable current ripple depending on its capacitance and size as shown in Table 1.4, and its life is shortened when the ripple current exceeds its allowable ripple current capacity. Therefore, it is necessary to analyze the component and magnitude of the ripple current to determine the capacitance and configuration of the series/parallel combination of capacitors in the design process. It should also be noted that the current ripple that can be tolerated on a capacitor is different for each frequency as shown in Table 1.5. A neutral-point voltage ripple reduction technique for reducing the capacitance of a three-level converter and increasing the lifetime of three-level converter is described in Chap. 4 of this book.

1.4.3 Leakage Current Reduction

Leakage current refers to current flowing to ground. This book describes leakage current that can occur during steady-state operation, not the leakage current caused by accidents. Many recently developed products use transformerless topology which is economical in terms of size and cost. Three-level converters have been applied without the transformer in many fields such as photovoltaic power generation, wind power generation and ESS (Energy Storage System). However, a power conversion apparatus of the transformerless type may cause leakage current. The magnitude of the leakage current is determined by the size of the leakage capacitor and the PWM method of the power converter. The magnitude of the leakage capacitor varies depending on environmental factors such as temperature and humidity. Large leakage currents can result in performance degradation, such as increasing the current THD of the power converter. In addition, since the size of leakage current is clearly regulated to ensure the safe operation of power conversion apparatus, reduction measures are required for authentication and product development. The most common solution is to connect additional filters to reduce leakage current. However, this causes problems such as increases in the price and size of products. In addition, it is possible to reduce leakage current by changing the PWM method of the power converter. Methods of reduce the leakage current of a three-level converter are described in Chap. 5 of this book. These methods do not require additional elements or devices and they reduce the leakage current by changing the PWM method of three-level converters.

References

1. Rodriguez, J., J.S. Lai, and F.Z. Peng. 2002. Multilevel inverters: A survey of topologies, controls, and applications. *IEEE Transactions on Industrial Electronics* 49 (4): 724–738.
2. Malinowski, M., K. Gopakumar, J. Rodriguez, and M.A. Pérez. 2010. A survey on cascaded multilevel inverters. *IEEE Transactions on Industrial Electronics* 57 (7): 2197–2206.
3. Rodriguez, J., S. Bernet, P.K. Steimer, and I.E. Lizama. 2010. A survey on neutral-point-clamped inverters. *IEEE Transactions on Industrial Electronics* 57 (7): 2219–2230.
4. Schweizer, M., I. Lizamay, T. Friedli, and J.W. Kolar. 2010. Comparison of the chip area usage of 2-level and 3-level voltage source converter topologies. In *Proceeding of IECON*, pp. 391–396, Nov. 2010.
5. Schweizer, M., and J.W. Kolar. 2012. Design and implementation of a highly efficient three-level T-type converter for low-voltage applications. *IEEE Transactions on Power Electronics* 28 (2): 899–907.
6. Soeiro, T.B., and J.W. Kolar. 2013. The new high-efficiency hybrid neutral point-clamped converter. *IEEE Transaction on Industrial Electronics* 60 (5): 1919–1935.
7. Soeiro, T.B., and J.W. Kolar. 2013. Analysis of high-efficiency three-phase two- and three-level unidirectional hybrid rectifiers. *IEEE Transaction on Industrial Electronics* 60 (9): 3589–3601.
8. Woo, H.M. 2014. Applications for next generation power semiconductor of SiC and GaN. *The Korean Institute of Electrical Engineers (KIEE)* 63 (5): 5.
9. Schweizer, M., T. Friedli, and J.W. Kolar. 2013. Comparative evaluation of advanced three-phase three-level inverter/converter topologies against two-level systems. *IEEE Transaction on Industrial Electronics* 60 (12): 5515–5527.
10. Wo, B. *High-power converter and AC drives*. Hoboken: Wiley
11. Lee, K.B., S.H. Huh, J.Y. Yoo, and F. Blaabjerg. 2005. Performance improvement of DTC for induction motor-fed by three-level inverter with an uncertainty observer using RBFN. *IEEE Transactions on Energy Conversion* 20 (2): 276–283.
12. Choi, U.M., F. Blaabjerg, and K.B. Lee. 2015. Reliability improvement of a T-type three-level inverter with fault-tolerant control strategy. *IEEE Transactions on Power Electronics* 30 (5): 2660–2673.

Chapter 2
Open-Circuit Fault Detection Methods for Three-Level Converters

2.1 Switch Fault Case and Reasons

IGBT modules are typically classified into two types: wire-bonded IGBT modules and press-pack IGBT modules. They have different characteristics which are a tradeoff between cost and performance, as shown in Table 2.1.

IGBT modules have a larger thermal resistance, a lower power density, and a higher failure rate because of the soldering and bond-wire connections of the internal chips. Press-pack packaging technology shown in Fig. 2.1 improves the connections between chips by direct press-pack contact. Therefore, press-pack IGBT modules have better reliability, a higher power density, and better cooling capability. However, their cost is also higher than that of conventional IGBT modules. Therefore, wire-bonded power device modules are still widely used in power electronics. Therefore, the failure mechanisms of conventional wire-bond IGBT modules are presented.

A wire-bonded IGBT is made up of several layers consisting of different materials. A silicon (Si) chip is soldered to a direct copper bonded (DCB) ceramic substrate. The DCB substrate insulates the Si chip from the baseplate and conducts the heat dissipated by the chip to the cooling system. Finally, the top side of the Si chip is contacted by aluminum (Al) bond wires. Figure 2.2 shows the structure of a wire-bond IGBT modules package.

The weak points in wire-bond IGBT modules are the wire bond and silicon interconnection, the silicon and DCB substrate solder joint, and the DCB substrate and baseplate solder joint. The primary failure mechanism in IGBT modules is the wire-bond liftoff as shown in Fig. 2.3. Al and Si material have different coefficients of thermal expansion (CTE), and thermal cycling causes repeated cooling and heating. This allows the disparately joined materials to expand and shrink at different rates when stress is applied at the point of contact. This CTE mismatch with temperature swings leads to wire-bond liftoff failures.

Another dominant failure mechanism is solder joint fatigue as shown in Fig. 2.4. Two solder joints are present in a standard IGBT module. They are between the Si

K.-B. Lee and J.-S. Lee, *Reliability Improvement Technology for Power Converters*, Power Systems, DOI 10.1007/978-981-10-4992-7_2

Table 2.1 Wire-bond IGBT modules versus press-pack IGBT modules

Characteristic	Wire-bond module	Press-pack IGBT module
Power density	Moderate	High
Reliability	Moderate	High
Cost	Moderate	High
Failure mode	Open circuit	Short circuit
Ease of maintenance	Better	Worse
Insulation of heat sink	Yes	No
Thermal resistance	Moderate	Small
Switching loss	Low	Low
Conduction loss	High	High

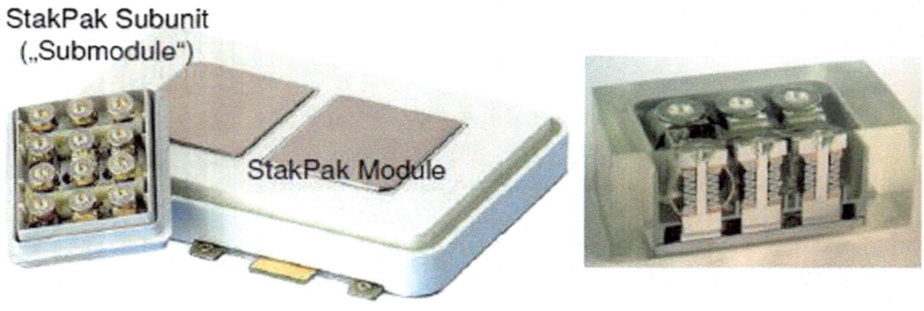

Fig. 2.1 Press-pack IGBT module

and the DCB and between the DCB and the baseplate. The CTE mismatch between the DCB substrate and the baseplate is higher than that between the DCB substrate and the Si. Therefore, the possibility of failure is higher in the solder joint between the DCB and the baseplate. This solder joint degradation increases the thermal resistance of the power device module. Therefore, the temperature in the module increases, which accelerates the lifting of the wire-bond.

The faults of IGBTs are classified into short-switch faults and open-switch faults. Short-switch faults may have serious impact on systems such as breakdown, as shown in Fig. 2.5, because a large current (which is above the rated current) flows through the short-circuited switch. Therefore, the system has to stop to ensure the safety of the other parts and the fuse, or additional devices, as shown in Fig. 2.6 [2], are used for preventing explosion from an abnormal large current.

On the other hand, open-switch faults (which are generated by thermal cycling, excessive collector current, and gate driver faults) cause input current distortions in NPC and T-type three-level converters [9, 10]. For the input side of the rectifier, this current distortion increases the total harmonic distortion (THD) and adversely affects the grid. Moreover, the power factor of the rectifier decreases, and the DC-link ripple increases. If an open-switch fault is not detected for a long time, other problems are generated because the distorted current is used to control the

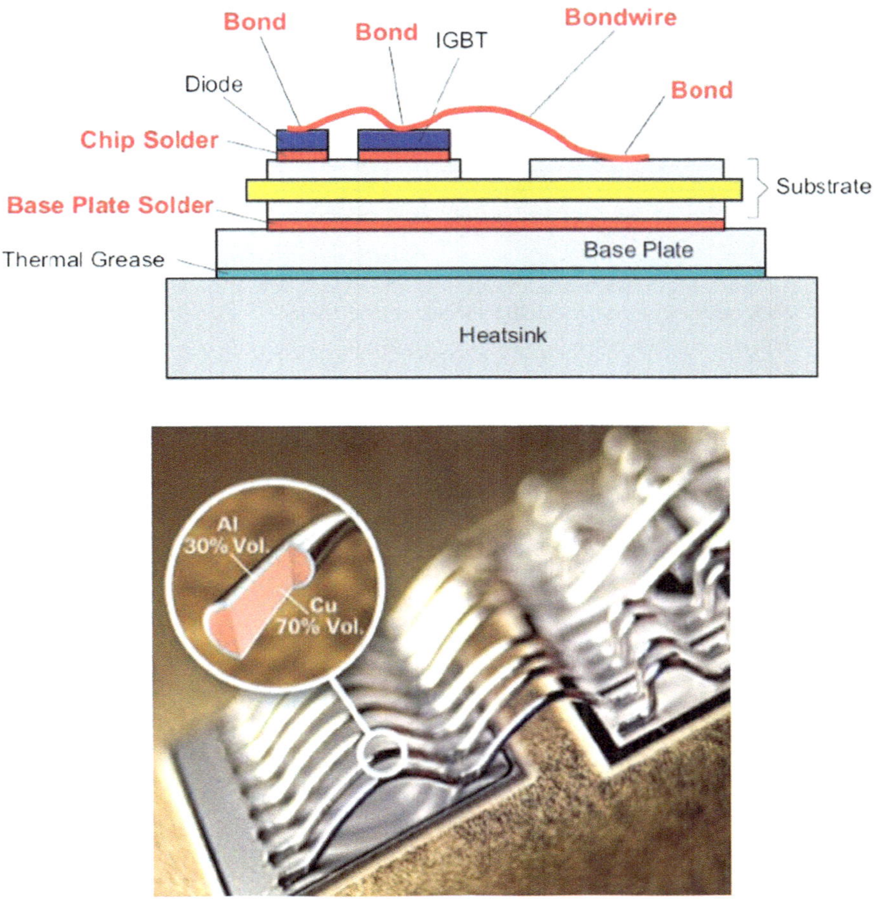

Fig. 2.2 Wire-bond IGBT module

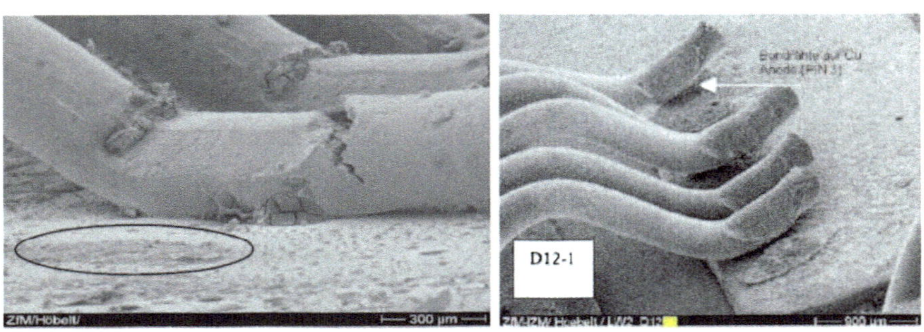

Fig. 2.3 Fault cases of a wire-bond IGBT module (wire fatigue)

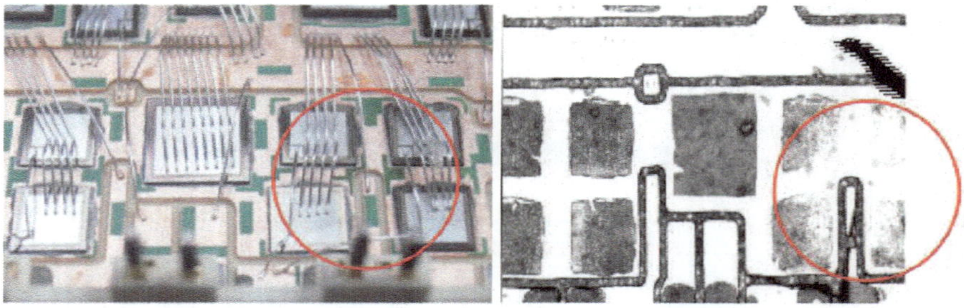

Fig. 2.4 Fault cases of a wire-bond IGBT module (solder fatigue)

Fig. 2.5 Breakdown case

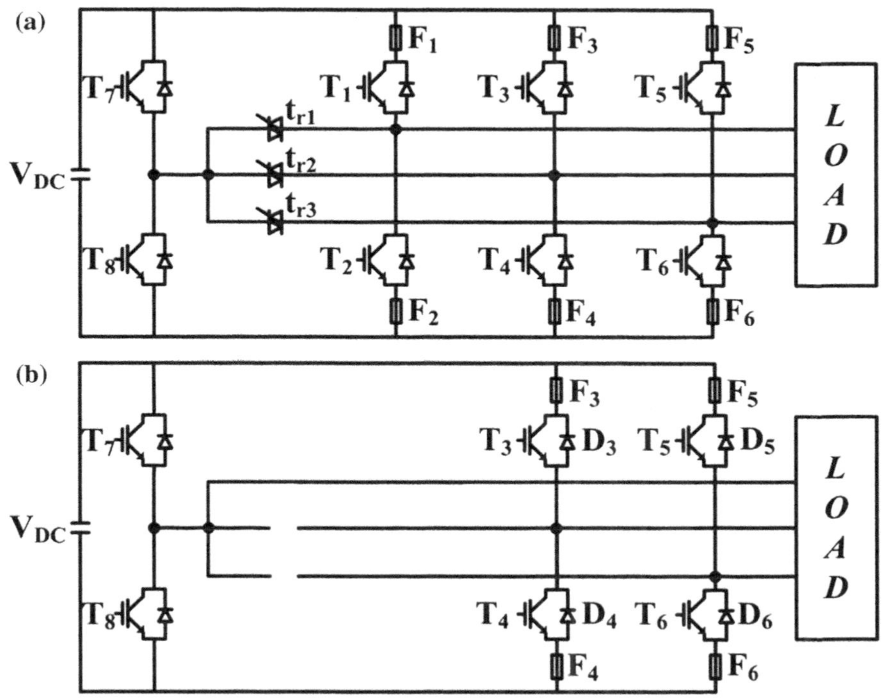

Fig. 2.6 Detection and tolerant control method using additional devices for switch short-circuit fault: **a** fault detection method using a fuse, **b** tolerant control method with an additional leg (drawn from Ref. [2], Fig. 1)

converter. Consequently, open-switch faults should be detected quickly, and tolerance control should be applied if possible.

Research into open-switch fault detection methods for three-level converters has been presented in a lot of papers. Some of these methods include an additional device for open-switch fault detection. Owing to this, current shape is considered in a lot of papers. Detection methods using currents are often based on the fact that the current pattern of the normal condition is different from the current pattern of an open-switch fault.

In this book, detection methods without additional devices for the open-circuit fault of three-level converters will be explained. Since the current patterns caused by a fault switch are different depending on the topology and the position of the switch, a detection method for each converter is required. In this chapter, the effects of the current under switch open-circuit faults will be introduced and the operation characteristic according to the operation mode (inverter mode and rectifier mode) will be explained. Finally, based on these results, several detection methods will be demonstrated.

2.2 Switch Open-Circuit Fault Detection Methods for NPC Three-Level Inverters

The three-level converter is used as a power conversion system in motor drives and renewable energy generation. Inverter operation means that the energy stored in the DC-link is transferred to the output of the inverter. This chapter shows switch open-circuit fault detection methods for NPC three-level inverters.

2.2.1 Switch Open-Circuit Fault Detection Methods Using Additional Devices

A. Switch open-circuit fault detection method using voltage sensors

Figure 2.7 shows a two-level inverter used in motor drive applications [1]. There are voltage sensors for detecting switch open-circuit faults. This method uses the fact that the output voltage of the three legs cannot make the desired output voltage

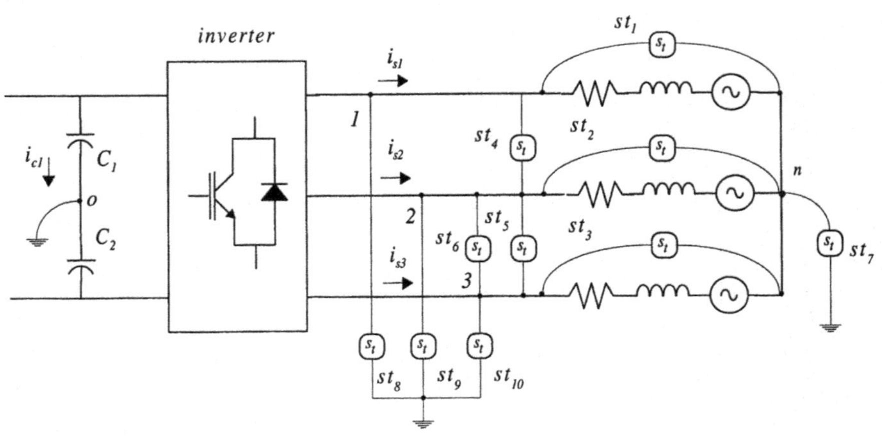

Fig. 2.7 Switch open-circuit fault detection method using voltage sensors (reprinted from Ref. [1], Fig. 4)

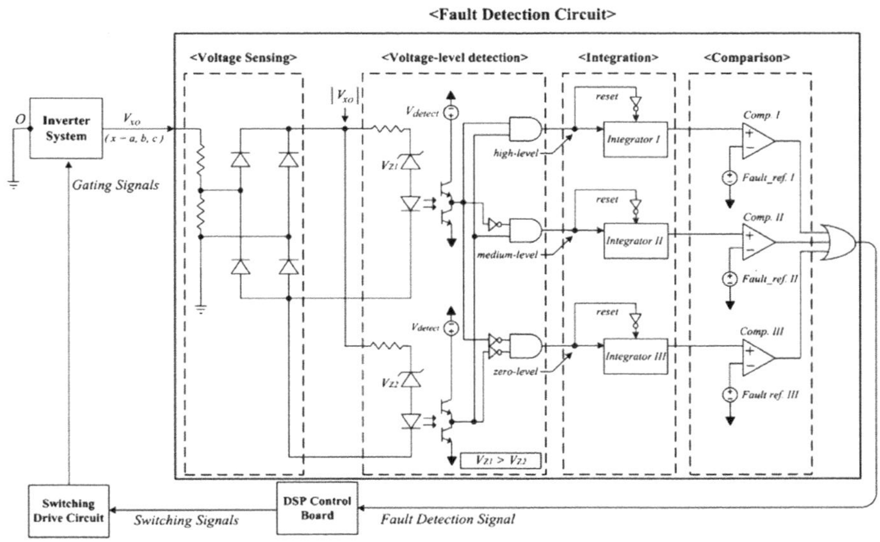

Fig. 2.8 Switch open-circuit fault detection method using additional detection circuits (reprinted from [3], Fig. 9)

in the presence of a switch open-circuit fault. In this method, the detection signal is generated by comparing two voltages in the normal condition and in the switch fault condition.

B. Switch open-circuit fault detection method using additional detection circuits

Figure 2.8 shows a detection circuits for switch open-circuit faults [3]. The inputs of the detection circuits are the pole voltage (V_{xo}, $x = a, b, c$) of each leg and the fault signal is generated through the Voltage-level detection circuit, the Integration circuit, and the Comparison circuit in order. This method can identify a fault state exactly. However, it cannot distinguish the position of the switch fault.

2.2.2 *Switch Open-Circuit Fault Detection Method Using Current Distortion*

The detection methods using additional devices have big drawbacks due to their increases in terms of size and cost. Therefore, a lot of papers have proposed detection methods using current distortion without additional devices.

The abnormal voltage caused by a switch open-circuit fault leads to current distortion. The output currents which are measured by current sensors are used to control the system and to obtain useful information for detecting fault switches.

A. The effects on current from switch open-circuit faults

The basic operation of a NPC three-level converter is shown in Chap. 1. The switching states are divided into three states P, O, and N. By considering the current direction, six current paths are expressed as shown in Fig. 2.9. When the switching state is P, there are two current paths depending on the current direction. In one path, the current flows through the two switches and the current direction is from the output to the DC-link. In the other path, the current direction is from the

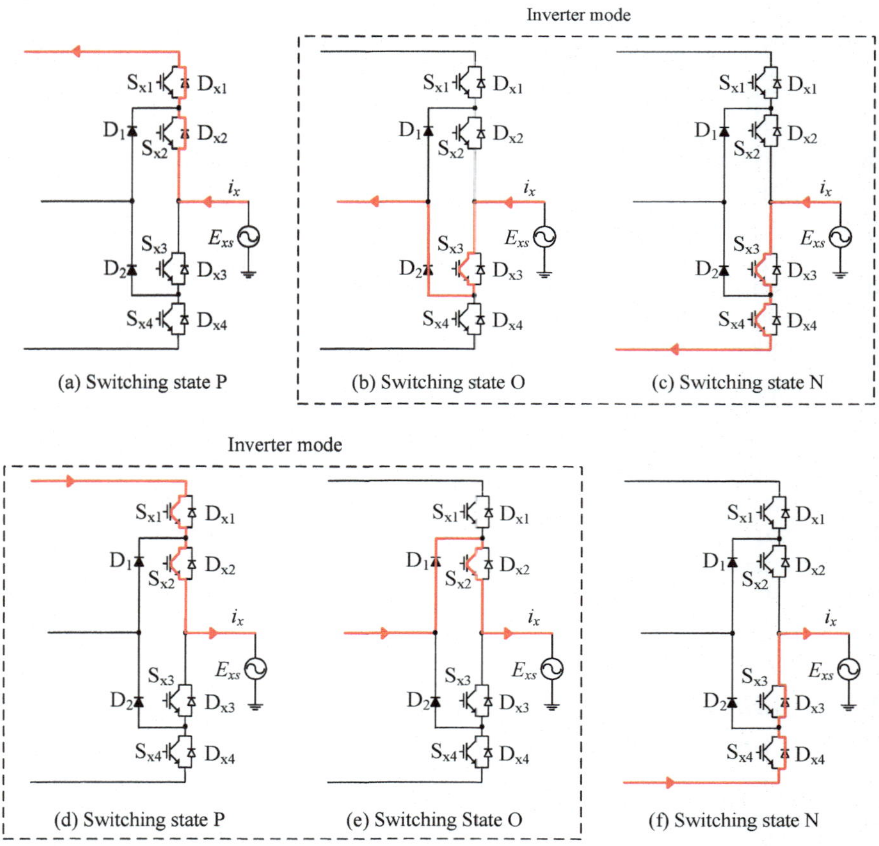

Fig. 2.9 Current paths of the NPC topology for the current directions and switching states

DC-link to the output and the current flows through two diodes. Based on the results, the switching states O and N can be analyzed.

When the NPC three-level converter operates in the inverter mode with unity power factor, only four current paths are used because the sign of the output voltage is the same as that of the output current. A positive output voltage is generated by the P- and O-switching states and the current flows from the DC-link to the output. The N- and O-switching states lead to a negative output voltage and the current flows from the output to the DC-link. The P- (or N-) switching state is a valid switching state which causes the current to flow in the inverter mode. The O-switching state act as a freewheeling path.

- S_{x1} open-circuit fault

When the open-switch fault occurs in switch S_{x1} while the switching state is "P," the phase current path in the grid-connected system is different from that under the RL load condition. Because of the grid phase voltage E_{xs}, the diode D_1 becomes reverse-biased. The grid phase voltage E_{xs} is the voltage between the grid and ground. The current path is opened, and the positive phase current becomes zero as shown in Fig. 2.10a. If E_{xs} is higher than the dc-link voltage ($V_{DC}/2$), the current path is formed through diodes D_{x1} and D_{x2}. Since this period is very short and the potential difference is very small, the negative phase current is almost zero.

- S_{x2} open-circuit fault

When the open-switch fault occurs in switch S_{x2} while the switching state is "P," the phase current does not flow as shown in Fig. 2.10b, because the phase current path is opened. Even when the switching state is "O," the phase current path is not formed, because the switch S_{x2} is opened. If E_{xs} is higher than V_{xz}, the phase current is formed through switch S_{x3} and diode D_2. However, the negative phase current does not flow, because this period is very short.

- S_{x3} open-circuit fault

When the switching state is "N," the phase current path is opened because of the occurrence of the open-switch fault in switch S_{x3}. The negative phase current becomes zero as shown in Fig. 2.10c.

Even when the switching state is "O," the phase current path is not formed, because switch S_{x3} is opened. The path is formed through diode D_1 and switch S_{x2} when V_{xz} is higher than E_{xs}. However, the positive phase current does not flow because of the abovementioned reasons.

Table 2.2 Feasible and infeasible switching state of a T-type three-level converter depending on the position of the open-switch fault

Position of open-switch fault	Feasible switching state	Infeasible switching state
S_{x1}	N (Fig. 2.9c, f) O (Fig. 2.9b, e) P (Fig. 2.9a]	P (Fig. 2.9d)
S_{x2}	N (Fig. 2.9c, f) O (Fig. 2.9b) P (Fig. 2.9a)	P (Fig. 2.9d) O (Fig. 2.9e)
S_{x3}	N (Fig. 2.9f) O (Fig. 2.9e) P (Fig. 2.9a, d)	N (Fig. 2.9c) O (Fig. 2.9b)
S_{x4}	N (Fig. 2.9f) O (Fig. 2.9b, e) P (Fig. 2.9a, d)	N (Fig. 2.9c)

- S_{x4} open-circuit fault

When the open-switch fault occurs in switch S_{x4} while the switching state is "N," diode D_2 becomes reverse biased. The current path follows the open circuit, and the phase current becomes zero as shown in Fig. 2.10d.

The feasible and infeasible switching states of NPC three-level converter depending on the position of the open-switch fault are summarized in Table 2.2.

B. Analysis of the current distortion caused by open-circuit faults

The measured three-phase currents (I_a, I_b, I_c) can be transformed into two dimensions (I_{ds}, I_{qs}) through the following equations:

$$\begin{aligned} I_{ds} &= I_a \\ I_{qs} &= \frac{1}{\sqrt{3}}(I_b - I_c) \end{aligned} \tag{2.1}$$

The current patterns that indicate the location of a faulty switch can be classified into 12 patterns. The shape of the current pattern under a healthy condition is a circle [4, 5].

In the occurrence of an open-switch fault, there is a change in the phase current value at the location where the fault has occurred. Therefore, a change in the circle shape and angle represents the occurrence of an open fault condition and is an indicator of the location of the faulty switching device, as shown in Table 2.3 and Fig. 2.11 under RL load. The magnitude and semicircle angle are defined as:

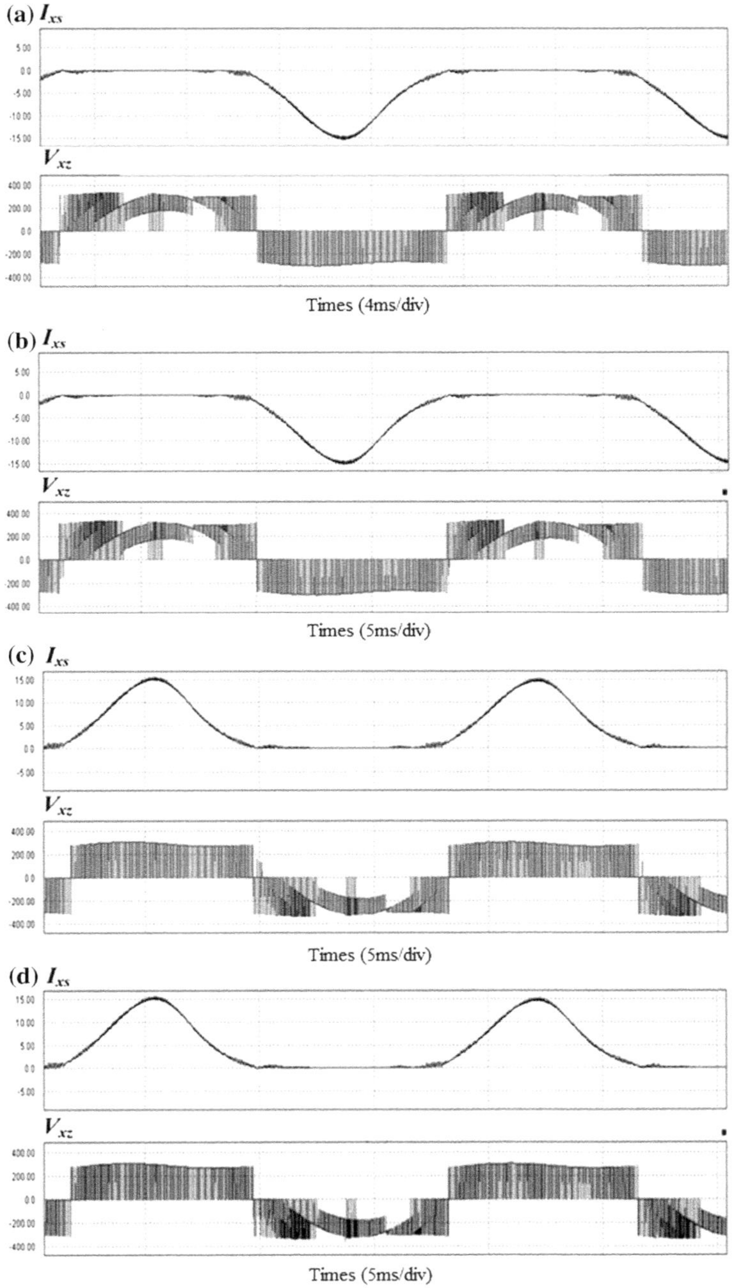

Fig. 2.10 Grid-connected NPC inverter characteristics: **a** S_{x1}, **b** S_{x2}, **c** S_{x3}, **d** S_{x4} (reprinted from [4], Figs. 10–13)

Table 2.3 Angle of semicircles during the occurrence of an open-switch fault

θ	Fault switch
30°–90°	S_{c1} or S_{c2} fault
90°–150°	S_{b3} or S_{b4} fault
150°–210°	S_{a1} or S_{a2} fault
210°–270°	S_{c3} or S_{c4} fault
270°–330°	S_{b1} or S_{b2} fault
330°–30°	S_{a3} or S_{a4} fault

$$I_{mag} = \sqrt{I_{qs}^2 + I_{ds}^2} \tag{2.2}$$

$$\theta = \tan^{-1}\left(\frac{I_{qs}}{I_{ds}}\right) \tag{2.3}$$

The open-switch fault detection steps under RL load are as follows

(1) First, calculate the magnitude of the current vector using (2.2).
(2) Second, compare I_{mag} with the magnitude of the current reference.

$$I_{mag,ref} = \sqrt{I_{qe,ref}^2 + I_{de,ref}^2} \tag{2.4}$$

(3) Third, if the difference of (2.2) exceeds the defined value, determine the angle of (2.3).
(4) Fourth, determine the position of the open-circuit fault by Table 2.3.

In the third step, by dividing the detection level into two levels, S_{x1} and S_{x2} (or S_{x3} and S_{x4}) the faults can be classified.

Figure 2.12 shows the current patterns of a grid-connected NPC inverter system under the fault condition. As in the previous analysis, the phase currents are the same in the two situations when either the S_{x1} switch or the S_{x2} switch are broken down. The phase currents when a fault occurs in either S_{x3} or S_{x4} are also equal. It is impossible to distinguish a fault in switch S_{x1} from that in switch S_{x2}, or a fault in switch S_{x3} from that in switch S_{x4} using the method explained in the previous section.

C. Experimental result of switch open-circuit fault detection in NPC three-level inverters

- S_{x1} and S_{x2} (S_{x3} and S_{x4}) classification using the pulse of the O-switching state

The main difference between a fault in switch S_{x1} and that in switch S_{x2} or between a fault in switch S_{x3} and that in switch S_{x4} is the possibility of switching state O. However, E_{xs} nullifies positive phase current and negative phase current,

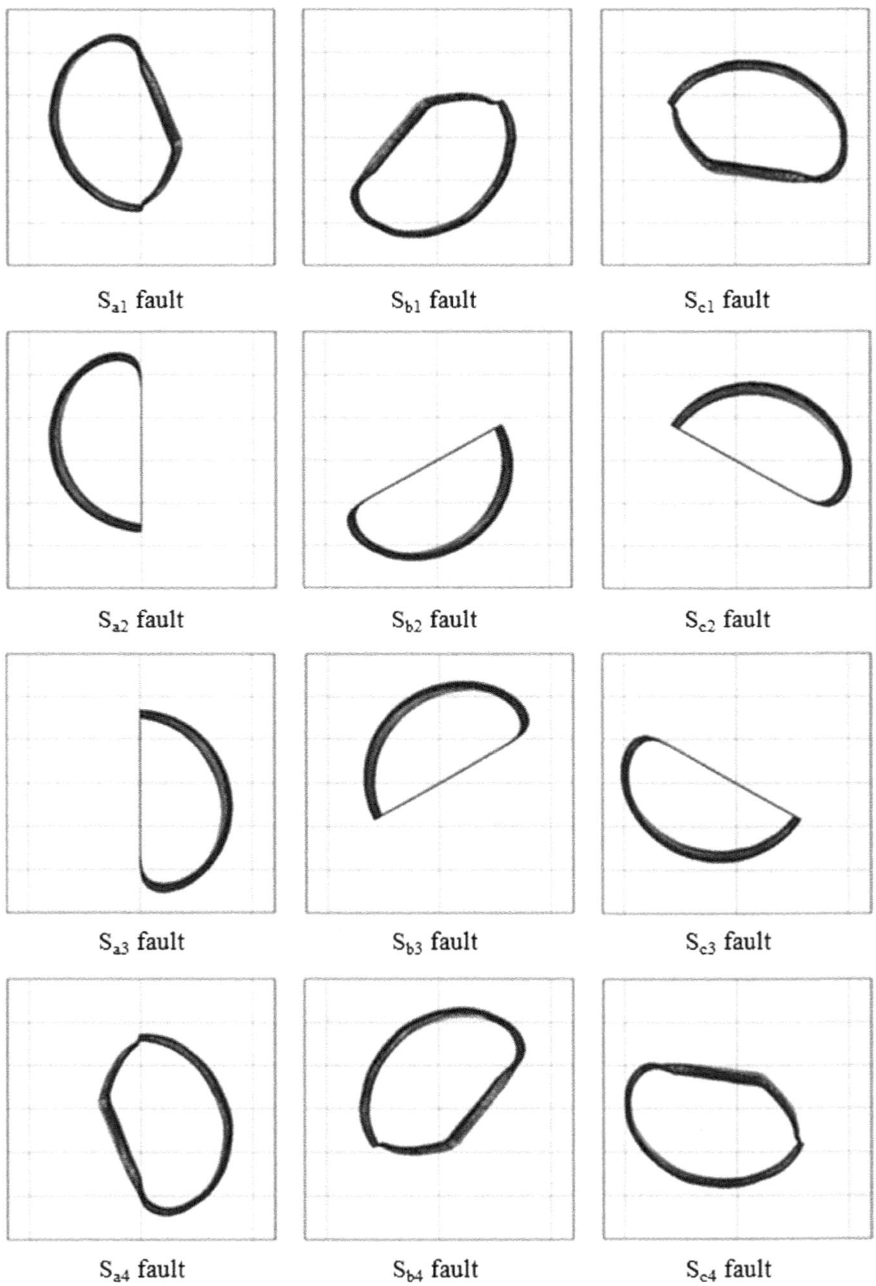

Fig. 2.11 Simulated current patterns during the occurrence of an open-switch fault under RL load (reprinted from [4], Fig. 9)

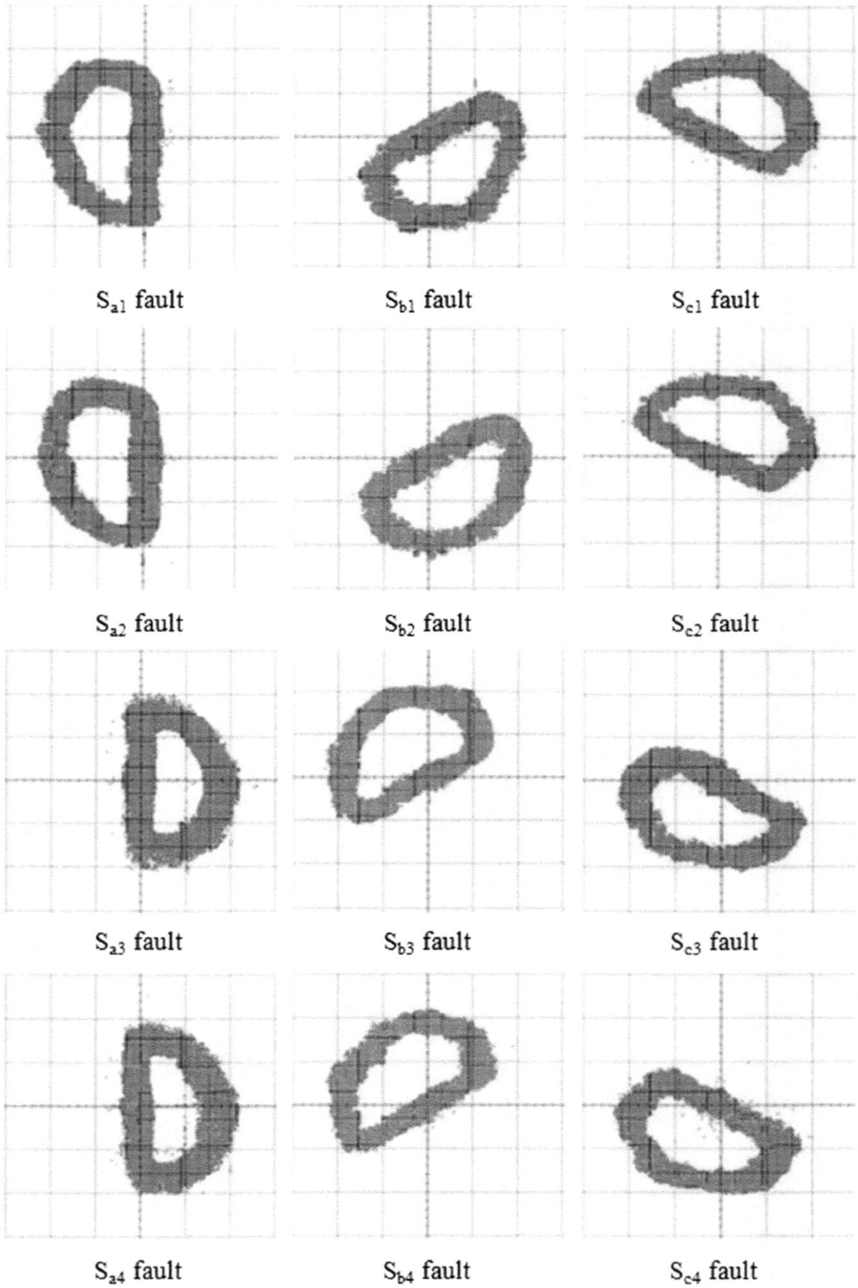

Fig. 2.12 Measured current patterns during the occurrence of an open-switch fault in a grid-connected system (reprinted from [4], Fig. 30)

regardless of the possibility of the O-switching state. It is difficult to determine the effects of the O-switching state on the current patterns. It is only possible to determine whether an open-switch fault occurs in one of the two upper switches (S_{x1}, S_{x2}) or in one of the two lower switches (S_{x3}, S_{x4}), on the basis of the current patterns. The location of a faulty switch either between the two upper switches or between the two lower switches in a grid-connected NPC inverter system can be identified by adding a simple switching scheme to the conventional method.

(5.1) Current generation identification using the pulse of the O-switching state [4]

If an open-switch fault occurs in one of the two upper switches, the O-switching state is applied for a very short period when E_{xs} is negative. If positive phase current flows for a short period, then it can be inferred that an open-switch fault has occurred in switch S_{x1}. If positive current does not flow, then it can be inferred that an open-switch fault has occurred in switch S_{x2}.

If an open-switch fault occurs in one of the two lower switches, the O-switching state is applied for a very short period when E_{xs} is positive. If negative phase current flows for a short period, then it can be inferred that an open-switch fault has occurred in switch S_{x4}. If negative current does not flow, it can be inferred that an open-switch fault has occurred in switch S_{x3}.

Figure 2.13a shows experimental results under the parameters of Table 2.4 when an open-switch fault occurs in switch S_{x1}. It is possible to detect the open-switch fault within half of a fundamental period and to identify the location of the faulty switch within two fundamental periods. Figure 2.13b shows the fault signal, phase angle, and phase current when an open-switch fault occurs in switch S_{x2}. Similarly, Fig. 2.13c, d show the phase angle, fault signal, and phase current when an open-switch fault occurs in switches S_{x3} and S_{x4}, respectively. Open switch fault detection is also possible within half of a fundamental period, and identification of the faulty switch is possible within two fundamental periods. It is noted here that a satisfactory diagnosis is obtained. It can also be seen from the experimental results that the location of the faulty switch can be identified with the proposed algorithm by using a simple switching control.

- S_{x1} and S_{x2} (S_{x3} and S_{x4}) classification using reactive current injection [11]

Usually, grid-connected inverters transfer electric power to the grid with a unity Power Factor (PF). Therefore, the polarities of the output phase current and the output voltage are almost the same, as shown in Fig. 2.14a. If an under excited reactive power is injected, the phase current leads the output pole voltage and grid voltage. Therefore, regions that have the different polarities between the phase

Table 2.4 Experimental parameters

DC-link voltage	600 V	Line-to-line voltage	380 Vrms
DC-link capacitor	1230 μF	Control period	160 μs
Switching frequency	6.25 kHz	Injection time for the O switching state	1–2 ms

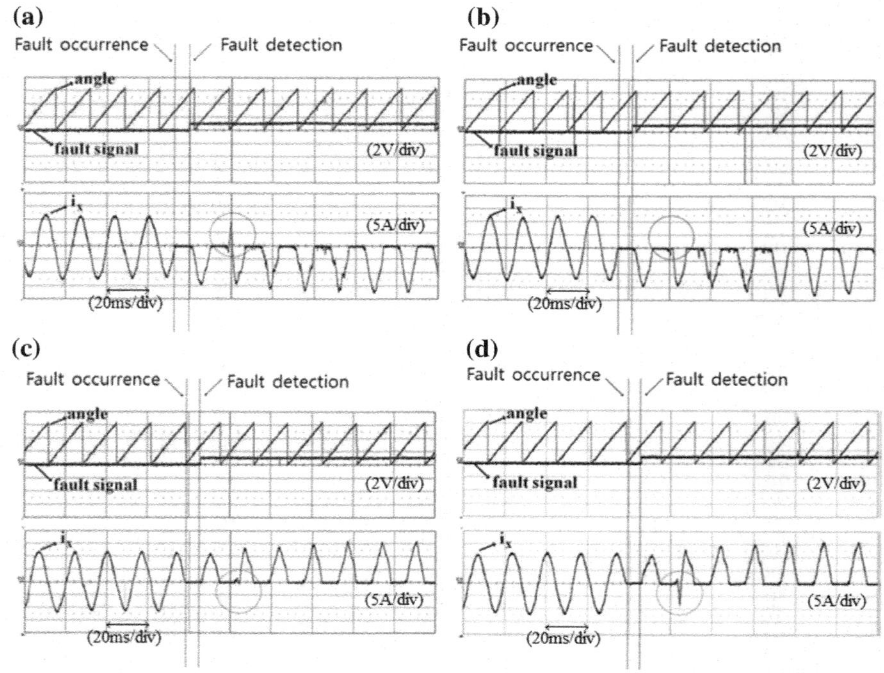

Fig. 2.13 Measured results of fault diagnosis when an open-switch fault occurs: **a** S_{a1}, **b** S_{a2}, **c** S_{a3}, **d** S_{a4} (reprinted from Ref. [4], Figs. 31–34)

current and the output voltage occur as shown in Fig. 2.14b. In this section, region 1 means that the phase current is positive and that the output pole voltage is negative, while region 2 means that the phase current is negative and that the output voltage is positive. By making regions 1 and 2 through under excited reactive power injection, the possibility of O-switching state can be determined and the faulty switch can be identified. Whether or not the current in regions 1 or 2 flows depending on the faulty switch.

(5.2) Current generation identification using reactive current injection [11]

First detection steps 1–4 are performed. Then reactive current is injected. If current is generated, the fault switch is S_{x4} (or S_{x1}), if it is not generated, the switch is S_{x2} (or S_{x3}).

Figure 2.15 shows experimental results under the parameters of Table 2.5. In the case of a S_{a1} fault, positive phase current flows in the region where under excited current is injected as shown in Fig. 2.15a. However, if an open-circuit fault occurs in S_{a2}, positive current does not flow as shown in Fig. 2.15b. From this result, the

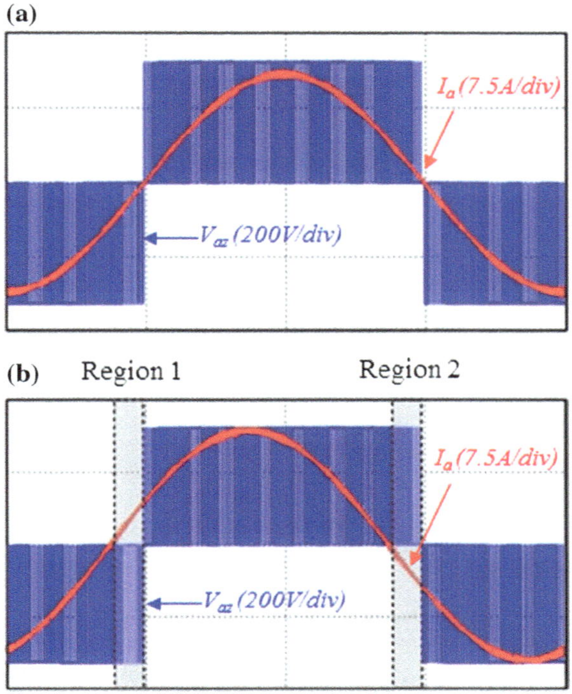

Fig. 2.14 Phase currents I_a and pole voltage V_{az}: **a** before and **b** after the injection of under excited reactive current (reprinted from [11], Fig. 6)

Table 2.5 Experiment parameters

DC-link voltage	600 V	Line-to-line voltage	380 Vrms
DC-link capacitor	2200 µF	Control period	100 µs
Switching frequency	10 kHz	Injection time for O switching state	2 A

faulty switch between the upper two switches can be correctly identified. Figure 2.15c, d show experimental results obtained with the discussed fault detection method when Type-A open-circuit faults have occurred in S_{a3} and S_{a4}, respectively. To determine the fault location from between the two lower switches (S_{a3} and S_{a4}), under excited reactive current has been injected. Negative current does not flow under an open-circuit fault in S_{a3}. However, negative current flows in the case of an open-circuit fault in S_{a4}. The location of the faulty switch is identified precisely by the discussed method.

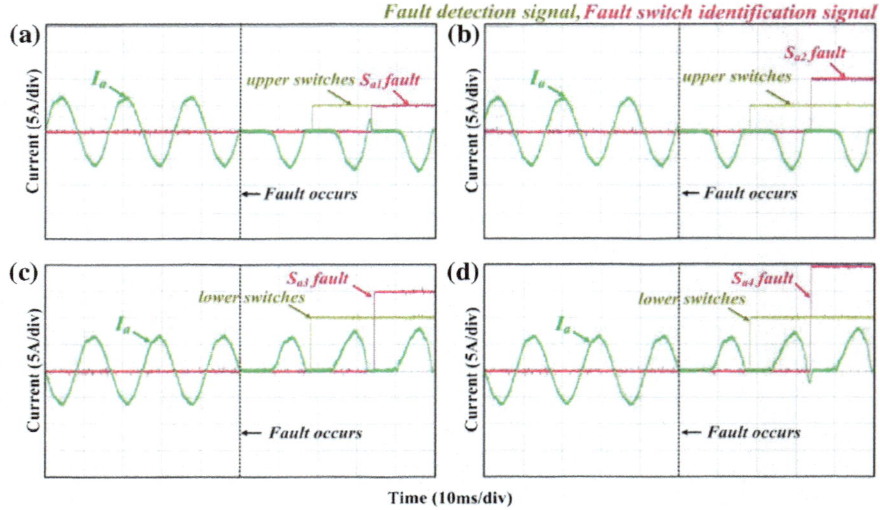

Fig. 2.15 Diagnosis of a faulty switch under a type-A open-circuit fault in: **a** S_{a1}, **b** S_{a2}, **c** S_{a3}, **d** S_{a4} by the injection of under excited reactive current (reprinted from [11], Fig. 20)

2.3 Switch Open-Circuit Fault Detection Methods for NPC Three-Level Rectifiers

2.3.1 Switch Open-Circuit Fault Detection Method Using Additional Devices

Switch open-circuit fault detection methods using additional devices can detect the position of an open-circuit fault regardless of the operation mode (inverter mode or rectifier mode). The methods mentioned in Sect. 2.2.1 can be used for detecting switch faults in NPC-type three-level rectifiers.

2.3.2 Switch Open-Circuit Fault Detection Method Using Current Distortion

A. The effects on current from switch open-circuit faults

The operating status of the switches and the pole voltage in NPC inverters can be represented by switching states. The switching state P indicates that the two upper switches S_{x1} and S_{x2} are ON and that the pole voltage V_{xz}, which is the voltage at terminal x ($x = a, b, c$) with respect to the neutral-point voltage Z, is $+V_{dc}/2$. The

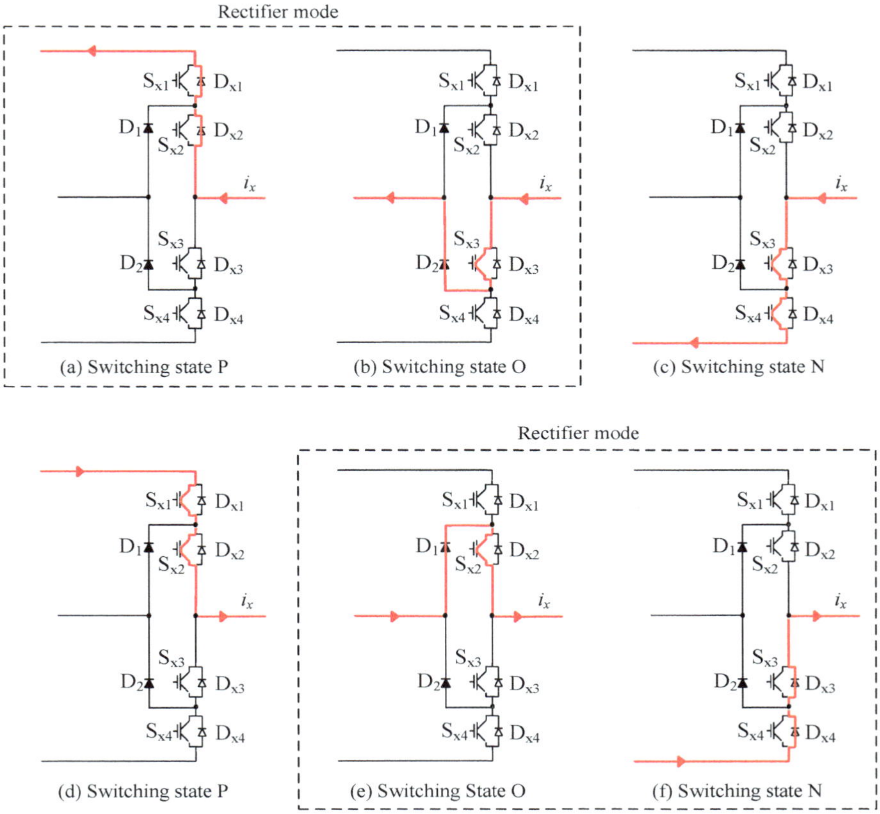

Fig. 2.16 Current paths of the NPC topology for the current directions and switching states

switching state O denotes that the two inner switches S_{x2} and S_{x3} are ON and that V_{xz} is zero. The switching state N signifies that the two lower switches S_{x3} and S_{x4} are ON and that V_{xz} is $-V_{dc}/2$.

Figure 2.16 shows the current paths according to the switching state and the phase current direction. The phase current includes information on the switching states in the NPC inverter. If an open-switch fault occurs, the switching state does not reach the desired state. This causes a change in V_{xz} and the phase current i_x.

B. S_{x1} and S_{x4} open-circuit faults of the three-level converters used in WTSs [8, 12]

Figure 2.17 shows a back-to-back power converter using a neutral-point-clamped (NPC) topology in a Wind Turbine Systems (WTS). The voltage difference ΔV between the input voltage, provided by the NPC rectifier, and the back electromotive force (EMF) of the PMSG generates input current, which has a different phase angle than the voltage difference ΔV because of the inductance and resistance of the PMSG. The magnitude of the input voltage is almost the same as

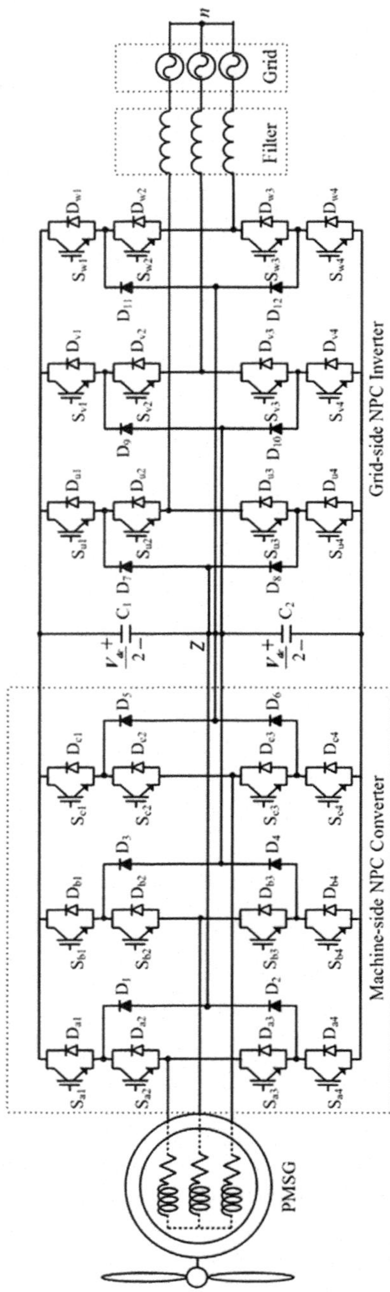

Fig. 2.17 Back-to-back converter using the NPC topology for a WTS (reprinted from [8], Fig. 1)

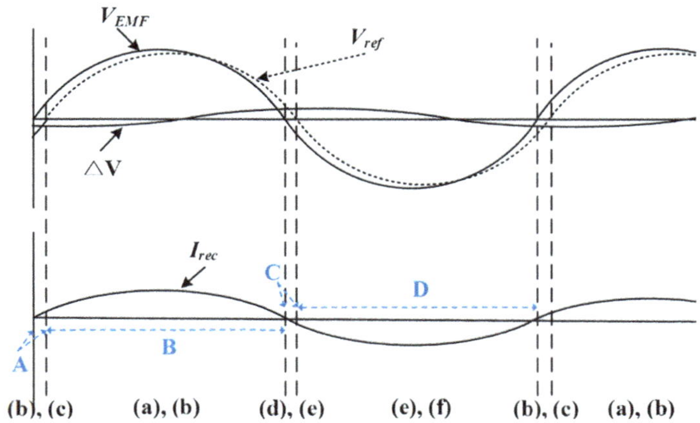

Fig. 2.18 Normal current flow between a NPC rectifier and a PMSG (reprinted from [12], Fig. 4)

(or smaller than) the back EMF. If the two voltage magnitudes are almost the same, the current magnitude is determined by the phase angle difference between the input voltage and the back EMF. Zero current means that the two voltages are the same and that the phase angle of the input voltage matches that of the back EMF. For unity power factor, the input voltage is controlled to correspond to the phase of the current with the same phase as the back EMF, as shown in Fig. 2.18.

If the time intervals containing the paths in Fig. 2.16c, d are not long enough to distort the current, S_{x2} and S_{x3} open-switch faults (which are related to the paths of Fig. 2.16b, e, excluding S_{x1} and S_{x4} open-switch faults) can be considered because there is no current through the IGBTs in the paths of Fig. 2.16a, f. Therefore, the switching state O, which is shown in Fig. 2.16b, e, is a valid switching state that generates current in the rectifier.

However, the time intervals of [(b), (c)] and [(d), (e)] are influenced by the current magnitude and modulation index. There are two Cases causing wide intervals [(b), (c)] and [(d), (e)]:

- Case 1: When V_{EMF} is small and a large I_{rec} is required
- Case 2: When reactive current is injected for the IPMSG

Figure 2.19 shows the current flow when a NPC rectifier generates a large current with a small modulation index. To create a large current flow, the phase angle difference is expanded. As a result, the time intervals of [(b), (c)] and [(d), (e)] increase. In this case, current distortion is generated by S_{x1} and S_{x4} open-switch faults. Consequently, all of the open-switch faults in the rectifier must be considered.

C. Open-circuit faults detection of the three-level converters used in WTSs

The current pattern is divided into four patterns depending on the location of the open-switch fault. An S_{x2} open-switch fault prevents the generation of negative

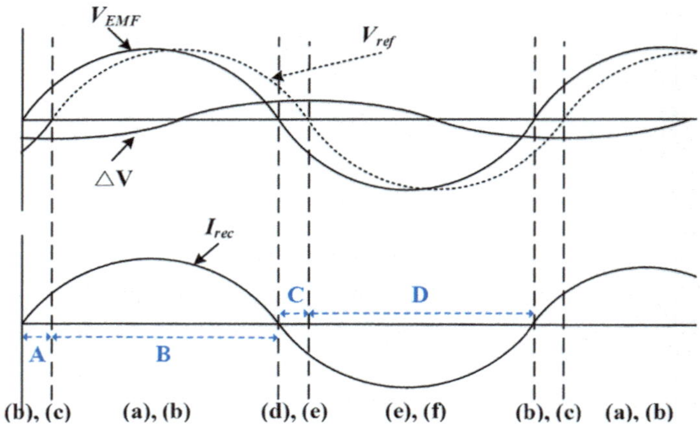

Fig. 2.19 Current flow process with a small modulation index and a large current (reprinted from [12], Fig. 5)

current, and an S_{x3} open-switch fault prevents the generation of positive current. An S_{x1} open-switch fault makes the current zero for some duration in the negative current, while the current becomes zero for some duration in the positive current when an S_{x4} open-switch fault occurs.

The range where the current distortion occurs can be defined as shown in Fig. 2.20. In each range, a switch open-circuit fault makes the current zero.

In the discussed detection method, the current can be divided into three ranges, as shown in Fig. 2.21, to separate the four patterns. The time of the zero range $T_{Z,Ix}$ is used to detect the location of an open-switch fault. The current belongs in one of the three ranges. If an S_{x2} open-switch fault occurs, no current flows in the negative range. An S_{x3} open-switch fault prevents current from flowing in the positive range, and the current is in the zero range for a longer time.

If the previous sign of the current was positive and if the current is in the zero range for half of a period, this indicates an S_{x2} open-switch fault, as shown in Fig. 2.21c. In the case of an S_{x3} open-switch fault, the previous sign of the input current is negative, and all of the other characteristics are the same as those in the S_{x2} open-switch fault detection method.

The detection method for the $T_{Z,Ix}$ of S_{x2} and S_{x3} open-switch faults is expressed as:

$$D_{23} = R_{23} T_{half} = R_{23} \frac{1}{2f_s} < T_Z \quad \begin{cases} if\ C_{pre,sign} = Positive, S_{x2}\ fault \\ if\ C_{pre,sign} = Negative, S_{x3}\ fault \end{cases} \tag{2.5}$$

where R_{23} is the detection parameter, $C_{pre,sign,Ix}$ is the previous sign of the current, T_{half} is half of the current period, and f_s is the angular frequency of the PMSG.

The previous sign of the current is determined when the input current enters the zero range and the time of the zero range is calculated, as shown in Fig. 2.22.

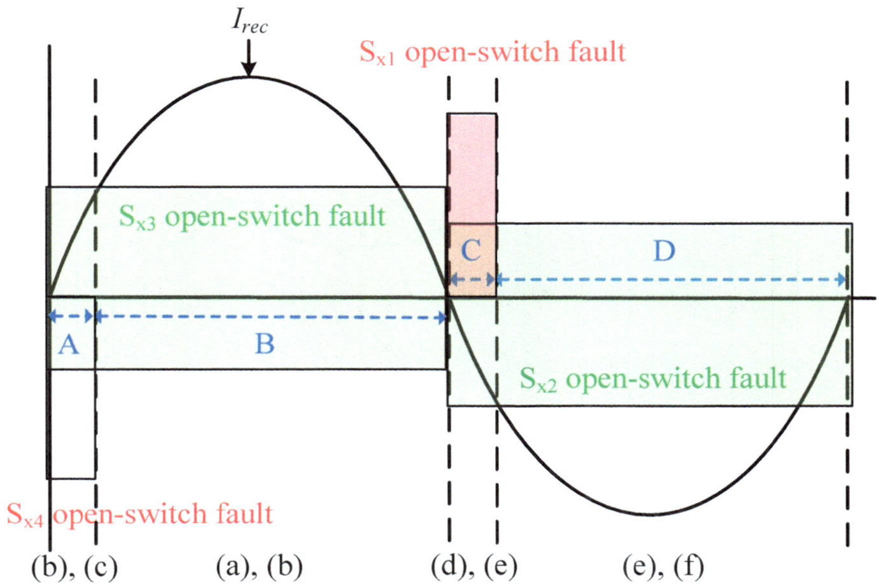

Fig. 2.20 Current distortion range depending on the position of a switch open-circuit fault (reprinted from [12], Fig. 6)

S_{x1} and S_{x4} open-switch faults are also detected by calculating $T_{Z,Ix}$. When the current is sinusoidal, T_P and T_N are almost the same, and $T_{Z,Ix}$ is small. However, an S_{x1} open-switch fault leads to a decrease in T_N and an increase in $T_{Z,Ix}$, as shown in Fig. 2.21b, and an S_{x4} open-switch fault decreases T_P and increases $T_{Z,Ix}$. Therefore, the detection method for S_{x1} and S_{x4} open-switch faults is expressed as:

$$D_{14} = R_{14}T_{half} = R_{14}\frac{1}{2f_s} < T_Z \quad \begin{cases} if\ C_{pre,sign} = Positive, S_{x1} fault \\ if\ C_{pre,sign} = Negative, S_{x4} fault \end{cases} \tag{2.6}$$

where R_{14} is the detection parameter.

To determine the zero range value, the current ripple is considered. 10% of the current is used in this study. T_{half} for open-switch fault detection is changed depending on the angular frequency because the period of the current is proportional to the speed of the PMSG.

- Determination of the R_{23} and R_{14} values

In the determination of the R_{23} and R_{14} values, the zero range value due to each open-switch fault should be considered. When an S_{x2} or S_{x3} open-switch fault occurs, $T_{Z,Ix}$ is almost the same as a half period of f_s (1/2 f_s) regardless of the system parameters. Therefore, if R_{23} is lower than 1, it is possible to detect S_{x2} and S_{x3}

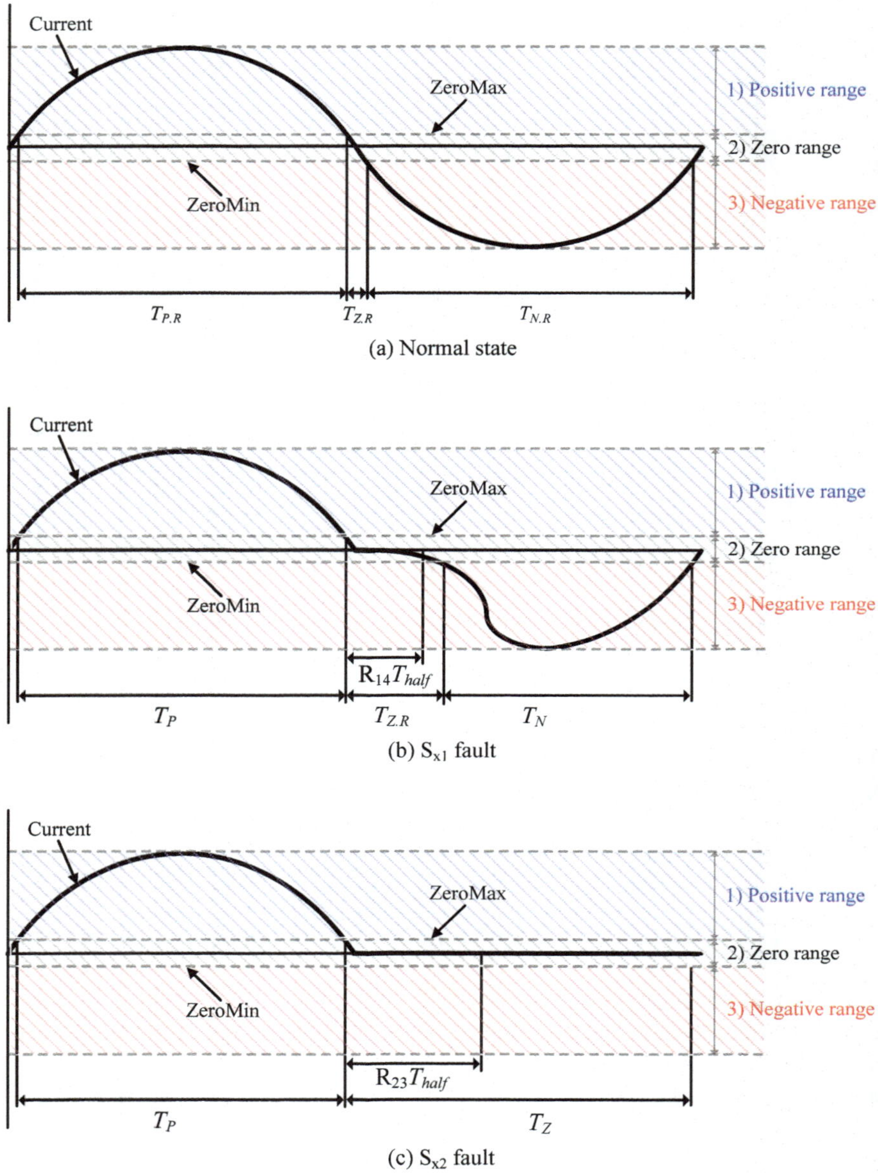

Fig. 2.21 Concept of the open switch fault detection method (reprinted from [8], Figs. 9, 10, 12)

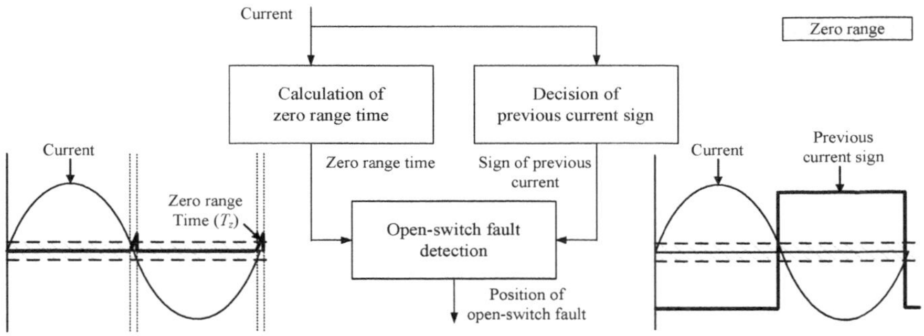

Fig. 2.22 Block diagram for the detection of the previous current sign and the calculation of the zero range duration (reprinted from Ref. [8], Fig. 11)

open-switch faults, and the detection performance can be improved by decreasing R_{23} close to 0.

To selecting R_{14}, the zero range value $T_{Z,Ix}$ caused by an S_{x1} or S_{x4} open-switch fault should be considered. Regardless of the open-switch fault, 10% of the current is contained in the zero range, and its range θ_n is expressed as:

$$\theta_n = 2 \times \sin^{-1}\left(\frac{1}{10}\right) = 11.5^\circ \tag{2.7}$$

To consider the time of the zero range $T_{Z,Ix}$ caused by an S_{x1} or S_{x4} open-switch fault, the system parameters shown in Fig. 2.23 are used. The voltage–current equation is expressed as

$$I_{x,rec} = \frac{BackEMF - V_{ref}\angle\theta}{R_s + j2\pi f_s L_s} \tag{2.8}$$

where V_{ref} is the rectifier input voltage and θ is the phase angle difference between the input voltage and the Back EMF.

θ is represented as:

$$\theta = \frac{-I_{x,rec} \times 2\pi f_s L_s}{BackEMF - I_{x,rec} R_s} \frac{360^\circ}{\pi} \tag{2.9}$$

Fig. 2.23 PMSG circuit model (reprinted from [8], Fig. 13)

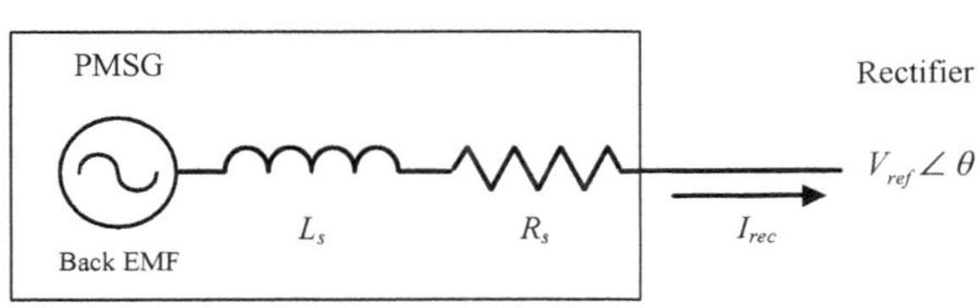

When an S_{x1} or S_{x4} open-switch fault occurs, the range of the zero current θ_{fault} is expressed as:

$$\theta_{fault} = \theta + \frac{\theta_n}{2} \tag{2.10}$$

For selecting R_{14}, the requirement can be expressed as:

$$\frac{\theta_n}{180^\circ}\frac{1}{2f_s} < R_{14}\frac{1}{2f_s} < \frac{\theta_{fault}}{180^\circ}\frac{1}{2f_s} \tag{2.11}$$

Consequently, the values of R_{23} and R_{14} are in the range of 0–1, and R_{23} should be larger than R_{14} because the $T_{Z,Ix}$ for S_{x2} and S_{x3} open-switch faults is longer than that for S_{x1} and S_{x4} open-switch faults. R_{14} should satisfy (2.11).

The location of an open-switch fault in a rectifier can be detected using (2.5) and (2.6). S_{x2} and S_{x3} open-switch faults satisfy both (2.5) and (2.6). However, S_{x1} and S_{x4} open-switch faults satisfy only (2.6). Therefore, an open-switch fault is detected using (2.6), Then (2.6) is used to determine the location of the open-switch fault. The proposed detection method is applied to three-phase currents.

- Detection principle for open-switch faults in NPC inverters

There are two current patterns caused by open-switch faults in an inverter with a unity power factor. These are the same as those in the cases of S_{x2} and S_{x3} open-switch faults in a rectifier. In the non-unity power factor case, although S_{x1} and S_{x2} open-switch faults create different current patterns, these faults generate the same pattern for at least one-quarter of a period of the current (from the starting point of the zero current), as shown in Fig. 2.24.

Therefore, (2.6) can be used to detect an open-switch faults in an inverter. The previous sign of the current and the duration of the zero range are determined in the same procedure of the rectifier, as shown in Fig. 2.22. The detection equation for open-switch faults of an inverter is expressed as:

$$\begin{gathered} D_{\mathrm{up,low}} = I_{up,low}T_{half} = I_{\mathrm{up,low}}\frac{1}{2\times 60} > T_{Z,Ix} \\ \begin{cases} if\ C_{pre,sign,Ix} = Positive, S_{x1} fault \text{ or } S_{x2} fault \\ if\ C_{pre,sign,Ix} = Negative, S_{x3} fault \text{ or } S_{x4} fault \end{cases} \end{gathered} \tag{2.12}$$

where $I_{up,low}$ is the detection parameter. A grid frequency of 60 Hz is used for f_s, and $I_{up,low}$ should be in the range of 0–0.5 due to the zero power factor.

The location of an open-switch fault can be classified into an upper-side (S_{x1} and S_{x2}) fault or a lower-side (S_{x3} and S_{x4}) fault using this detection method. When the open-switch fault detection method using the duration of the zero current is applied to a back-to-back converter using the NPC topology, a block diagram of the open-switch fault detection method for the back-to-back converter is shown in Fig. 2.25.

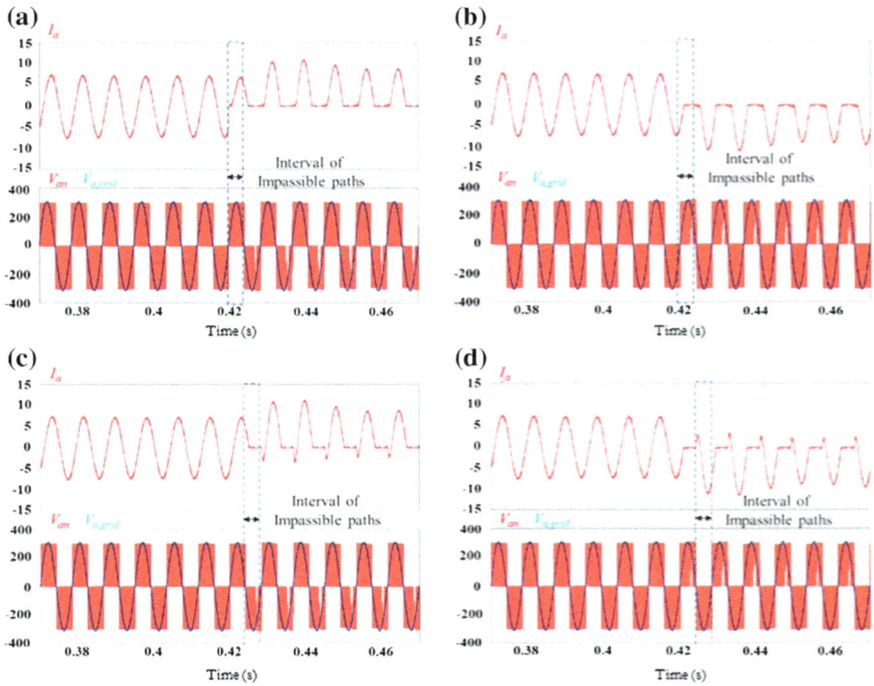

Fig. 2.24 Current distortion and a-phase pole voltage of open-switch faults for a power factor of 0.5: **a** S_{a2}, **b** S_{a1}, **c** S_{a3}, **d** S_{a4} (reprinted from [8], Fig. 8)

C. Experimental results of the switch open-circuit fault detection in NPC three-level rectifiers/inverters

The steps for switch fault detection in NPC three-level rectifiers/inverters is as follows:

(1) Calculation of the current sign ($C_{pre,sign}$) as shown in Fig. 2.22
(2) Calculation of the zero range using (2.5) and (2.6)
(3) If it exceeds only D_{14}, an S_{x1} or S_{x4} open-circuit fault occurs. Then the final position of the open-circuit fault is determined by $C_{pre,sign}$. If it exceeds D_{14} *and* D_{23}, an S_{x2} or S_{x3} open-circuit fault occurs. Then final position of the open-circuit fault is decided by $C_{pre,sign}$.

Figure 2.26 shows the experimental setup and the simulation and experimental parameters are shown in Table 2.6.

Figure 2.27 shows distorted currents caused by open-switch faults in a rectifier at 600 rpm. The negative current remains zero for some time after the zero crossing point when an S_{a1} open-switch fault occurs, as shown in Fig. 2.27a. An S_{a4} open-switch fault prevents the flow of positive a-phase current for some time after

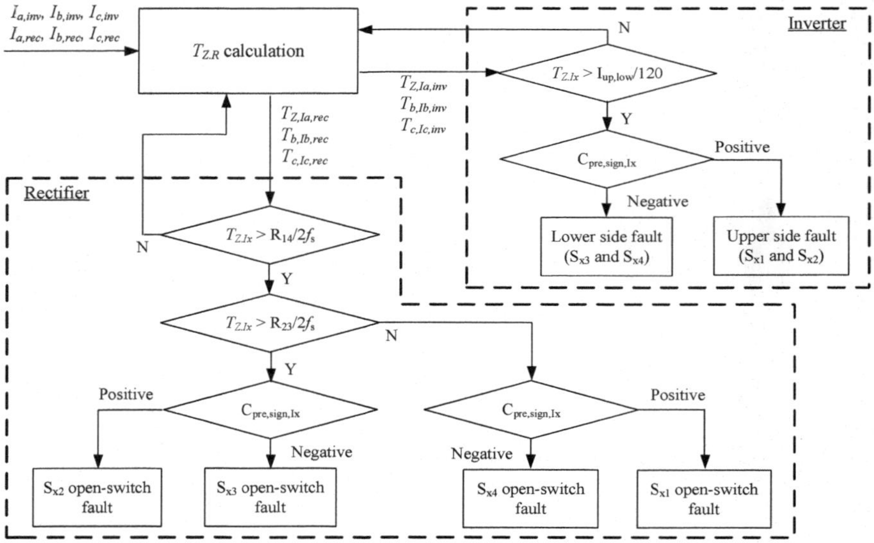

Fig. 2.25 Block diagram of the open-switch fault detection method in a back-to-back converter (reprinted from [8], Fig. 14)

Table 2.6 Simulation and experimental parameters

Generator			
Rated power	11 kW	Rated power	191 Vrms
pole	6	pole	1450 rpm
Stator resistance	0.099 Ω	Stator resistance	4.07/4.65 mH
System parameter			
DC-link voltage	600 V	DC-link voltage	380 Vrms
DC-link capacitance	1100 μF	DC-link capacitance	60 Hz
Switching frequency	10 kHz	Switching frequency	100 μs
Detection parameter			
R_{14}	0.075	$R_{23}/I_{up,low}$	0.3

the zero crossing point, as shown in Fig. 2.27b. The negative current in the a-phase becomes zero when an S_{a2} open-switch fault occurs, as shown in Fig. 2.27c. An S_{a3} open-switch fault prevents the flow of positive current in the a-phase, as shown in Fig. 2.27d. This characteristic of open-switch faults, which is verified by experimental results, is the same as that of the simulated results.

Figure 2.28 shows the performance of the open-switch fault detection method in a rectifier at 600 rpm. $T_{Z,Ix}$ is always smaller than D_{14} for normal operation. However, $T_{Z,Ix}$ increases when the current is zero because of an open-switch fault. Then, an open-switch fault is detected when $T_{Z,Ix}$ exceeds D_X (x = 14 and 23). The detection signal, which is indicated by the red line, represents the location of the

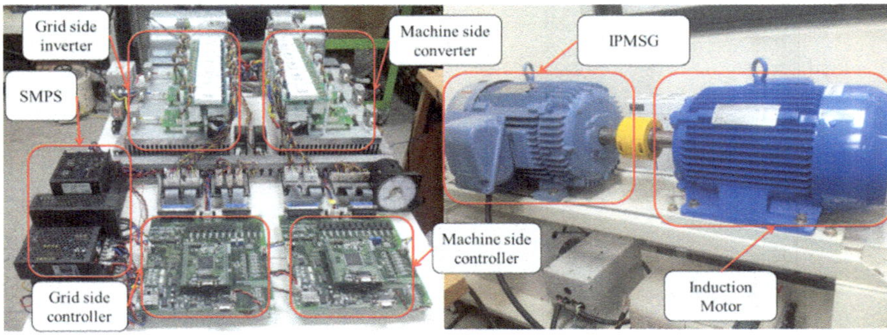

Fig. 2.26 Experimental setup of a back-to-back converter using a NPC three-level topology (reprinted from [8], Fig. 15)

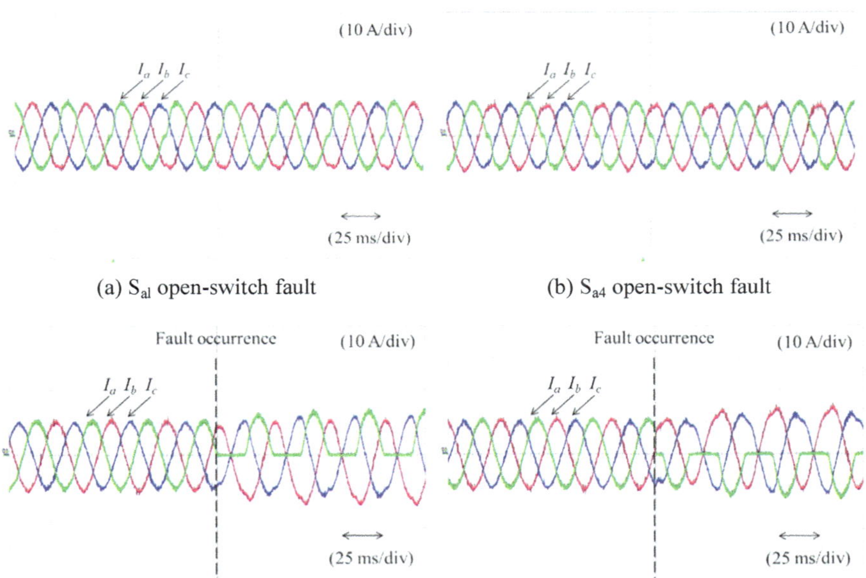

(a) S_{a1} open-switch fault

(b) S_{a4} open-switch fault

(c) S_{a2} open-switch fault

(d) S_{a3} open-switch fault

Fig. 2.27 Current distortion caused by open-switch faults in the rectifier of a PMSG at 600 rpm (reprinted from [8], Fig. 16)

open-switch fault. The open-switch fault detection method determines the location of the open-switch fault using $T_{Z,Ix}$ and the previous sign of the current, as shown in Fig. 2.28. Each fault detection signal is sent within one period of the current. Figure 2.29 shows the performance of the open-switch fault detection method in a rectifier at 1200 rpm. Because an angular frequency of 1200 rpm is twice that of

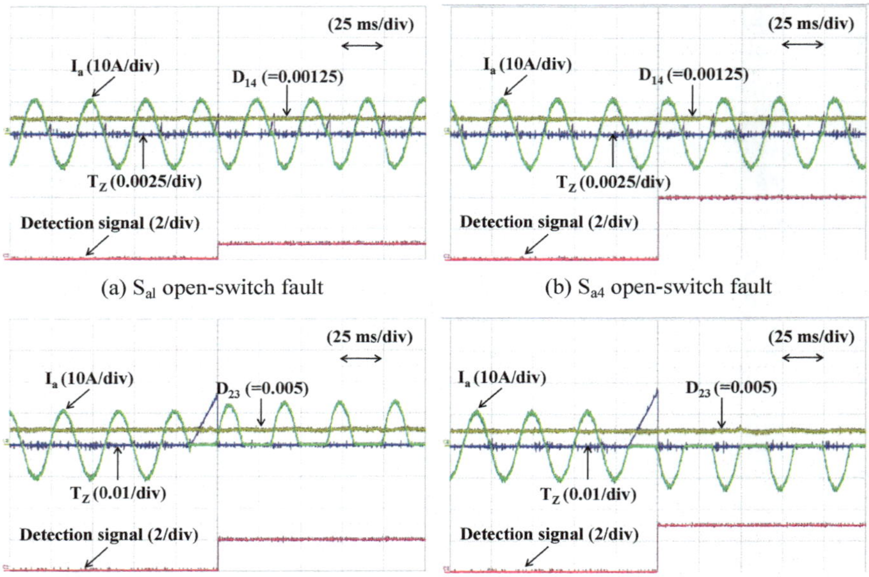

(a) S_{a1} open-switch fault (b) S_{a4} open-switch fault

(c) S_{a2} open-switch fault (d) S_{a3} open-switch fault

Fig. 2.28 Experimental results for the proposed open-switch fault detection in the rectifier of a PMSG at 600 rpm (reprinted from [8], Fig. 17)

600 rpm, D_{14} and D_{23}, which are calculated using (1) and (2), decrease in magnitude by half, and their values are shown in Fig. 2.29. Similar to the case in Fig. 2.28, $T_{Z,Ix}$ increases after an open-switch fault occurs, and the location of the open-switch fault is determined within one period of the current.

Figure 2.30 shows the performance of the open-switch fault detection method in an inverter at a unity power factor. In the inverter, S_{x1} and S_{x2} (S_{x3} and S_{x4}) open-switch faults lead to the same current pattern shown in Fig. 2.30. An open-switch fault is detected when $T_{Z,Ix}$ exceeds $D_{up,low}$. A value of 1 for the red line, which is the detection signal, indicates that there is one S_{x1} and S_{x2} open-switch fault. Meanwhile, a value of 2 for the red line indicates that there is one S_{x3} and S_{x4} open-switch fault. The open-switch fault detection method determines the location of the open-switch fault, as shown in Fig. 2.30. Each fault detection signal is sent within one period of the current.

Figure 2.31 shows the performance of the open-switch fault detection method in an inverter at a power factor of 0.5. In this case, S_{x1} and S_{x2} (S_{x3} and S_{x4}) open-switch faults lead to different current patterns. An S_{x2} (S_{x3}) open-switch fault leads to a zero current for a longer duration than an S_{x1} (S_{x4}) open-switch fault, as shown in Fig. 2.31. Although the power factor is not unity, the open-switch fault detection method accurately determines the location of the open-switch fault. Each fault detection signal is sent within one period of the current.

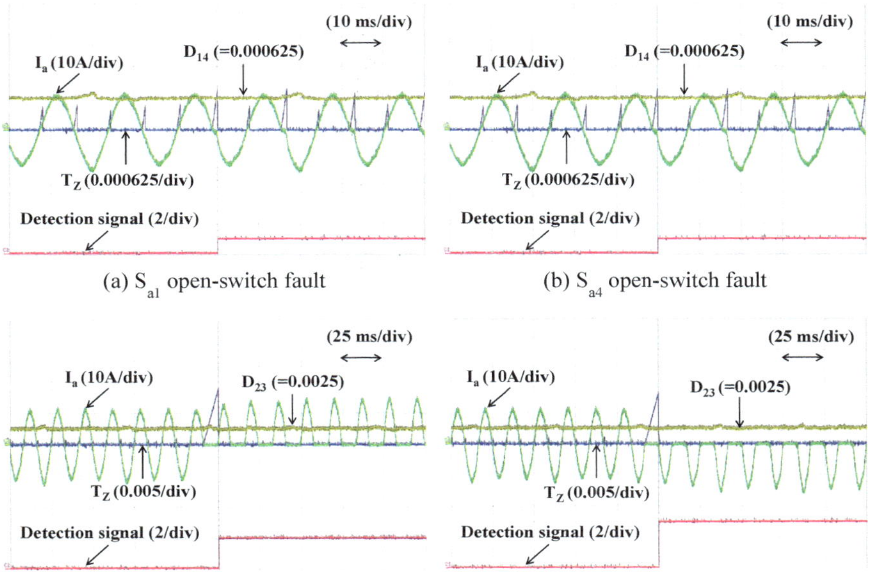

(a) S_{a1} open-switch fault (b) S_{a4} open-switch fault

(c) S_{a2} open-switch fault (d) S_{a3} open-switch fault

Fig. 2.29 Experimental results for the proposed open-switch fault detection in the rectifier of a PMSG at 1200 rpm (reprinted from [8], Fig. 18)

2.4 Switch Open-Circuit Fault Detection Method for T-Type Three-Level Inverter

There are three switching states for three-level topologies. The first switching state is named P, and this means that S_{x1} and S_{x2} are ON and that S_{x3} and S_{x4} are OFF. The next switching state is named O, and this means that S_{x2} and S_{x3} are ON and that S_{x1} and S_{x4} are OFF. The switching state N means that S_{x1} and S_{x2} are OFF and that S_{x3} and S_{x4} are ON. Depending on the current direction and switching state, there are six current paths in the T-type topology, as shown in Fig. 3.32.

2.4.1 Switch Open-Circuit Fault Detection Method Using Current Distortion

A. The effects on current depending on the position of an open-circuit fault

The current paths of inverter and rectifier operations are different. For inverter operation with a unity power factor, most of the positive current flows through the paths of Fig. 2.32b, c, and most of the negative current flows through the paths of

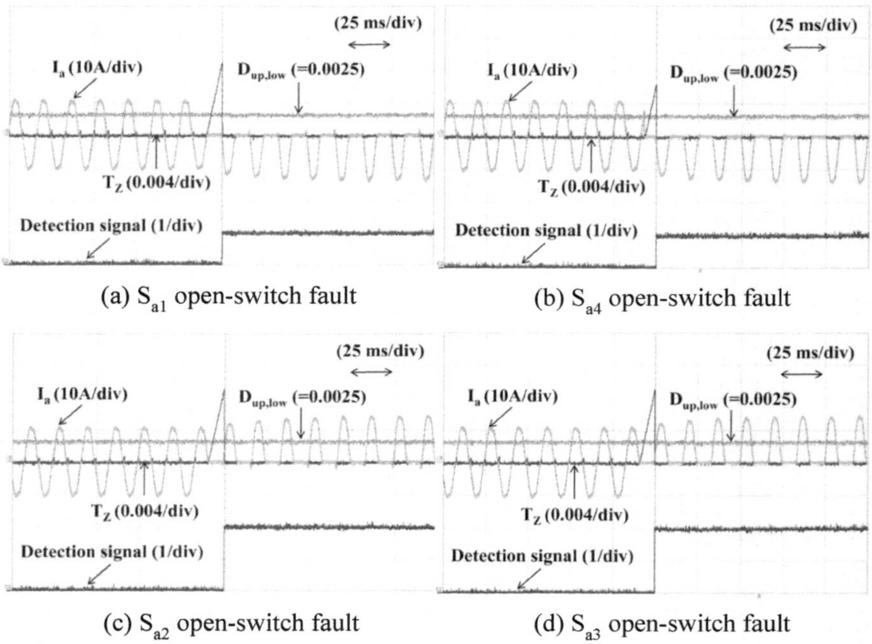

(a) S_{a1} open-switch fault (b) S_{a4} open-switch fault

(c) S_{a2} open-switch fault (d) S_{a3} open-switch fault

Fig. 2.30 Experimental results for the open-switch fault detection in an inverter at a unity power factor (reprinted from [8], Fig. 19)

Fig. 2.32d, e. Therefore, in a NPC inverter, the P-switching state of Fig. 2.32d is made infeasible by an S_{x1} or S_{x2} open-switch fault regardless of the position of the open-switch fault. This means that NPC inverters have a structural limitation to realizing tolerance control without additional devices. On the other hand, in rectifier operation with a unity power factor, most of the positive current flows through the paths of Fig. 2.32a, b, and most of the negative current flows through the paths of Fig. 2.32e, f. These paths do not include the current flowing through S_{x1} and S_{x4}.

- S_{x1} open-circuit faults

In T-type three-level inverters, S_{x1} open-circuit fault makes the P-switching state infeasible. This means that the current path of Fig. 2.32d cannot be generated. If R-L load is connected to the output of the inverter, an S_{x1} open-circuit fault leads to variations of the load voltage, and an undesired current flows through the O-switching state instead of the P-switching state. However, the grid voltage and back EMF of the motor drive do not generate any current due to the impossibility of the P-switching state.

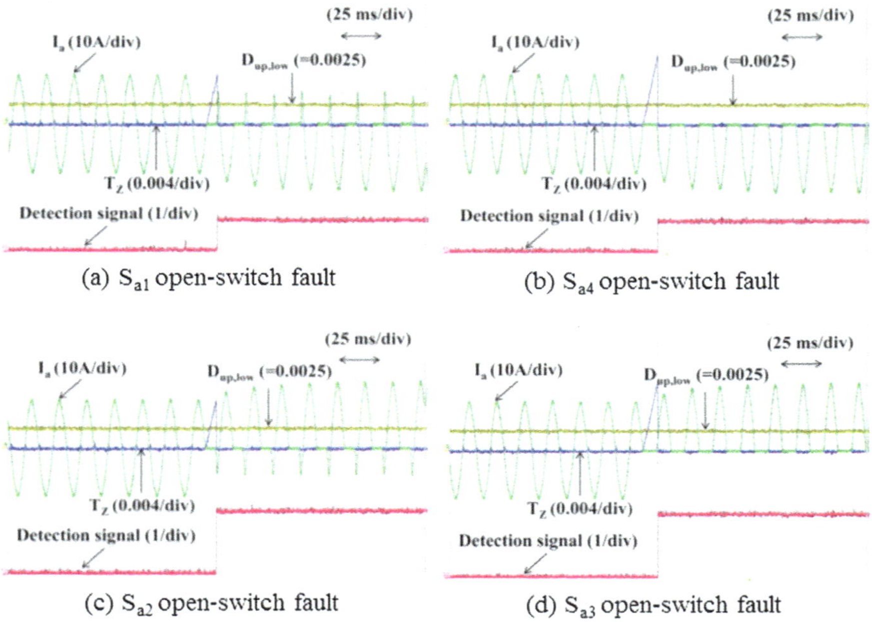

Fig. 2.31 Experimental results for the open-switch fault detection in an inverter at a power factor of 0.5 (reprinted from [8], Fig. 20)

- S_{x2} open-circuit faults

In T-type three-level inverters, an S_{x2} open-circuit fault influences the O-switching state. Therefore, the current path of Fig. 2.32e cannot be generated.

- S_{x3} open-circuit faults

In T-type three-level inverters, an S_{x3} open-circuit fault influences the O-switching state. This is related to the current path of Fig. 2.32b which cannot be generated.

- S_{x4} open-circuit fault

S_{x4} open-circuit faults lead to the impossibility of the N switching state. As a result, the current path of Fig. 2.32c cannot be generated. Similar to S_{x1} open-circuit faults, in the R-L load condition, current is generated by the feasible O-switching state. On the other hand, there is no current in grid-connected and motor drive applications when an S_{x4} open-circuit fault occurs regardless of the validity of the O-switching state.

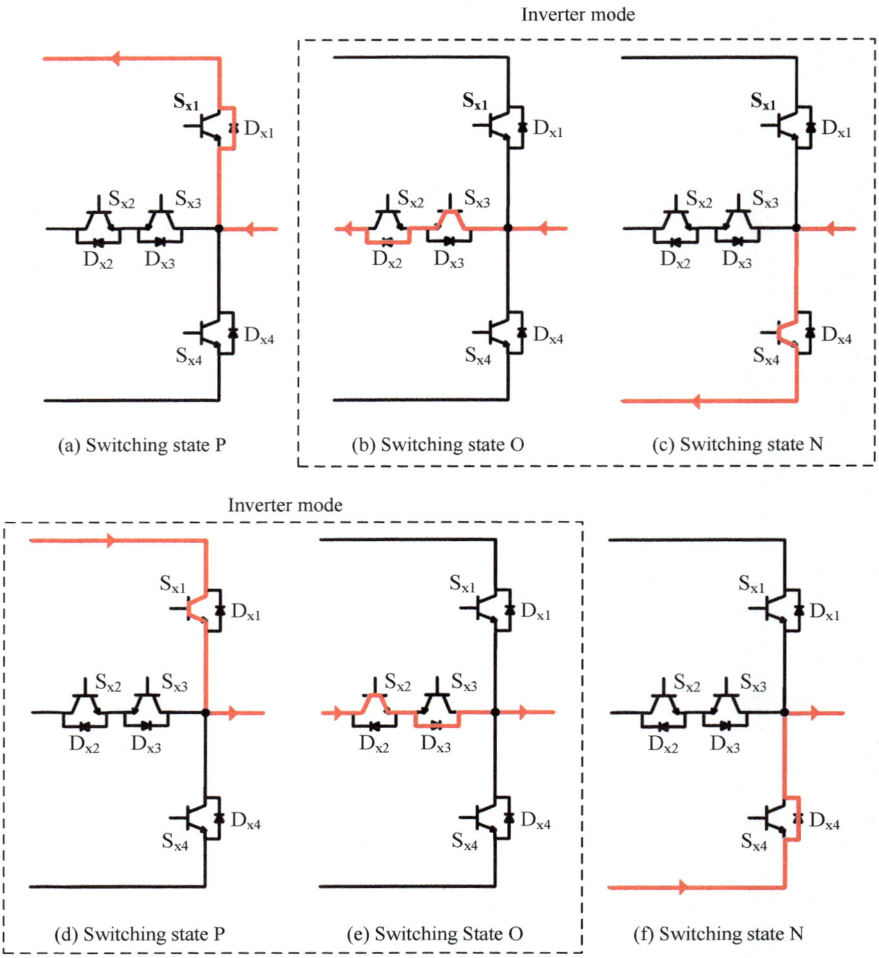

Fig. 2.32 Current paths depending on the current direction and operating state

The feasible and infeasible switching states of T-type three-level converter depending on the position of the open-switch fault are summarized in Table 2.7.

B. Switch open-circuit fault detection method for T-type converters [6]

- The classification of upper switch (S_{x1}, S_{x2}) and lower switch (S_{x3}, S_{x4}) open-circuit faults

Under normal conditions, the average of the phase current is zero. However, in faulty cases, the value of the phase current's average is changed. If an open-switch fault occurs in S_{x1} or S_{x2}, the average of the phase current is a negative value because the positive phase current is distorted as shown in Fig. 2.33. If an

Table 2.7 Feasible and infeasible switching states of a T-type three-level converter depending on the position of an open-switch fault

Position of open-switch fault	Feasible switching state	Infeasible switching state
S_{x1}	N (Fig. 2.32c, f) O (Fig. 2.32b, e) P (Fig. 2.32a)	P (Fig. 2.32d)
S_{x2}	N (Fig. 2.32c, f) O (Fig. 2.32b) P (Fig. 2.32a, d)	O (Fig. 2.32e)
S_{x3}	N (Fig. 2.32c, f) O (Fig. 2.32e) P (Fig. 2.32a, d)	O (Fig. 2.32b)
S_{x4}	N (Fig. 2.32f) O (Fig. 2.32b, e) P (Fig. 2.32a, d)	N (Fig. 2.32c)

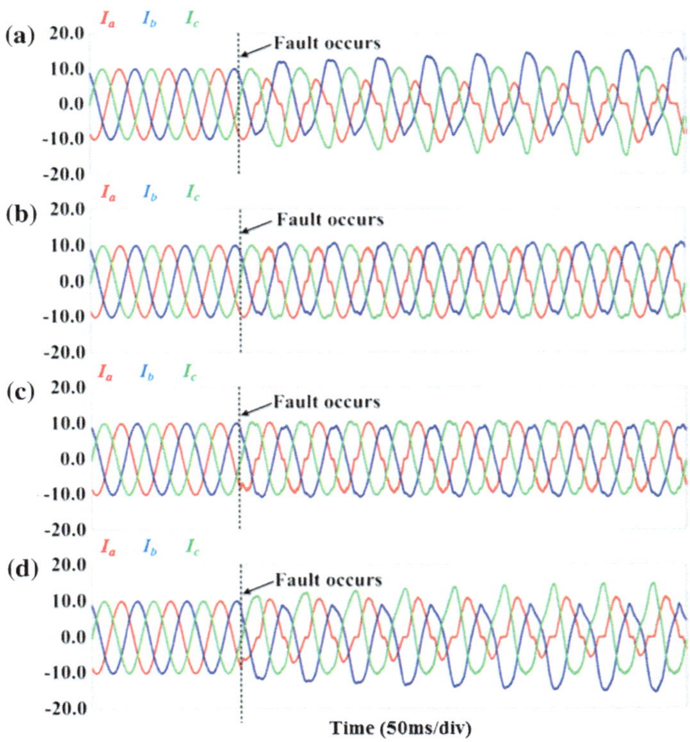

Fig. 2.33 Phase current in a T-type three-level inverter with RL load depending on the position of an open-circuit fault: **a** S_{x1}, **b** S_{x2}, **c** S_{x3}, **d** S_{x4} (reprinted from [6], Figs. 6–9)

open-switch fault occurs in S_{x3} or S_{x4}, the average of the phase current has a positive value because the negative phase current is distorted in the faulty leg.

The averages of the other phase currents are also changed due to distortions of the phase current in the faulty leg. However, if these values are used directly, they may lead to a false alarm when harmonic components exist in the phase current or when this system is notably under fast transients. To improve the accuracy of detection, the normalized current method is employed. The normalized current can be obtained by the following equation:

$$\begin{aligned} I_s &= \sqrt{I_{qs}^2 + I_{ds}^2} \\ I_{xN} &= \frac{I_x}{I_s} \end{aligned} \tag{2.13}$$

where $x = a, b, c$, and I_{ds} and I_{qs} are the reactive and active currents in the stationary reference frame, respectively.

The average of the phase current ($I_{xN,avg}$) for one-fundamental period can be expressed as:

$$I_{xN,avg} = \int_0^{2\pi} I_{xN} \tag{2.14}$$

Using these values, it is possible to identify the faulty leg and whether the open-switch faults occur in one of the two switches S_{x1} and S_{x2} or in one of the two switches S_{x3} and S_{x4}. To improve accuracy, the threshold value I_{thr} is set. It is summarized as:

$$\begin{aligned} I_{xN,avg} > I_{thr} &: \quad U_x = 1 \\ I_{xN,avg} < -I_{thr} &: \quad U_x = -1 \\ -I_{thr} < I_{xN,avg} < I_{thr} &: \quad U_x = 0 \end{aligned} \tag{2.15}$$

- An analysis of the characteristic difference between upper switch (S_{x1}, S_{x2}) and lower switch (S_{x3}, S_{x4}) open-circuit faults

As mentioned above, by using the average phase current ($I_{xN,avg}$), it is possible to divide faults into upper switch (S_{x1}, S_{x2}) and lower switch (S_{x3}, S_{x4}) open-circuit faults. However, the positions should be classified as S_{x1} and S_{x2} (S_{x3} and S_{x4}) in detail.

To do this, it is necessary to analyze T-type three-level converter operation in the case of a switch fault. The operation conditions can be divided into the R-L load condition and the voltage condition. The T-type three-level converter operates as an inverter, and the P-switching state, which is a valid switching state, leads to the current flow. In the case of the RL load condition, when the P-switching state is infeasible, current flows through S_{x2} as shown in Fig. 2.34. This circuit is the same as that of the O-switching state. Although the O-switching state is not a valid

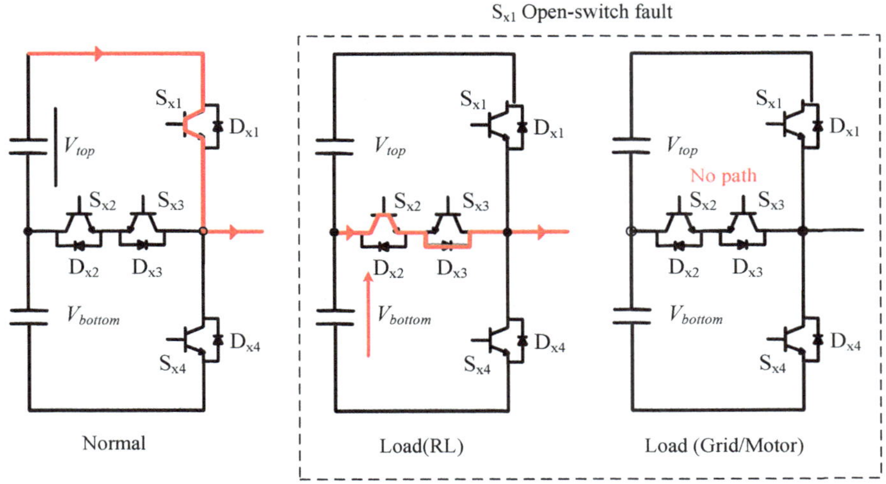

Fig. 2.34 Current path of a T-type three-level converter with an S_{x1} open-circuit fault during the P switching state

switching state, current flows because the current is determined by the output voltage in the R-L load condition. However, in the case of the grid-connected condition or the motor drive condition, since the current is affected by both the output voltage and the grid voltage or the back EMF of the motor, current does not flow if the P-switching state is infeasible as shown in Fig. 2.34.

In addition, two DC-link voltages are changed owing to the current variation of an S_{x1} open-circuit fault as shown in Fig. 2.34. The P-switching state transfers power from two DC-link capacitors to the output under normal operation. However, an S_{x1} open-circuit fault in the R-L load condition leads the current flow through the neutral-point and it flows from the bottom capacitor. This means that S_{x1} open-circuit faults cause different voltage variations in both of the DC-link capacitors.

S_{x2} open-circuit faults make the O-switching state infeasible which is not a valid switching state. Therefore, it generates current directly and just operates as a freewheeling path for the current. The current generated by a valid switching state flows continuously through D_{x4} when an S_{x2} open-circuit fault occurs. S_{x2} open-circuit faults makes the same current flow, as shown in Fig. 2.35, regardless of the condition of the load (R-L or voltage). DC-link voltage variation is also generated by S_{x2} open-circuit faults. The current flows through the neutral-point and it flows to the bottom capacitor which is the opposite situation when compared to S_{x1} open-circuit faults.

Using the change of the two capacitor voltages, a faulty switch can be identified between switch S_{x1} and switch S_{x2} or between switch S_{x3} and switch S_{x4}. If V_{dc1} is bigger than V_{dc2}, the open-switch fault occurs in switch S_{x1} or switch S_{x3}. Reversely, if V_{dc2} is larger than V_{dc1}, the open-switch fault occurs in switch S_{x2} or switch S_{x4}.

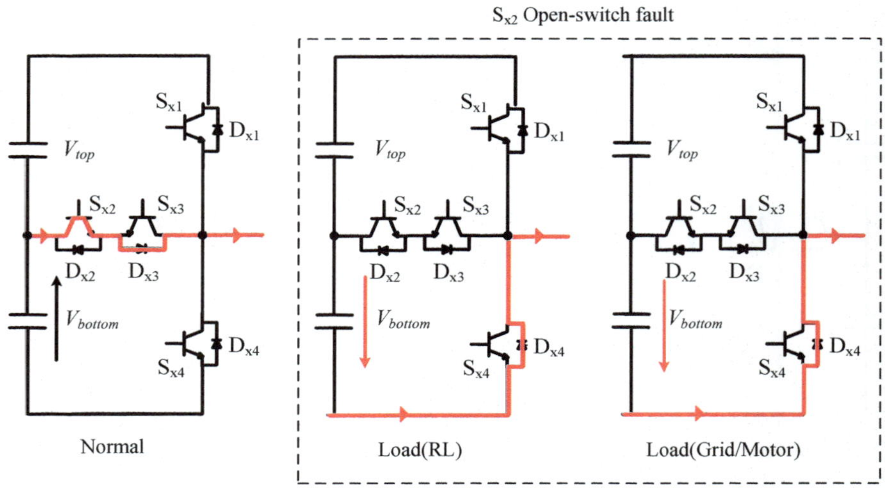

Fig. 2.35 Current path of a T-type three-level converter with an S_{x2} open-circuit fault during the O switching state

To improve accuracy, the threshold value V_{thr} is set. It is expressed as:

$$\begin{aligned} V_{top} - V_{bottom} > V_{thr}: &\quad V_x = 1 \\ V_{top} - V_{bottom} < -V_{thr}: &\quad V_x = -1 \\ -V_{thr} < V_{top} - V_{bottom} < V_{thr}: &\quad V_x = 0 \end{aligned} \tag{2.16}$$

C. Experimental results of switch open-circuit fault detection in T-type three-level inverters

The steps for the fault diagnosis of T-type three-level inverters is as follows.

(1) Calculate the average of the normalized phase currents.
(2) Using these values, identify the faulty leg and whether the open-switch faults occur in one of the two switches S_{x1} and S_{x2} or in one of the two switches S_{x3} and S_{x4}.
(3) Identify the location of the faulty switch between S_{x1} and S_{x2} or between S_{x3} and S_{x4} using the change of the two capacitor voltages. Values of the diagnostics variables are shown in Table 2.9.

Figure 2.36 shows experimental results of the proposed fault diagnosis method under the parameters of Table 2.8 when an open-switch fault occurs in switches S_{a1},

Table 2.8 Experimental parameters

DC-link voltage	200 V	Fundamental frequency	60 Hz
Control period	100 μs	Load	10 mH, 10 Ω
Switching period	10 kHz	I_{thr}, V_{thr}	0.08, 10 V

Table 2.9 Values of the diagnostics variables

Fault position	Diagnosis variables			
	U_a	U_b	U_c	V_x
S_{a1}	−1	+1	–	+1
S_{a2}	−1	+1	–	−1
S_{a3}	+1	−1	–	+1
S_{a4}	+1	−1	–	−1
S_{b1}	–	−1	+1	+1
S_{b2}	–	−1	+1	−1
S_{b3}	–	+1	−1	+1
S_{b4}	–	+1	−1	−1
S_{c1}	+1	–	−1	+1
S_{c2}	+1	–	−1	−1
S_{c3}	−1	–	+1	+1
S_{c4}	−1	–	+1	−1

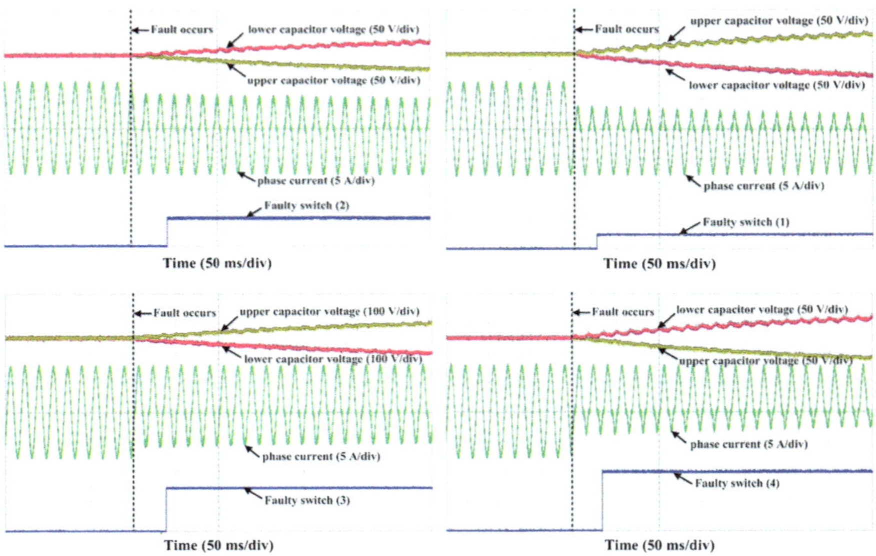

Fig. 2.36 Experiment results of the proposed fault diagnosis method (reprinted from [6], Figs. 26–29)

S_{a2}, S_{a3}, and S_{a4}, respectively. After the fault, the phase current is distorted, and the upper capacitor voltage becomes larger than the lower capacitor voltage as discussed in the previous analysis. The faulty switch is detected within 30 ms. In all of the cases, the faulty switch is identified accurately in about 50 ms. More time is needed to detect the fault switch when an open-switch fault occurs in neutral-point switches (S_{x2} or S_{x3}) than when an open-switch fault occurs in half-bridge switches (S_{x1} or S_{x4}). This is because the phase currents are less distorted in the neutral-switch fault case than those of the half-bridge switch fault case. This diagnosis time is closely related to the magnitude of the phase current, the size of the capacitor, and the threshold value. Experimental results show that the proposed fault diagnosis method precisely detects the faulty switch.

2.5 Switch Open-Circuit Fault Detection Method for T-Type Three-Level Rectifiers

2.5.1 Switch Open-Circuit Fault Detection Method Using Current Distortion

A. Analysis of T-type rectifiers in the presence of an open-switch fault.

Depending on the operating state and current direction, there are six paths as shown in Fig. 2.37. In the rectifier, almost all of the current is generated when the operating state is O, and the current continuously flows through a diode if the operating state is changed to P or N. This means that the infeasibility of the O-operating state leads to the zero current regardless of the P-operating state.

There are three paths when the current is positive. In the positive current part, most of the current of the rectifier which has a unity power factor, flows through paths (a) and (b). Therefore, the range of a-path (c) is small compared to ranges of paths (a) and (b). Hence, an S_{x3} open-switch fault in the O-operating state is fatal to rectifier operation and causes current distortion. On the other hand, an S_{x1} open-switch fault does not cause current distortion in the P-operating state because the current flows through a diode. In the N-operating state, the current flows through S_{x4}. An S_{x4} open-switch fault has little effect on the current when the rectifier is operated with a unity power factor. In addition, S_{x1} also does not affect the current in the negative current part.

B. Current distortion of T-type three-level rectifiers in the grid-connected condition [7]

Figure 2.38a shows the current, DC-link voltage, and pole voltage when an S_{a2} open-switch fault occurs. The S_{a2} open-switch fault has a fatal effect on the input current. In the negative current part, the valid O-operating state, which is shown in Fig. 2.37e, becomes impossible. Therefore, the whole negative current cannot be generated. Moreover, the DC-link voltage is largely changed. The negative current of Fig. 2.38a is caused by the decreased DC-link voltage. When the DC-link voltage is lower than the line-to-line peak voltage of the grid, the current can flow into the DC-link capacitor from the grid through the diodes of the rectifier.

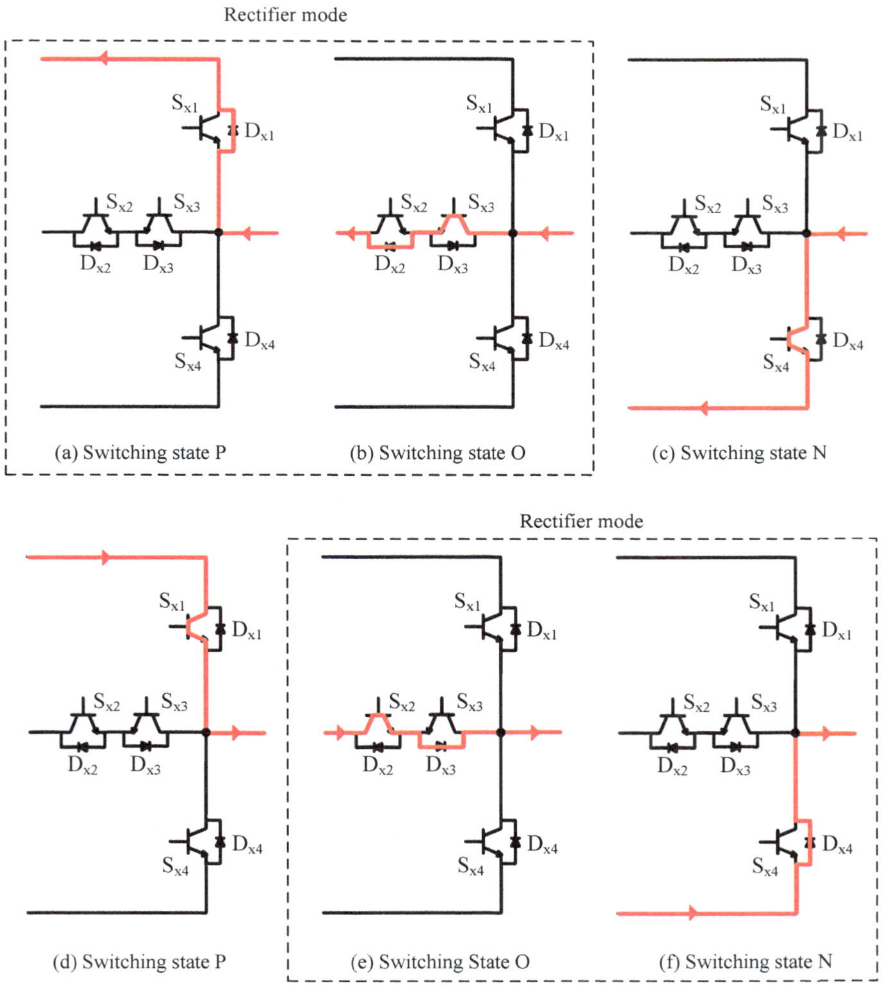

Fig. 2.37 Current paths depending on the current direction and operating state

Although the P-operating state is possible, current does not flow because P is not a valid operating state. The pole voltage of Fig. 2.38a does not have meaning in the negative current part. This is because, when the input voltage is a valid operating state O, the rectifier inner circuit can be an open circuit by an S_{a2} open-switch fault Fig. 2.40.

Figure 2.38b shows the current, DC-link voltage and pole voltage when an S_{a3} open-switch fault occurs. Like an S_{a2} open-switch fault, a valid operating state is impossible. Therefore, an S_{a3} open-switch fault makes the positive current zero for some parts and causes DC-link voltage ripple. Open-switch faults of S_{a2} and S_{a3} in

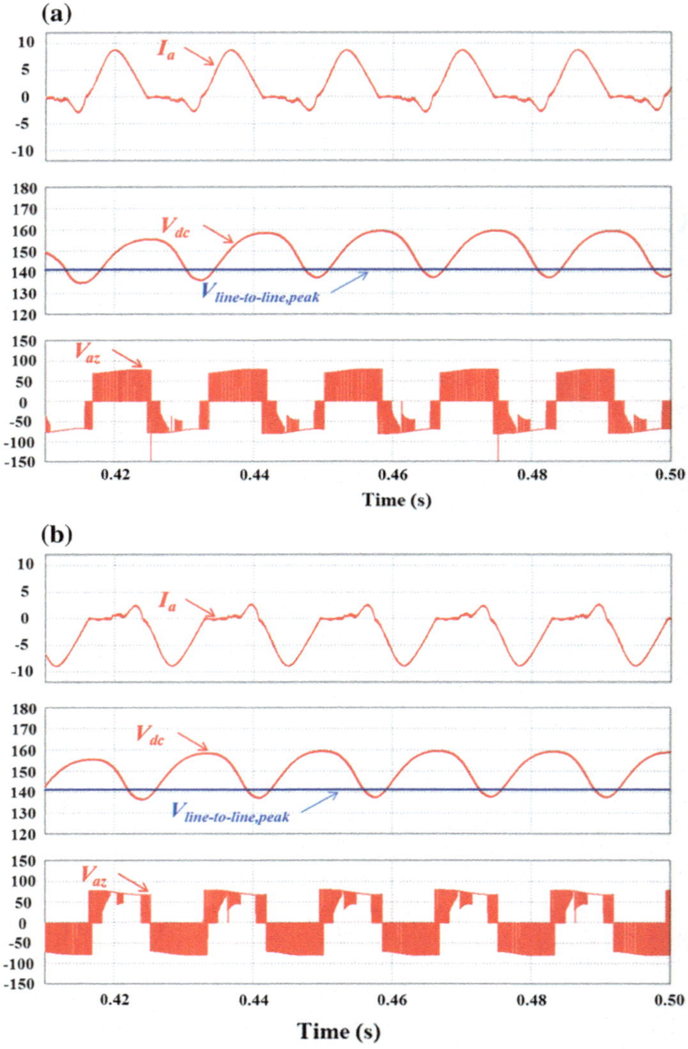

Fig. 2.38 Rectifier waveforms: I_a, V_{dc}, $V_{line\text{-}to\text{-}line,peak}$, and V_{az}: **a** S_{x2}, **b** S_{x3} (reprinted from [7], Figs. 5, 6)

the b-phase and c-phase legs cause the same current distortion as those of the a-phase leg.

To detect all of the open-switch faults without additional devices, current shapes should be considered. The input current angle is the same as the grid voltage angle with a unity power factor. Therefore, the grid voltage angle is used to detect open-switch faults.

By using the measured grid voltages, the d-axis and q-axis voltages are calculated as:

$$\begin{aligned} E_{ds} &= \frac{2E_a - E_b - E_c}{3} \\ E_{qs} &= \frac{E_b + E_c}{\sqrt{3}} \end{aligned} \tag{2.17}$$

By using the d-axis and q-axis voltages, the grid voltage angle is calculated as:

$$\theta = \tan^{-1}\left(\frac{E_{qs}}{E_{ds}}\right) \tag{2.18}$$

A calculated angle (θ) of (2.18) is more accurate than the angle from the grid current because the grid voltage always has a large value. Depending on an position of the open-switch fault, the angle value which makes the current zero is different. Figure 2.39 shows the angle range with zero current depending on the position of an open-switch fault. The zero current range is changed by the DC-link capacitor and load current. A large DC-link capacitor decreases the DC-link voltage change and keeps the DC-link voltage higher than the grid rectifier voltage for a long time. This means that the zero current range is expanded. On the other hand, a small DC-link capacitor reduces the zero current range. A large load current also reduces the zero current range because a large current flows of the load from the DC-link capacitor. However, the start angle value of the zero current range is always the same regardless of the DC-link capacitor value and load current. When an S_{a2} open-switch fault occurs, zero current is generated from the angle value $\pi/2$. In an S_{a3} open-switch fault, the current becomes zero from the angle value $3\pi/2$.

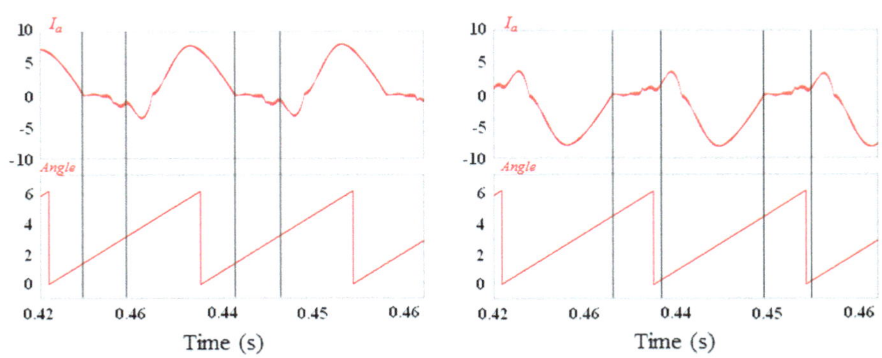

(a) S_{a2} open-switch fault

(b) S_{a3} open-switch fault

Fig. 2.39 Angle value with zero current according to the position of an open-switch fault in the a-phase leg (reprinted from [7], Fig. 7)

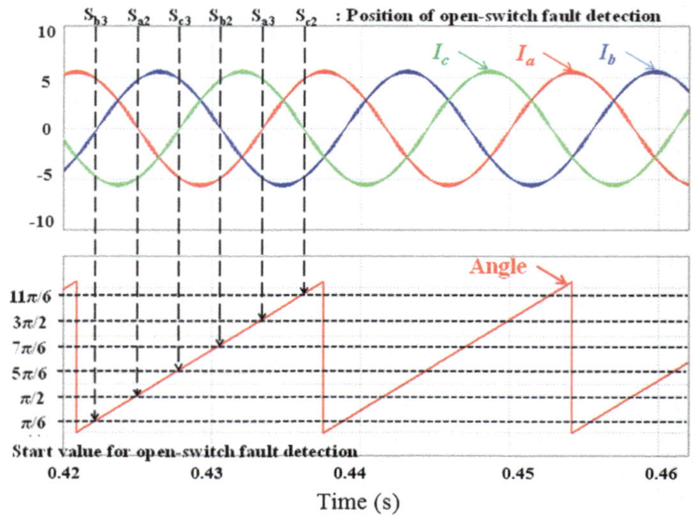

Fig. 2.40 Detection angle values according to the position of an open-switch fault in all of the phases (reprinted from [7], Fig. 8)

Through considerations of b-phase and c-phase current distortions, the start angle values which lead to zero current are defined as shown in Fig. 2.40. To detect the position of an open-switch fault, the angle value and differential equation for each of the currents are used, and the differential equation is defined as:

$$\Delta I = \frac{dI_x}{dt} = \frac{I_x[n] - I_x[n-1]}{T_{fault}}, \quad \begin{cases} x = a, & \theta = \pi/2 \sim 2\pi/3, 3\pi/2 \sim 5\pi/3 \\ x = b, & \theta = \pi/6 \sim \pi/2, 7\pi/6 \sim 4\pi/3 \\ x = c, & \theta = 5\pi/6 \sim \pi, 11\pi/6 \sim 2\pi \end{cases} \tag{2.19}$$

where T_{fault} is a period of the differential equation and I_x (x = a, b, c) is the measured current. The proposed detection method monitors open-switch faults for the π/6 range from each start angle value.

The derivative value of the current at the detection angle range becomes zero when an open-switch fault occurs because the slope of the current is steepest at the zero crossing point of the current. The derivative value of the current has its largest value at the zero crossing point of the normal current. Therefore, if an open-switch fault occurs, a dramatic change can be observed because the current and the derivative value of the current become zero at the zero crossing point of the current.

Consequently, the proposed open-switch detection method uses the current value and grid angle as shown in Table 2.10. Moreover, a block diagram of the summarized open-switch fault detection method is shown in Fig. 2.41. For the final decision determination of an open-switch fault, both of the fault alarms 1 and 2 have to be generated, which means that the input current has a smaller value than

Table 2.10 Open-switch fault detection table

Detection angle range	$\pi/6 \sim \pi/3$	$\pi/2 \sim 2\pi/3$	$5\pi/6 \sim \pi$	$7\pi/6 \sim 4\pi/3$	$3\pi/2 \sim 5\pi/3$	$11\pi/6 \sim 2\pi$
Detection boundary value	$\lvert I_b\rvert < A$, $\lvert dI_b/dt\rvert < B$	$\lvert I_a\rvert < A$, $\lvert dI_a/dt\rvert < B$	$\lvert I_c\rvert < A$, $\lvert dI_c/dt\rvert < B$	$\lvert I_b\rvert < A$, $\lvert dI_b/dt\rvert < B$	$\lvert I_a\rvert < A$, $\lvert dI_a/dt\rvert < B$	$\lvert I_c\rvert < A$, $\lvert dI_c/dt\rvert < B$
Detection switch	S_{b3}	S_{a2}	S_{c3}	S_{b2}	S_{a3}	S_{c2}

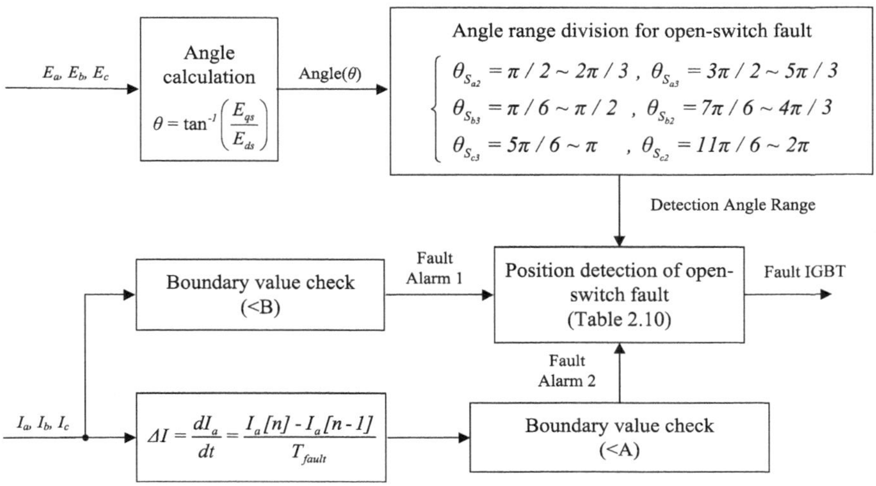

Fig. 2.41 Block diagram of the proposed detection method for a T-type rectifier (reprinted from [7], Fig. 9)

the boundary value (0.2), and the derivative value of the current is smaller than the boundary value (0.5).

In the discussed open-switch fault detection method, the boundary values of the current and the differential equation of the current are important factors in the performance of the open-switch fault detection. Ideally, an open-switch fault causes the result of (2.19) to be 0 for some part of the current. Therefore, an open-switch fault can be detected when the result of (2.19) is 0. In such a case, the boundary values can be chosen as 0. However, boundary values should be used, excluding 0, because currents contain ripple and noise components. Selecting boundary values

Table 2.11 Simulation and experimental results

DC-link voltage	150 V	Grid line-to-line voltage	150 Vrms
DC-link capacitor	550 μF	Grid frequency	60 Hz
Switching frequency	10 kHz	Load	33.3 Ω
Control period	100 us	A, B	0.5, 0.2

for currents should consider the current ripple, and this value should be larger than the current ripple. However, a very large boundary.

C. Experimental results of switch open-circuit fault detection in T-type three-level rectifiers

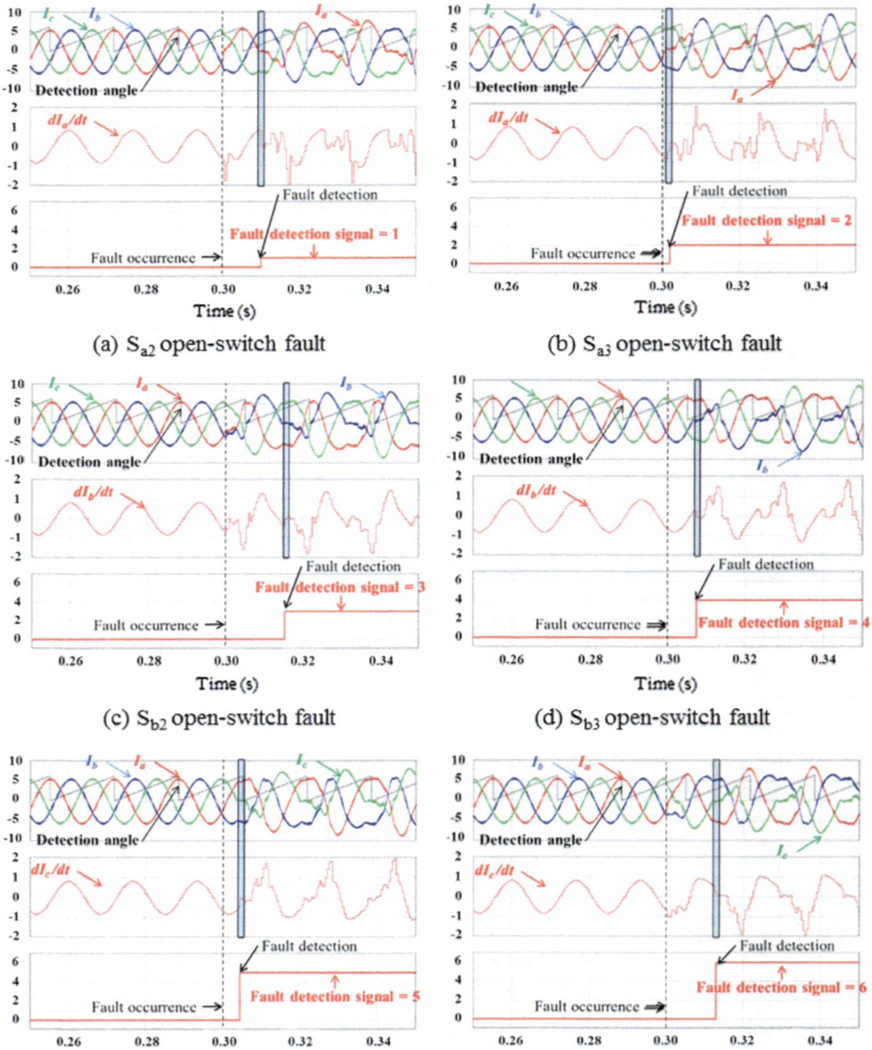

(a) S_{a2} open-switch fault
(b) S_{a3} open-switch fault
(c) S_{b2} open-switch fault
(d) S_{b3} open-switch fault
(e) S_{c2} open-switch fault
(f) S_{c3} open-switch fault

Fig. 2.42 Currents and fault detection signal depending on the open-switch fault position (reprinted from [7], Fig. 16)

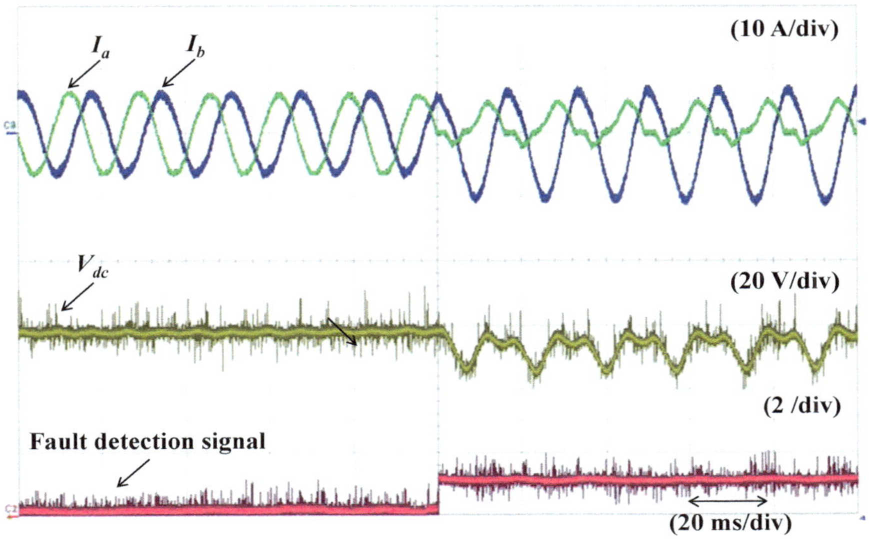

(a) S_{a2} open-switch fault

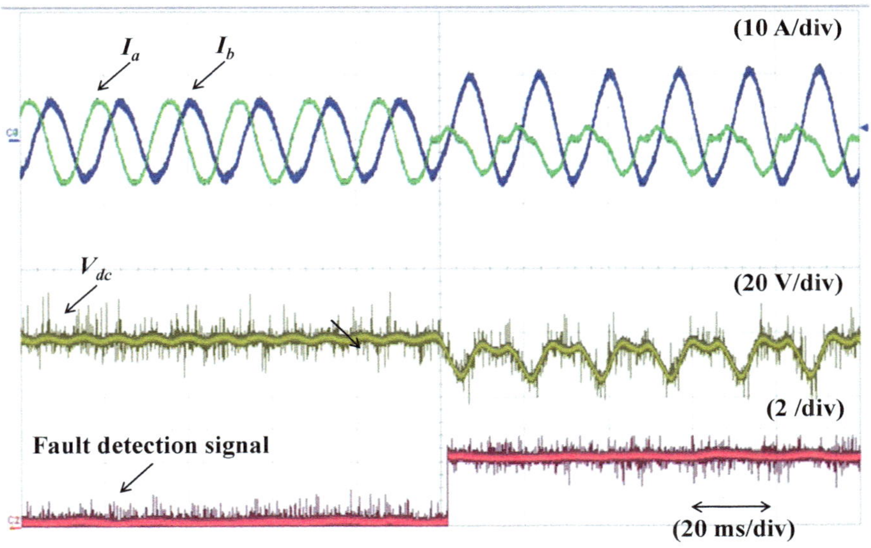

(b) S_{a3} open-switch fault

Fig. 2.43 Experimental results of open-switch fault detection (reprinted from [7], Fig. 21)

The steps for the fault diagnosis of T-type three-level rectifiers

(1) Calculate the grid angle using (2.18).
(2) During 60° from the six points shown in Fig. 2.40, calculate (2.19).
(3) Determine the position of the switch open-circuit fault with Table 2.10.

The simulation and experiment was implemented under the parameter of Table 2.11.

Figure 2.42 shows the input currents, fault detection angle, fault detection signal, and derivation of the current flowing through a leg that contains a fault switch. The open-switch fault occurs at 0.4 s, and the fault detection signal has a value of 1, 2, 3, 4, 5, and 6 depending on the position of the open-switch fault. The fault detection signal is generated by the rule of Table 2.10. In Fig. 2.42a, the fault detection signal becomes 1 when the a-phase current and the derivation of this current go into the boundary value at the detection angle $\pi/2$ of Table 2.10. The figures shown in Fig. 2.42b–f, also show the open-switch fault detection process. An open-switch fault of all the switches is detected in a period of the grid frequency (60 Hz) regardless of the position of the open-switch fault.

Figure 2.43a, b show the input currents (I_a and I_b), DC-link voltage, and fault detection signal when S_{a2} and S_{a3} open-switch faults occur. The a-phase input current I_a becomes zero at some part of the negative current due to the S_{a2} open-switch fault in Fig. 2.43. When an S_{a3} open-switch fault occurs, the a-phase input current I_a in some part of the positive current becomes zero. The DC-link voltage has a ripple, as shown in Fig. 2.43a, b, and this value is 16 V. The open-switch fault detection method identifies the open-switch fault and the position of the fault switch. The fault detection signal becomes 1, which means that the open-switch fault position is S_{a2}. The fault detection signal becomes 2 after an S_{a3} open-switch fault occurs as shown in Fig. 2.43b. These results are the same as those obtained in simulations.

References

1. Ribeiro, R.L.A., C.B. Jacobina, E.R.C. Silva, and A.M.N. Lima. 2003. Fault detection of open-switch damage in voltage-fed PWM motor drive systems. *IEEE Transaction on Power Electronics* 18 (2): 587–593.
2. Ribeiro, R.L.A., C.B. Jacobina, E.R.C. Silva, and A.M.N. Lima. 2004. Fault-tolerant voltage-fed PWM inverter AC motor drive systems. *IEEE Transaction on Industrial Electronics* 51 (2): 439–446.
3. Kim, T.J., W.C. Lee, and D.S. Hyun. 2009. Detection method for open-circuit fault in neutral-point-clamped inverter systems. *IEEE Transaction on Industrial Electronics* 56 (7): 2754–2763.
4. Choi, U.M., H.G. Jeong, K.B. Lee, and F. Blaabjerg. 2012. Method for detecting an open-switch fault in a grid-connected npc inverter system. *IEEE Transaction on Power Electronics* 27 (6): 2726–2739.

5. Ko, Y.J., K.B. Lee, D.C. Lee, and J.M. Kim. 2012. Fault diagnosis of three-parallel voltage-source converter for a high-power wind turbine. *IET Power Electronics* 5 (7): 1058–1067.
6. Choi, U.M., K.B. Lee, and F. Blaabjerg. 2014. Diagnosis and tolerant strategy of an open-switch fault for T-type three-level inverter systems. *IEEE Transactions on Industry Applications* 50 (1): 495–508.
7. Lee, J.S., and K.B. Lee. 2014. An open-switch fault detection method and tolerance controls based on SVM in a grid-connected T-type rectifier with unity power factor. *IEEE Transactions on Industrial Electronics* 61 (12): 7092–7104.
8. Lee, J.S., K.B. Lee., and F. Blaabjerg. 2015. Open-switch fault detection method of a back-to-back converter using NPC topology for wind turbine systems. *IEEE Transaction on Industrial Electronics* 51 (1): 325–335.
9. Choi, U.M., F. Blaabjerg, and K.B. Lee. 2015. Reliability improvement of a T-type three-level inverter with fault-tolerant control strategy. *IEEE Transaction on Power Electronics* 30 (5): 2660–2673.
10. Choi, U.M., F. Blaabjerg, and K.B. Lee. 2015. Study and handling methods of power IGBT module failures in power electronic converter systems. *IEEE Transaction on Power Electronics* 30 (5): 2517–2533.
11. Choi, U.M., J.S. Lee, F. Blaabjerg, and K.B. Lee. 2016. Open-circuit fault diagnosis and fault-tolerant control for a grid-connected NPC inverter. *IEEE Transactions on Power Electronics* 31 (10): 7234–7247.
12. Lee, J.S., and K.B. Lee. 2015. Open-switch fault tolerance control for a three-level NPC/T-type rectifier in wind turbine systems. *IEEE Transactions on Industrial Electronics* 62 (2): 1012–1021.

Chapter 3
Three-Level Converter Tolerant Control for Switch Faults

In power conversion systems, tolerant control can be defined as control for performing fundamental functions even if a fault occurs. Performance mitigation can be shown by tolerant control. However, the basic function of transferring power should be achieved. Managers and operators can prepare for secondary problems by using tolerant control. For example, blade accidents caused by the emergency stop of a WTS can be avoided.

Tolerant control for three-level converters can be divided into two cases which are those equipped with or without additional devices. Tolerant controls with additional switches can provide a solution for short-circuit faults or both short and open-circuit faults. Modified topologies adding an extra leg shown in Fig. 3.1 are typical among tolerant controls with additional switches, In these topologies, the extra leg operates instead of a leg containing a fault switch. However, adding switches or devices is a major drawback. It results in increases in both size and price, and requires additional circuits such as gate drivers and sensors.

On the other hand, tolerant controls for open-circuit faults of a switch can be realized without additional devices and they can ensure normal operation without the use of the fault switch.

The tolerant control for S_{x2} and S_{x3} open-circuit faults explained in chapter 3 can be applied to three-level converter regardless of the operation mode (rectifier mode or inverter mode). This is due to the fact that the infeasible vectors caused by switch open-circuit faults are the same in both operation modes.

In Sect. 3.1, tolerant controls with additional devices will be explained briefly. In Sect. 3.2, several tolerant controls, which do not use the additional devices, will be explained in detail.

K.-B. Lee and J.-S. Lee, *Reliability Improvement Technology for Power Converters*, Power Systems, DOI 10.1007/978-981-10-4992-7_3

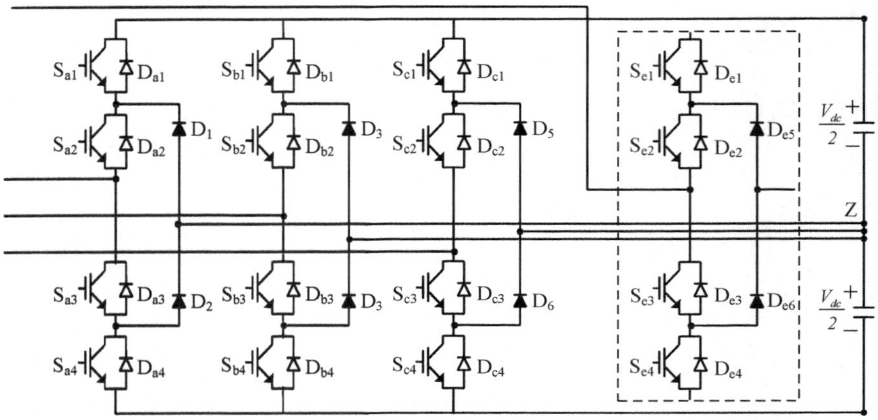

Fig. 3.1 Basic tolerant control topology with an addition leg

3.1 Three-Level Converter Tolerant Control in Topology Modifications

3.1.1 Tolerant Control with Six Thyristors [1]

Figure 3.2 shows a modified three-level converter topology with six thyristors. Two thyristors in parallel are located between the neutral-point and output point on each leg. The tolerant control can be implemented simply. If an open-circuit fault or short-circuit fault occurs, all of the IGBTs of the leg containing the fault switch do not operate and the two thyristors of the leg become ON. This means that the output of the leg only has the O-switching state. Since the P- or N-switching states cannot be generated in the tolerant control of the modified topology, the feasible output voltage range of the NPC three-level converter is reduced and limited to half. The limited output voltage range is a critical disadvantage in the tolerant control. If this topology is used in a grid-connected system, this topology cannot provide a solution for switch faults depending on the operating conditions although six thyristors are added to conventional NPC three-level topology.

3.1.2 Tolerant Control with an Additional Leg and Thyristors [2]

Figure 3.3 shows a modified three-level converter. This topology adds an additional leg to the conventional NPC three-level converter and six thyristors are added to

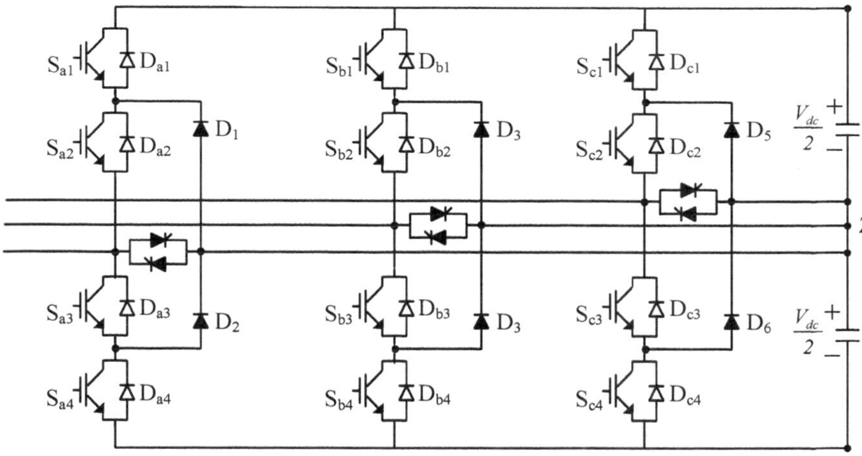

Fig. 3.2 The modified three-level converter topology with six thyristors (drawn from Ref. [1], Fig. 8)

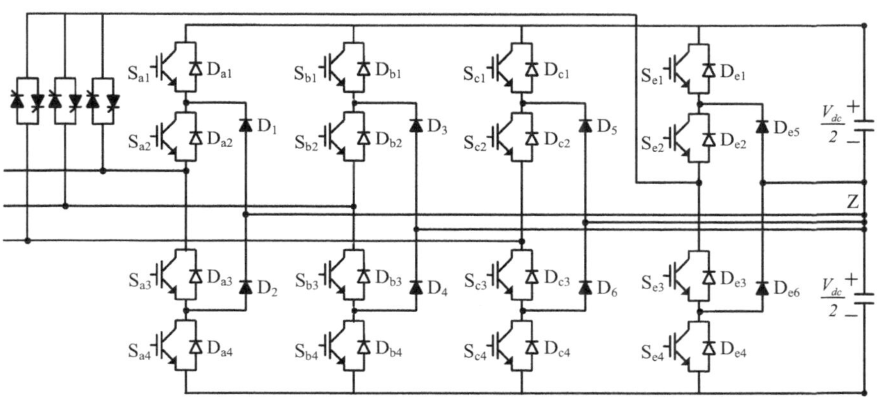

Fig. 3.3 The modified three-level converter topology with an additional leg and thyristors (drawn from Ref. [2], Fig. 1)

help the operation of the additional leg. The tolerant control of this topology is also simple. If an open-circuit fault or short-circuit fault occurs all of the IGBTs of the leg containing the fault switch do not operate. The two thyristors connected between the leg and the additional leg become ON and the additional leg operates instead of the leg containing the fault switch. Unlike the tolerant control shown in Fig. 3.2, this topology makes the whole output voltage range, which is the main benefit of the topology with an additional leg. However, several switching devices are used for configuring the additional leg and a lot of thyristors are used in this topology. Therefore, the large number of switching devices required in this topology are its main drawback.

3.1.3 Tolerant Control with Six Additional Switches [3]

Figure 3.4 shows a modified NPC three-level converter with six additional switches. Compared to the previous topologies, this topology only takes into account the short-circuit faults. In the tolerant control of this topology, the additional switches operate to prevent the occurrence of short circuits through DC-link and S_{x1}–S_{x3} (or S_{x2}–S_{x4}). If an S_{x1} or S_{x2} short-circuit fault occurs, S_{x6} turns OFF. If and S_{x3} or S_{x4} short-circuit fault occurs, S_{x5} turns OFF. By this tolerant control, the leg containing the fault switch is completely disconnected from the neutral-point. Therefore, the O-switching state cannot be generated and only the P- or N-switching state is used to make the output voltage. This tolerant control can operate like the normal operation. However, the output voltage/current THD is increased and the efficiency is decreased because of the absence of the O-switching state in the output voltage of the leg containing the fault switch.

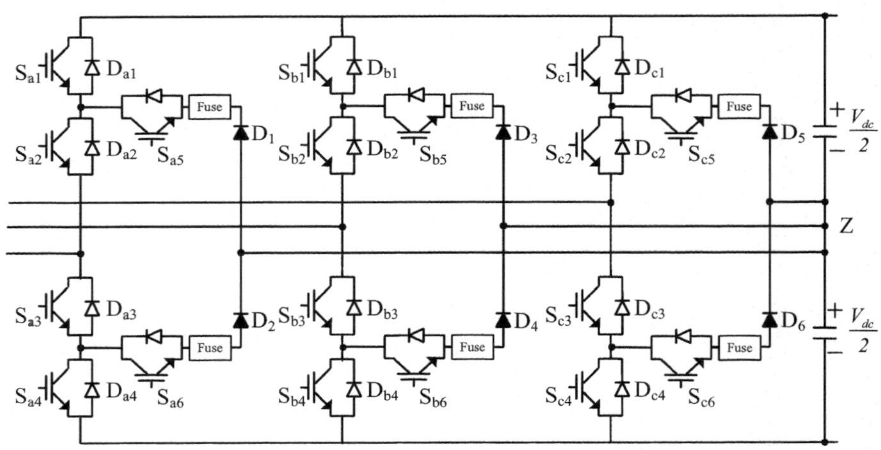

Fig. 3.4 The modified three-level converter topology with an additional six switches (drawn from Ref. [3], Fig. 5)

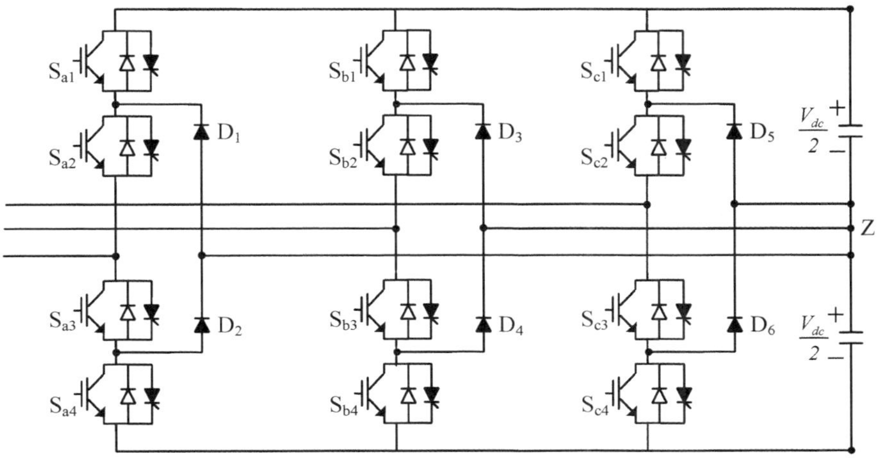

Fig. 3.5 Topology with twelve thyristors (drawn from Ref. [4], Fig. 4)

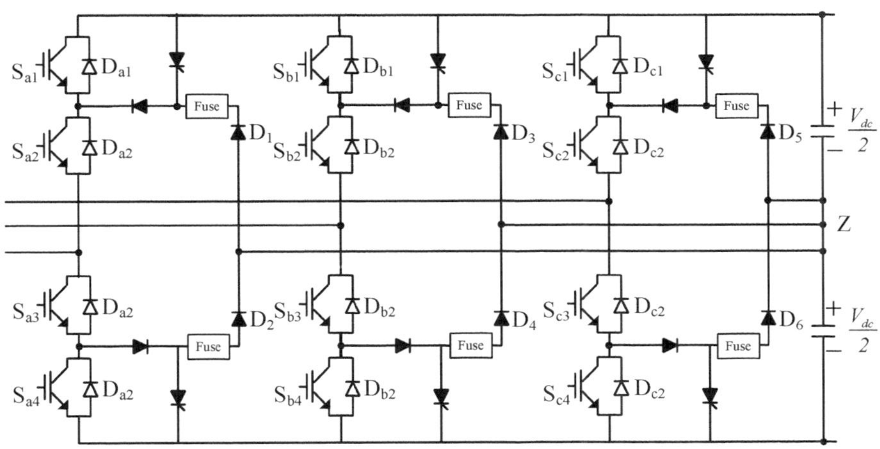

Fig. 3.6 Topology using additional diodes and thyristors (drawn from Ref. [3], Fig. 3)

3.1.4 Other Tolerant Control Topologies

Various topologies for tolerant control operation are shown in Figs. 3.5, 3.6, and 3.7 [3]–[5]. Like the previous topologies, the fault switch is isolated and additional devices operate to maintain normal operation.

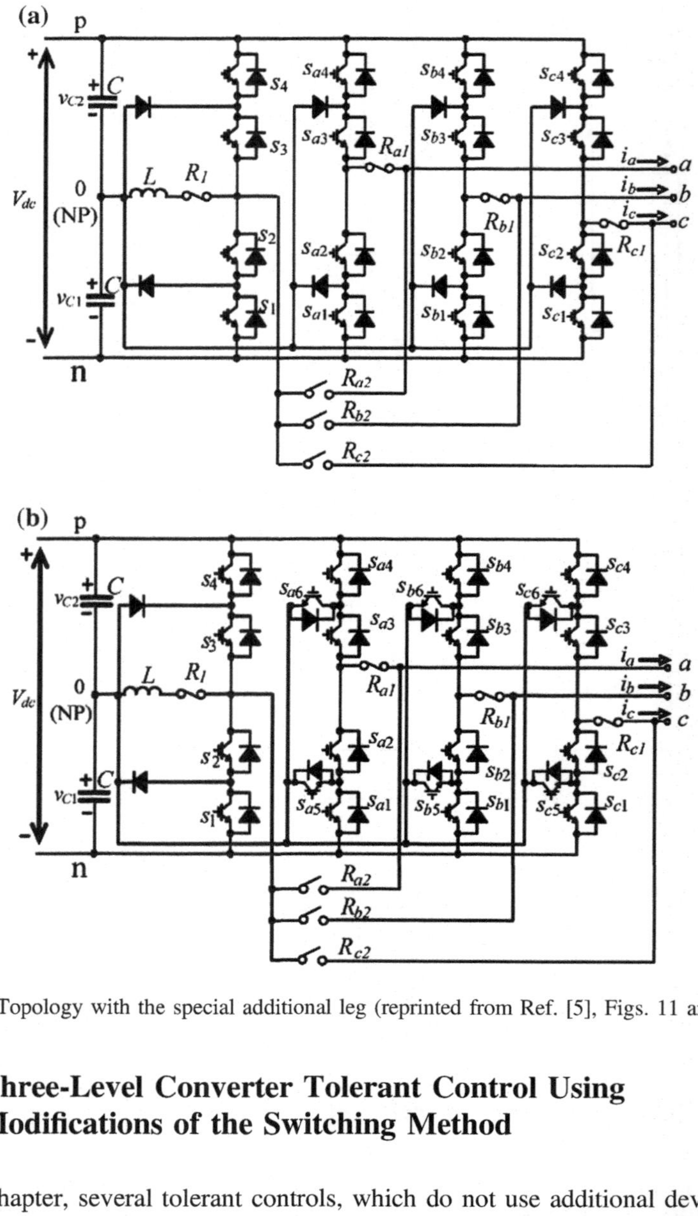

Fig. 3.7 Topology with the special additional leg (reprinted from Ref. [5], Figs. 11 and 12)

3.2 Three-Level Converter Tolerant Control Using Modifications of the Switching Method

In this chapter, several tolerant controls, which do not use additional devices, are introduced. Tolerant control should operate closely with a switch fault detection algorithm. Under the assumption that the switch fault detection algorithms introduced in previous chapter is used with the tolerant control, this chapter focuses on tolerant control for the open-switch faults.

3.2.1 Open-Circuit Faults in NPC Three-Level and T-Type Three-Level Converters

NPC three-level converters are vastly different from T-type three-level converters in terms of the structure. Depending on the position of an open-switch fault, the infeasible switching states are determined, and are shown in Table 3.1 [6, 7].

In both topologies, the infeasible switching state cases due to S_{x1} and S_{x4} open-circuit faults are the same. This means that the tolerant control for S_{x1} and S_{x4} open-circuit faults can be applied in both topologies. However, S_{x2} and S_{x3} open-circuit faults show different characteristic in both topologies. NPC three-level converters make the O- and P- (or N-) switching states infeasible. However, only O-switching state is infeasible in T-type three-level converters. This makes a big difference in developing tolerance control without additional devices. When the O- and P- (or N-) switching states are infeasible, like in the case of NPC three-level converters, the leg containing the open-circuit fault cannot make positive output voltage. This means that the tolerant control definitely needs additional devices. However, in the case of T-type three-level converters, S_{x2} and S_{x3} open-circuit faults just effect that O-switching state. Therefore, in the tolerant control, the output voltage is configured without the use of the O-switching state.

Consequently, some tolerant controls for three-level converters can be applied to both topologies. However, due to structural differences, unique tolerant control for only one topology can exist.

Table 3.1 Feasible and infeasible switching state of three-level converter depending on position of open-switch fault

Position of open-switch fault	NPC three-level converter		T-type three-level converter	
	Feasible switching state	Infeasible switching state	Feasible switching state	Infeasible switching state
S_{x1}	N (Fig. 2.9c, f) O (Fig. 2.9b, e) P (Fig. 2.9a)	P (Fig. 2.9d)	N (Fig. 2.32c, f) O (Fig. 2.32b, e) P (Fig. 2.32a)	P (Fig. 2.32d)
S_{x2}	N (Fig. 2.9c, f) O (Fig. 2.9b) P (Fig. 2.9a)	P (Fig. 2.9d) O (Fig. 2.9e)	N (Fig. 2.32c, f) O (Fig. 2.32b) P (Fig. 2.32a, d)	O (Fig. 2.32e)
S_{x3}	N (Fig. 2.9f) O (Fig. 2.9e) P (Fig. 2.9a, d)	N (Fig. 2.9c) O (Fig. 2.9b)	N (Fig. 2.32c, f) O (Fig. 2.32e) P (Fig. 2.32a, d)	O (Fig. 2.32b)
S_{x4}	N (Fig. 2.9f) O (Fig. 2.9b, e) P (Fig. 2.9a, d)	N (Fig. 2.9c)	N (Fig. 2.32f) O (Fig. 2.32b, e) P (Fig. 2.32a, d)	N (Fig. 2.32c)

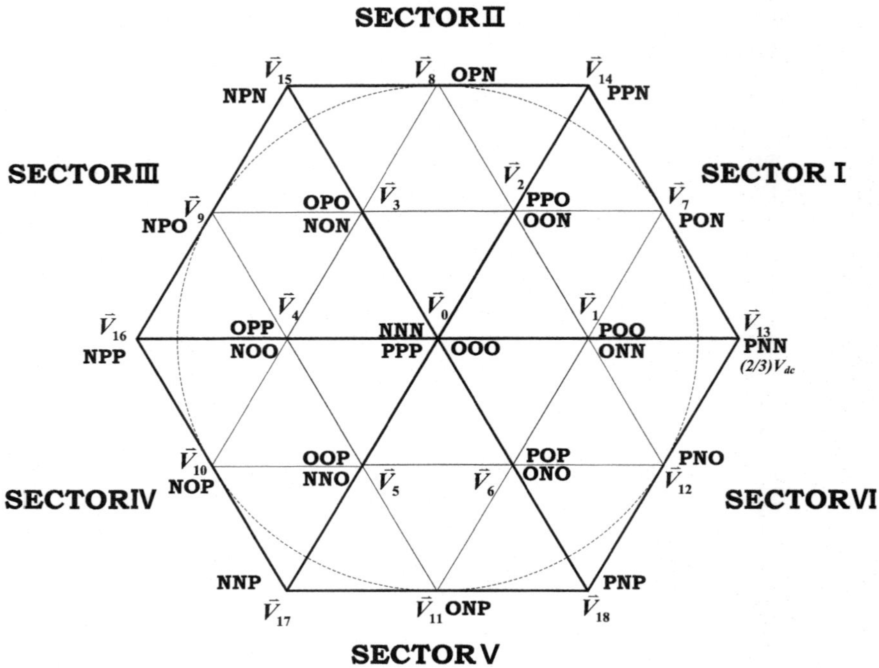

Fig. 3.8 Space vector diagram of three-level converter

3.2.2 Tolerant Control for S_{x2} and S_{x3} Open-Circuit Faults

As mentioned before, in NPC three-level converters, S_{x2} (or S_{x3}) open-circuit faults make both the O- and P- (or N-) switching states infeasible. Therefore, it is impossible to realize tolerant control without additional devices. On the other hand, T-type converters give the chance to apply tolerant control. In particular, when a T-type converter operates in the rectifier mode, S_{x1} and S_{x4} open-circuit faults cannot effect the output current. Therefore, only S_{x2} and S_{x3} open-circuit faults are normally taken into account in the tolerant control of T-type rectifiers.

Fig. 3.8 shows the space vector diagram where all of vectors are available in T-type three-level converters. S_{x2} open-circuit faults make the O-switching state infeasible. Although an S_{x2} open-circuit fault occurs, there is no effect on the output current if the current flows from the output to the DC-link (rectifier mode) because the current flows through D_{x2} not S_{x2}. Based on this result, Fig. 3.9 shows a space vector diagram of a three-level converter, which contains the infeasible vectors caused by an S_{a2} open-circuit fault. Vectors with the O-switching vector of the a-leg, where S_{a2} open-circuit fault has influence, cannot be generated. [OPN], [ONP], [OPO], [OPP], [OPO] and [OOO] cannot be generated due to the presence of an S_{a2} open-circuit fault. In the tolerant control of S_{x2} and S_{x2} open-circuit faults, the output voltage is configured without the infeasible voltage vectors of Fig. 3.9.

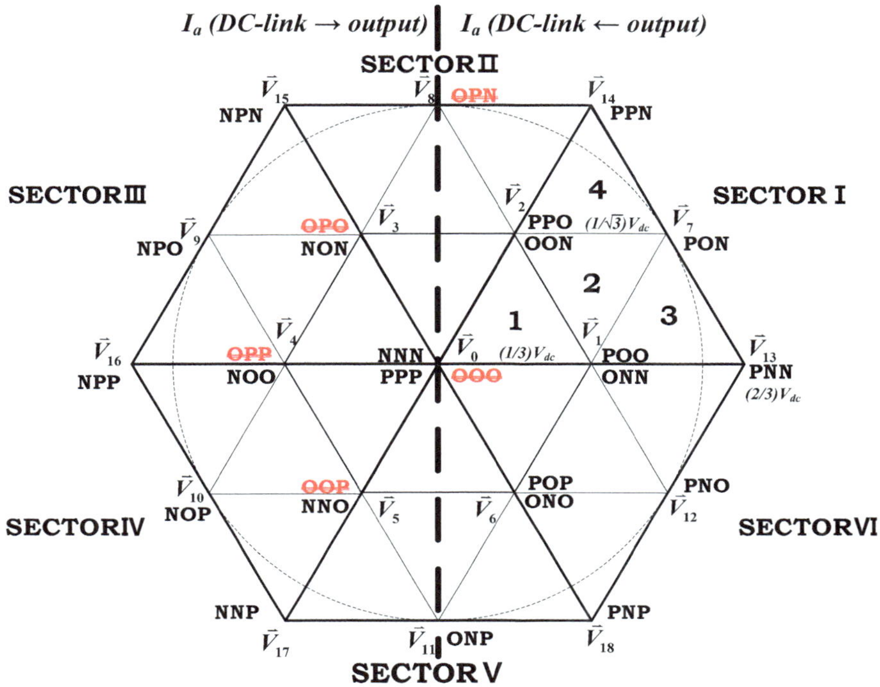

Fig. 3.9 Space vector diagram of three-level converter when S_{a2} open-circuit fault occurs

A. Replacement Two-Level Switching Tolerance Control (R2LS-TC) [8].

An S_{a2} open-switch fault causes two medium-vectors (OPN, ONP), three small-vectors (OPO, OPP, OOP), and one zero vector (OOO) to be infeasible when the a-phase current is negative. Therefore, in R2LS-TC, without using these infeasible vectors, the Two-Level Switching Method (2-LSM) using large vectors (NPN, NPP, NNP, PNP) which use the structure of three-level topology shown in Fig. 3.10 is implemented in sectors II–V as shown in the gray parts of Fig. 3.11 and Three-Level Switching Method (3-LSM) is conducted in the rest of sector. R2LS-TC replaces 3-LSM with 2-LSM in several sectors when an S_{a2} open-switch occurs. This tolerance control can be applied in the entire modulation index range. Moreover, consideration of neutral-point voltage balancing is not required because large vectors do not cause the neutral-point voltage unbalancing [23]. However, the input current ripple of each phase increases in the sectors, where the 2-LSM is applied, owing to the use of large vectors in 2-LSM. For an S_{a3} open-switch fault, 2-LSM is applied in sectors I, II, V, and VI.

The Replacement Two-Level Switching (R2LS) tolerance control for S_{x2} and S_{x3} (x = a, b, c) open-switch faults is shown in Table 3.2.

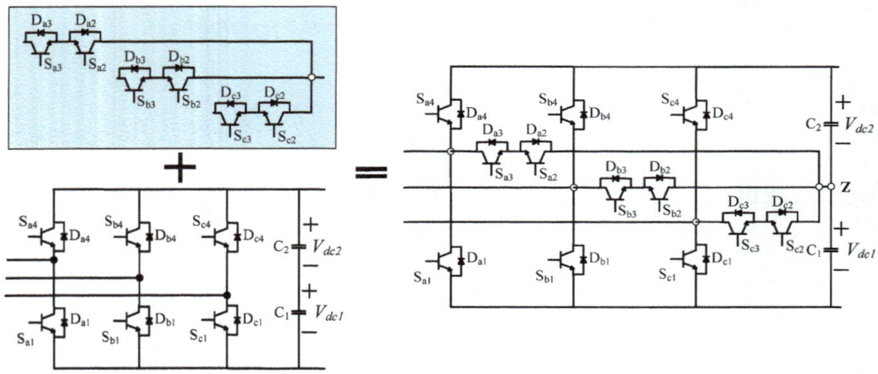

Fig. 3.10 Configuration of two-level converter and T-type three-level converter

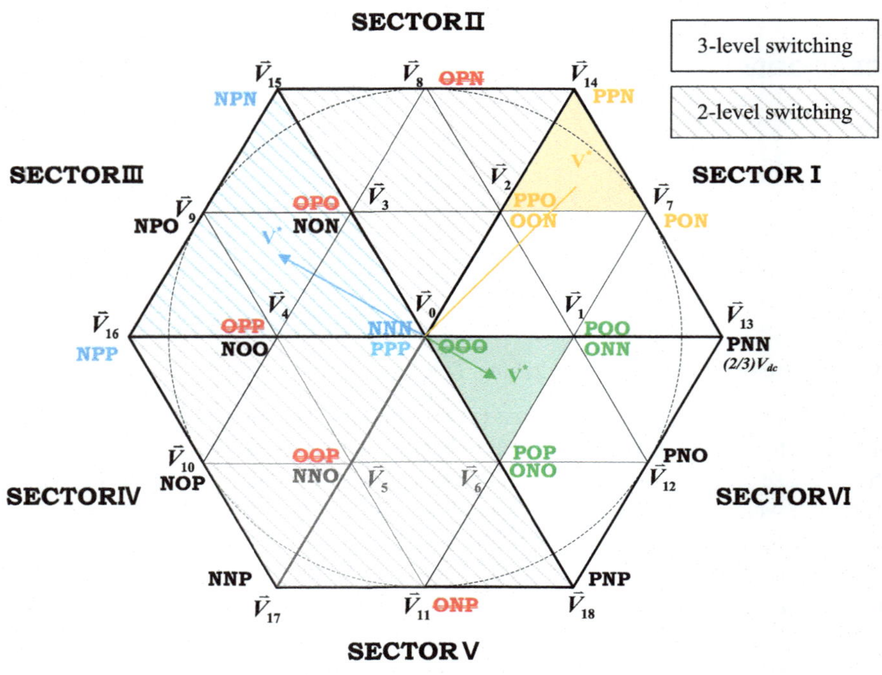

Fig. 3.11 R2LS-TC for S_{a2} open-switch fault (reprinted from Ref. [6], Fig. 7)

- Simulation and experimental results for R2LS-TC.

Simulation parameters are as follows: the rated power is 1.5 kW, the grid line-to-line voltage is 127 V_{rms}, the grid frequency is 60 Hz, the L-filter value is 1 mH, the DC-link voltage is 200 V, the DC-link capacitor is 1100 μF, the load is

Table 3.2 R2LS-TC depending on the position of open-switch fault

Position of open switch fault	Applied switching method	
	Sector for two-level switching	Sector for three-level switching
S_{a2}	Sector II, III, IV, V	Sector I, VI
S_{a3}	Sector I, II, V, VI	Sector III, IV
S_{b2}	Sector I, IV, V, VI	Sector II, III
S_{b3}	Sector I, II, III, IV	Sector V, VI
S_{c2}	Sector I, II, III, VI	Sector IV, V
S_{c3}	Sector III, IV, V, VI	Sector I, II

Table 3.3 Simulation and experiment parameters

DC-link voltage	200 V	Line-to-line voltage	127 Vrms
DC-link capacitance	1100 μF	Grid frequency	60 Hz
Switching frequency	10 kHz	L-filter	1 mH
Control period	100 μs	Load	33.3 Ω

33.3 Ω, the switching frequency is 10 kHz, and the control period is 100 us. They are listed in Table 3.3

Figure 3.12 shows the pole voltage and current waveforms of tolerance controls when an S_{a2} open-switch occurs. The R2LS-RC restores the distorted input currents after 0.5 s. The R2LS-TC makes the pole voltage of three-phase as two levels ($V_{dc}/2$ and $-V_{dc}/2$).

R2LS-TC replaces 3-LSM with 2-LSM in four Sectors as shown in the experimental result of Fig. 3.13. Therefore, the DC-link voltage ripple and the difference between the top voltage and the bottom voltage (V_{dif}), which are 0.5 V and 0.3 V, respectively, are small because 2-LSM does not cause any unbalance of the neutral-point voltage.

B. Maintenance Three-Level Switching Tolerance Control (M3LS-TC) [8].

S_{x2} and S_{x3} open-switch faults cause a large current distortion and DC-link voltage ripple. NPC-type rectifiers cannot remove current distortion completely due to structural limitations. However, T-type rectifier can restore the current distorted by an open-switch fault. Two methods for the tolerance control are proposed based on the SVM.

For an explanation of tolerance control, an S_{a3} open switch fault is considered as an example.

In the positive current part, the O-operating state is impossible in the presence of an S_{a3} open-switch fault. This means that OOO, OPN, OON, ONN, ONO, and ONP, which contain the O-operating state in the a-phase input voltage, in the right half plane of Fig. 3.14, cannot be made.

In sectors II and V, a two-level switching method is applied because the medium vectors OPN and ONP are impossible. A T-type rectifier can be operated as a

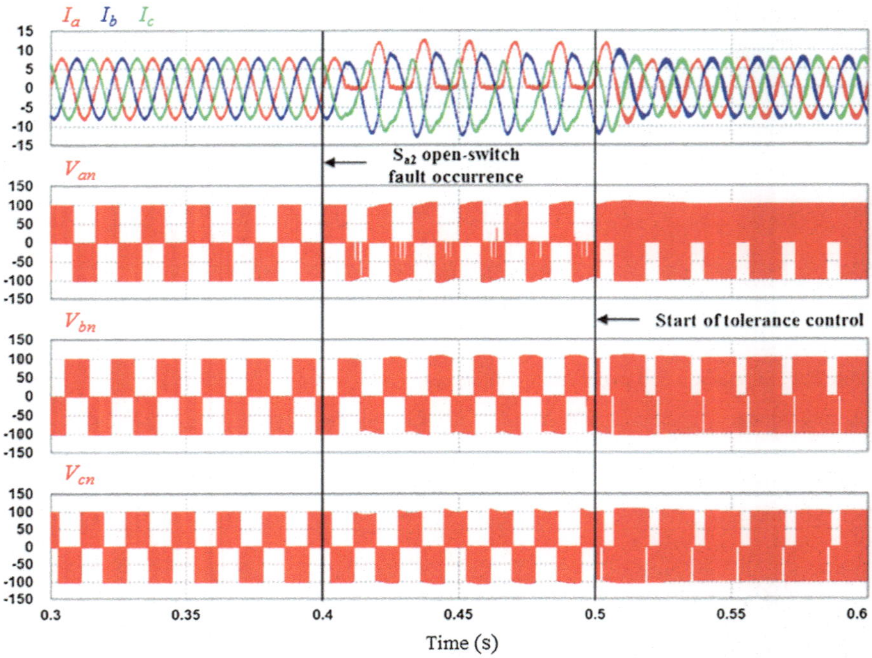

Fig. 3.12 Simulation results of R2LS-TC when S_{a2} open fault occurs (reprinted from Ref. [6], Fig. 13)

two-level rectifier by stopping the operation of S_{x2} and S_{x3}. To make the input voltage, NPN, PPN, NNN, and PPP are used in sector II. Similarly, in sector V, NNP, PNP, NNN, and PPP are used. The two-level switching method replaces the impossible O-operating state with the N-operating state.

In sectors I and VI, all of the impossible vectors are N-type small vectors. N-type small vectors can be replaced with P-type small vectors. There are four segments in sector I. Figure 3.15a shows the switching sequence in sector I-1 of Fig. 3.14 when the rectifier operates without any faults. This switching sequence consists of OON, PON, PPN, and PPO. The switching on-times ($T_{a,on}$, $T_{b,on}$, and $T_{c,on}$) and the switching off-times ($T_{a,off}$, $T_{b,off}$, and $T_{c,off}$) for the a-phase, b-phase, and c-phase are expressed as:

$$\begin{aligned} &T_{a,on}T_7 + T_{14} + \tfrac{T_2}{2}, \quad T_{b,on} = T_{14} + \tfrac{T_2}{2}, \quad T_{c,on} = \tfrac{T_2}{2} \\ &T_{a,off} = T_s - T_{a,on}, \quad T_{b,off} = T_s - T_{b,on}, \quad T_{c,off} = T_s - T_{c,on} \end{aligned} \tag{3.1}$$

where T_x (x = 1, 2, …, 18) denotes on-times V_x.

Due to S_{x3} open-switch faults, the N-type small vector, OON, is impossible in sector I-1. OON can be changed to PPO which is a P-type small vector as shown in Fig. 3.15b.

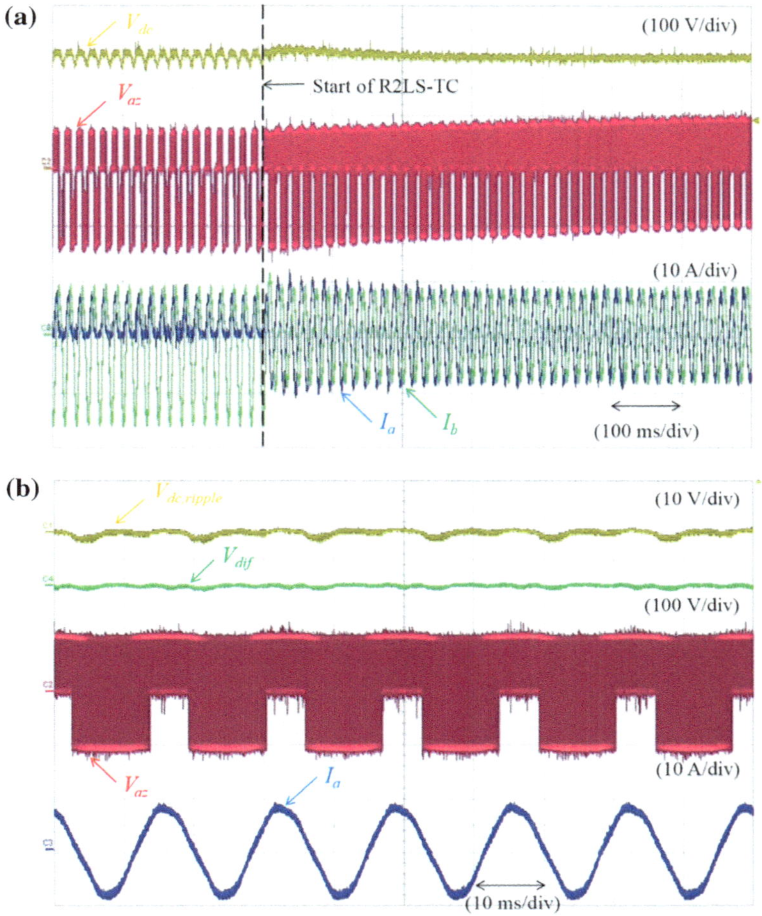

Fig. 3.13 Experimental results of R2LS-TC when S_{a2} open fault occurs (reprinted from Ref. [6], Figs. 20 and 21)

Switching on-times for the tolerance control ($T_{a,on,tolerance}$, $T_{b,on,tolerance}$, and $T_{c,on,tolerance}$) are simply calculated using the minimum on-time (T_{min}) among $T_{a,on}$, $T_{b,on}$, and $T_{c,on}$, and they are expressed as

$$
\begin{aligned}
T_{min} &= [\,T_{a,on}, \quad T_{b,on}, \quad T_{c,on}\,] \\
T_{offset} &= T_{min} \\
T_{x,on,tolerance} &= T_{x,on} + T_{offset},\ x = a, b, c
\end{aligned}
\tag{3.2}
$$

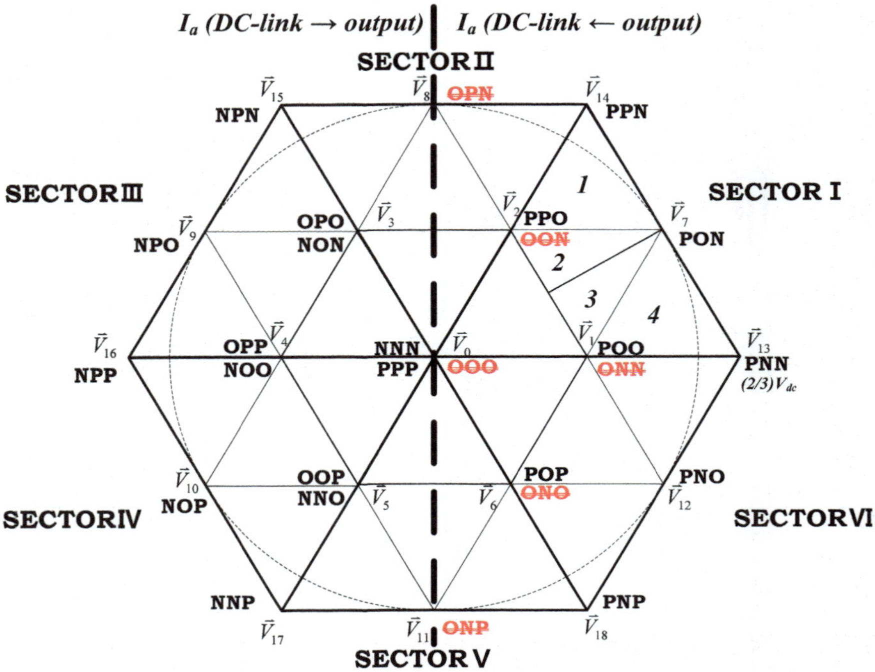

Fig. 3.14 Space vector diagram with the S_{a3} open-switch fault

Figure 3.16a shows the normal switching sequence in sector I-2. OON is impossible and can be changed to PPO as shown in Fig. 3.16b. $T_{a,on,tolerance}$, $T_{b,on,tolerance}$, and $T_{c,on,tolerance}$ are calculated using (3.2).

In sector I-3, there are two impossible vectors as shown in Fig. 3.17a. The N-type small vector, ONN, can be easily changed to the P-type small vector, PON, using (3.2). The calculated on-times ($T_{a,on,new1}$, $T_{b,on,new1}$, and $T_{c,on,new1}$) are shown in Fig. 3.17b. In addition, the N-type small vector, OON, should be changed to the P-type small vector, PPO. Then the switching order becomes PPO-PON-POO-PON-PPO as shown in Fig. 3.17c. Then the on-times are changed. $T_{a,on,tolerance}$, $T_{b,on,tolerance}$, and $T_{c,on,tolerance}$ are expressed as:

$$\begin{aligned} T_{max} &= \left[\, T_{a,on,new1}, \quad T_{b,on,new1}, \quad T_{c,on,new1} \,\right] \\ T_{offset2} &= T_s - T_{max} \\ T_{x,on,tolerance} &= T_{x,on,new1} + T_{offset2},\ x = a, b, c \end{aligned} \tag{3.3}$$

Changing a switching order is not necessary because each of the recalculated on-times $T_{b,on,tolerance}$ and $T_{c,on,tolerance}$ from (3.3) corresponds to the sums of the

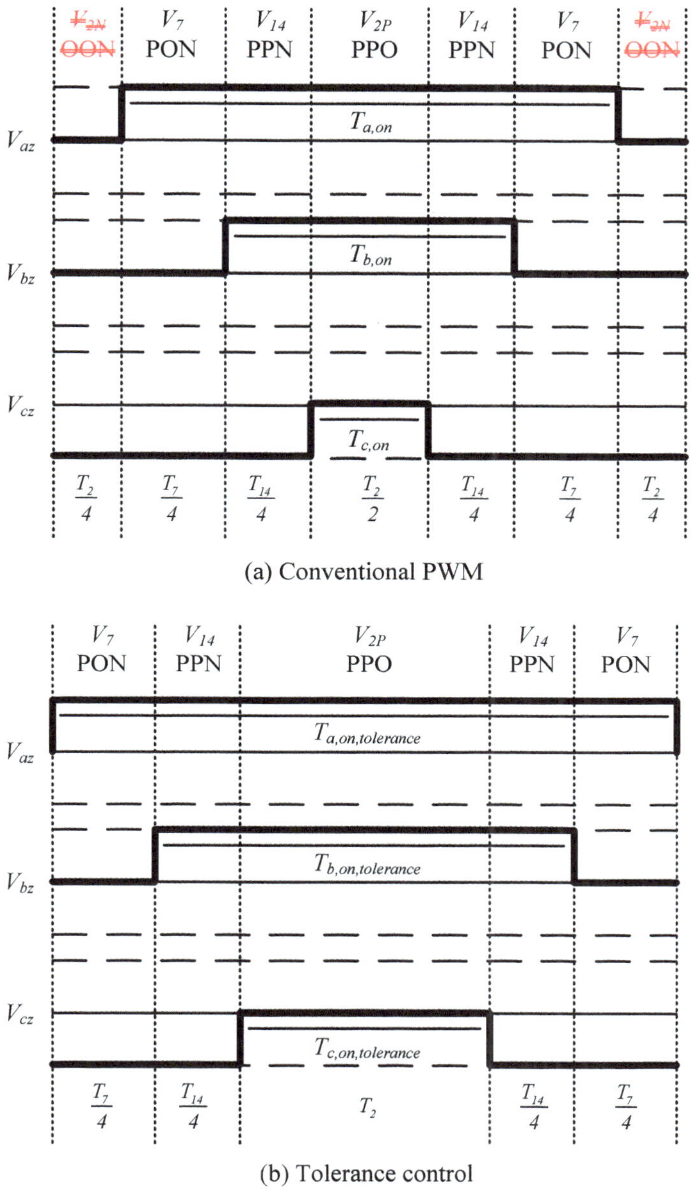

Fig. 3.15 Switching sequence in sector I-1 (reprinted from Ref. [7], Fig. 11)

b-phase separated times and the c-phase separated times in Fig. 3.17c. The recalculated on-times are relocated to the center of T_s. Therefore, the switching order becomes PON-POO-PPO-POO-PON.

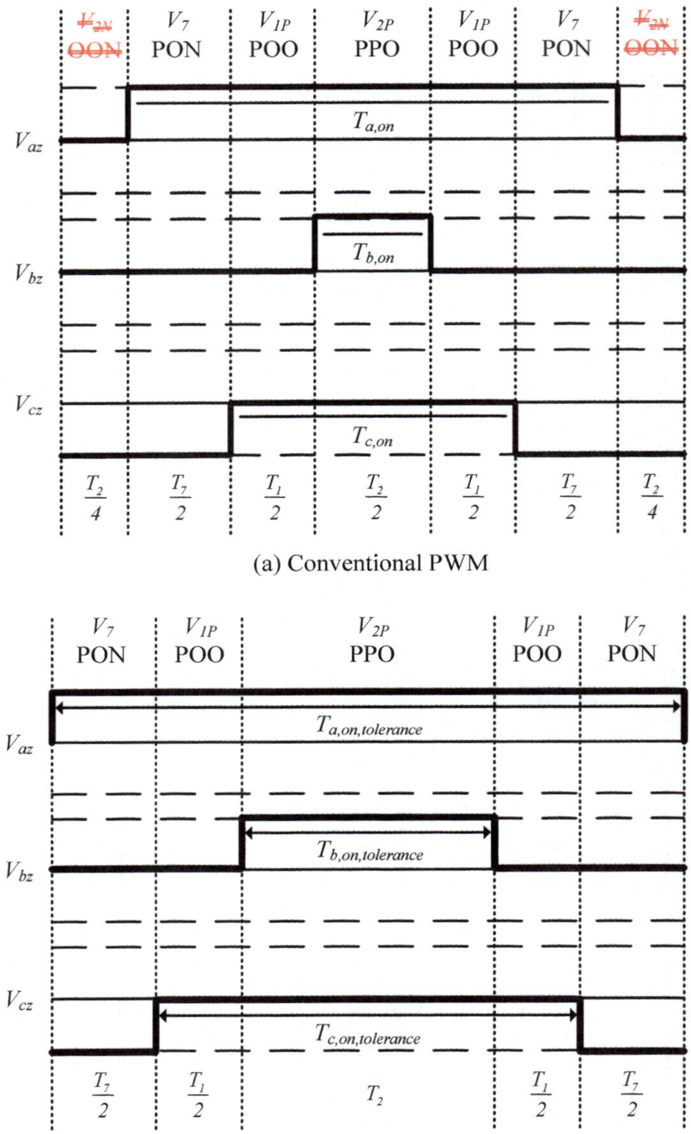

(b) Tolerance control

Fig. 3.16 Switching sequence in sector I-2 (reprinted from Ref. [7], Fig. 12)

In sector I-4, the N-type vector, ONN, is impossible. ONN can be changed to the P-type small vector, POO. $T_{a,on,tolerance}$, $T_{b,on,tolerance}$, and $T_{c,on,tolerance}$ are calculated using (3.2). The same switching methods used in sector I for the tolerance control are applied in sector VI.

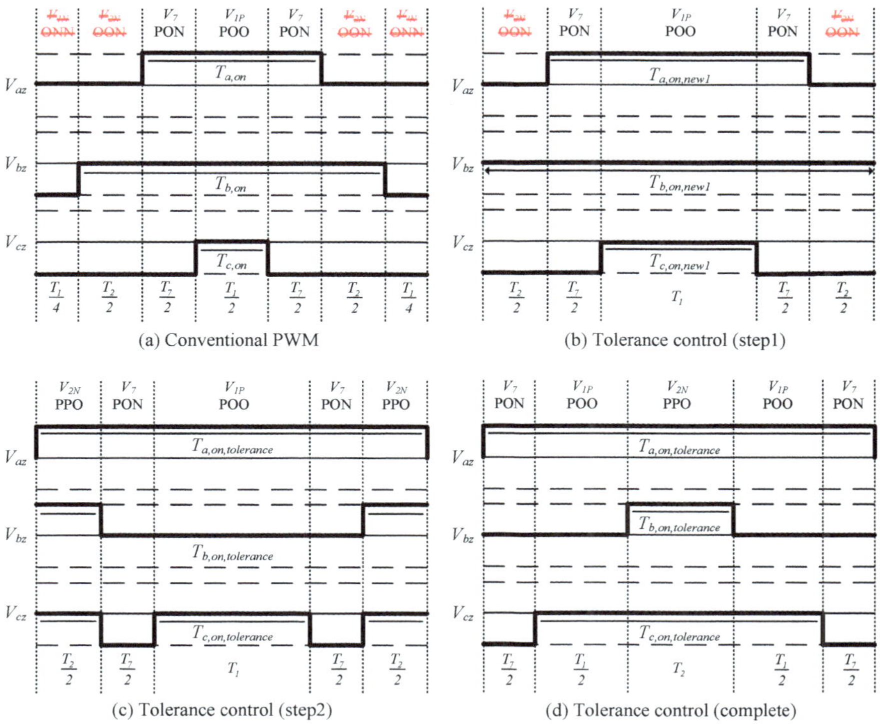

Fig. 3.17 Switching sequence in sector I-3 (reprinted from Ref. [7], Fig. 13)

Consequently, in sectors I and VI, all of the N-type small vectors are replaced with P-type small vectors. This causes a neutral-point voltage unbalance. Therefore, the P-type small vectors, OPO, OPP, and OOP, in sectors III and IV should be replaced with the N-type small vectors, NON, NOO, and NNO.

The tolerance control for S_{x2} and S_{x3} (x = a, b, c) open-switch faults is shown in Table 3.4. This tolerance control maintains the characteristic of three-level topology in four sectors as shown in Fig. 3.18. Therefore, the input current ripple is not greatly increased. However, neutral-point voltage ripple increases because the switching by the P-type or N-type small vectors is only continued for a half period of the output voltage.

- Simulation and experimental results of M3LS-TC.

The performance of the M3LS tolerance control is identified in Fig. 3.19. An S_{a2} open-switch fault occurs at 0.3 s as shown in Fig. 3.19a. As a result, the three phase currents are distorted, and the DC-link voltage has a ripple of 20 V. The M3LS tolerance control is applied at 0.4 s, and these distorted currents are restored to a sinusoidal current. Moreover, DC-link voltage ripple is also eliminated.

Table 3.4 M3LS-TC depending on the position of an open-switch fault

	Position of open-switch fault		
	S_{a2} or S_{a3}	S_{b2} or S_{b3}	S_{c2} or S_{c3}
Sector I	OON → PPO ONN → POO	Two-level switching method	OON → PPO ONN → POO
Sector II	Two-level switching method	OON → PPO OPO → NON	OON → PPO OPO → NON
Sector III	OPO → NON OPP → NOO	OPO → NON OPP → NOO	Two-level switching method
Sector IV	OPP → NOO OOP → NOO	Two-level switching method	OPP → NOO OOP → NNO
Sector V	Two-level switching method	OOP → NNO ONO → POP	OOP → NNO ONO → POP
Sector VI	ONN → POO ONO → POP	ONN → POO ONO → POP	Two-level switching method

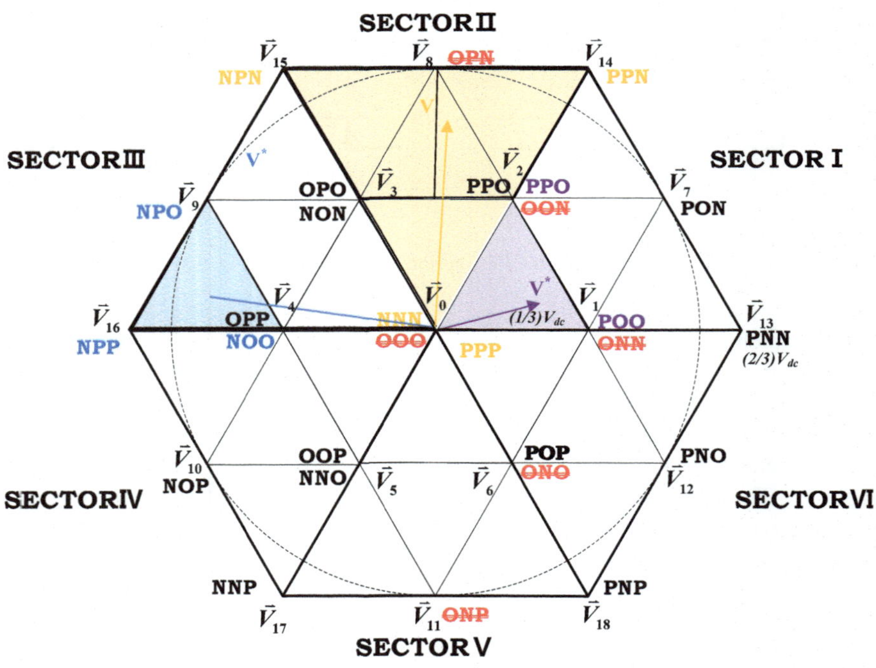

Fig. 3.18 M3LS-TC for an S_{a2} open-switch fault

Figure 3.19b shows the input currents and DC-link voltage when the M3LS tolerance control is applied. When an S_{a3} open-switch fault occurs, the switching levels of the input voltages in sectors I and VI, which are the shaded parts of Fig. 3.19b, have the characteristic of three-level topology.

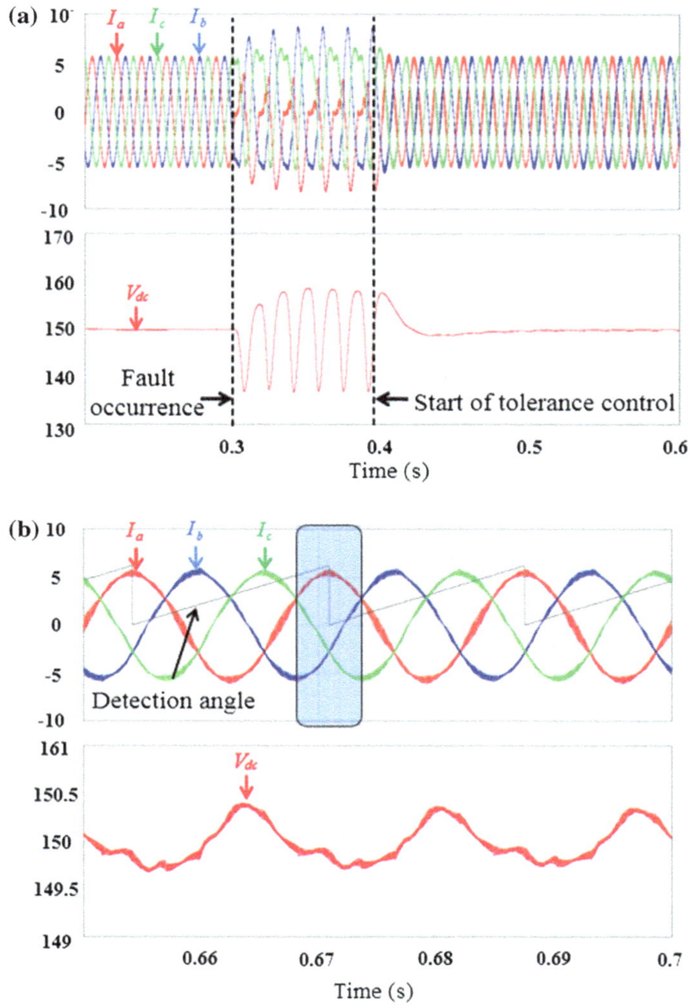

Fig. 3.19 Simulation results of M3LS-TC when an S_{a3} open fault occurs (reprinted from Ref. [7], Figs. 17 and 19)

Figure 3.20a shows the performance of the M3LS tolerance control when an S_{a3} open-switch fault occurs. The distorted input currents are restored by the M3LS tolerance control, and the ripple due to the open-switch fault decreases. The DC-link voltage of Fig. 3.20a has a ripple of 6.5 V, because the M3LS tolerance control considers neutral-point balancing. In comparison with the DC-link voltage ripple of Fig. 3.13, the DC-link voltage ripple of the R2LS tolerance control is smaller than that of the M3LS tolerance control since the R2LS tolerance control

(a)

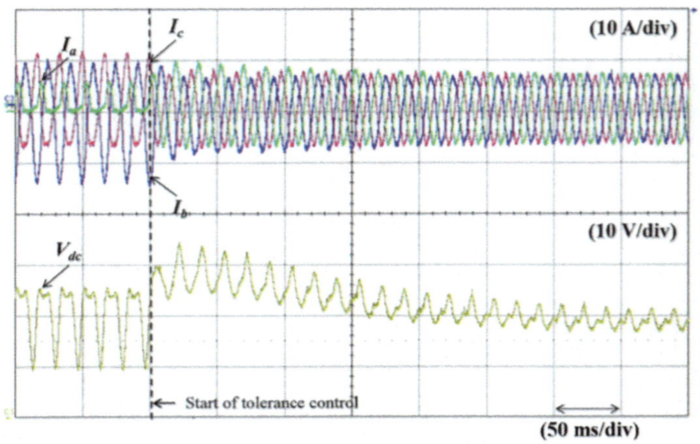

(b)

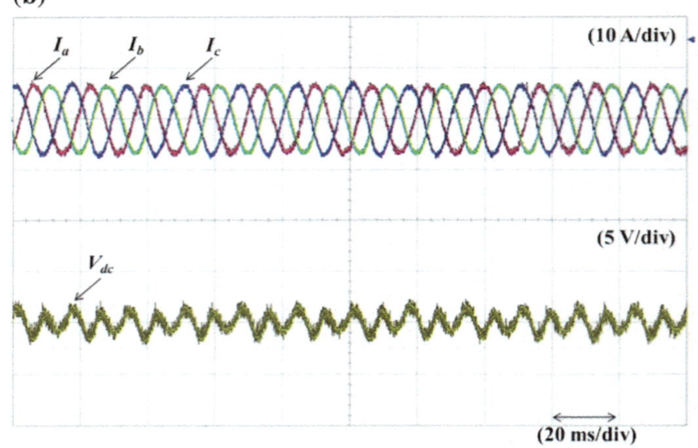

Fig. 3.20 Experimental results of M3LS-TC when an S_{a3} open fault occurs (reprinted from Ref. [7], Figs. 22 and 24)

does not need to consider neutral-point balancing. However, the input current THD of the R2LS tolerance control is larger than that of the M3LS tolerance control.

C. One Phase Two-Level Switching Tolerance Control (OP2LS-TC) [9].

For an S_{a2} open-switch fault, OP2LS-TC maintains three-level switching in the two phases that do not contain the open-switch fault, and the input voltage of the one phase with the open-switch fault has two levels by OP2LS-TC. In order to satisfy the operation of OP2LS-TC when an S_{a2} open-switch fault occurs, for all the sectors, the offset value is added to the switch on-time ($T_{x,on}$) of the one-phase

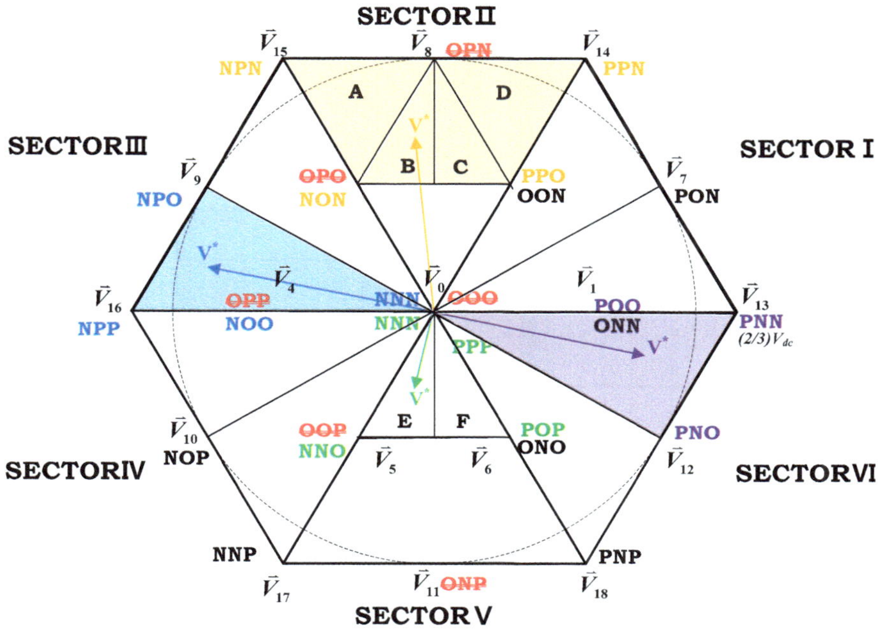

Fig. 3.21 OP2LS-TC for an S_{a2} open-switch fault (reprinted from Ref. [6], Fig. 12)

reference voltage, which contains the open-switch fault, to make its phase voltage level two. This means that each Sector is divided into two parts, as shown in Fig. 3.21, and the nearest vectors are used to generate input voltages.

To explain the modification of the switching on-times ($T_{a,on}$, $T_{b,on}$, $T_{c,on}$), the case of an S_{a2} open fault is used as an example.

In sector II-A, the switching sequence is shown in Fig. 3.22a. When an S_{a2} open fault occurs, there are infeasible [OPN] and [OPO] vectors. The switching on-time of the a-leg should be changed to apply the two-level switching method only to the a-leg. The averaged output voltage for a switching period has to be equal before and after the tolerant control is applied. As a result, the switching on-time in sector II-A can be expressed as:

$$T_{a,on,TC} = \frac{1}{2} T_{a,on} \tag{3.4}$$

By changing the switch on-time of the a-leg in OP2LS-TC, the a-leg does not contain the O switching state as shown in Fig. 3.22b. It only generates the P and N switching states which is the main characteristic of the two-level switching method. Nevertheless, the averaged output voltage is maintained at the same value by decreasing the switching on-time by half. In addition, we can see that there are no changes at the b-leg or c-leg.

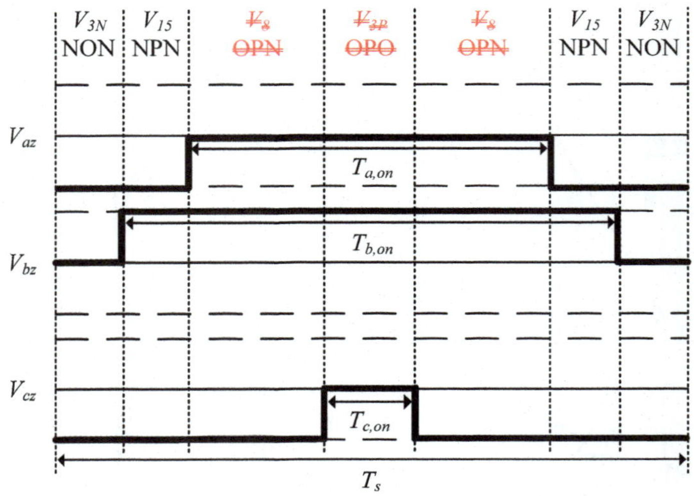

(a) Normal switching

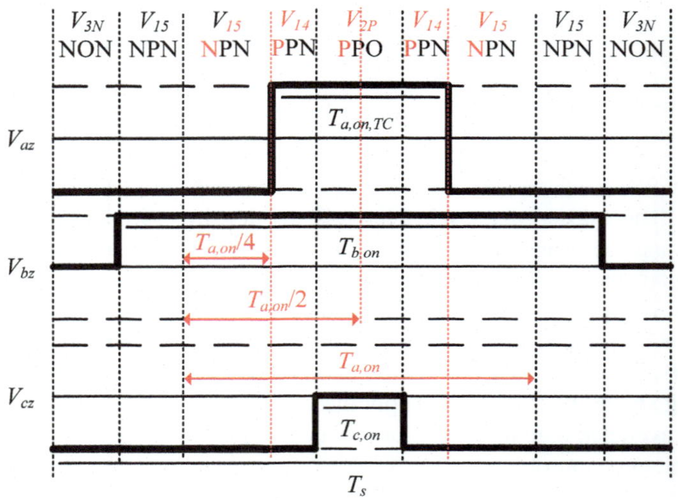

(b) Tolerance switching

Fig. 3.22 Switching sequence in sector II-A (reprinted from Ref. [6], Fig. 9)

The switching on-time ($T_{a,on}$) of the a-leg in sector II-B should be changed, as shown in Fig. 3.23, to apply the two-level switching method in the a-leg containing the open-switch fault. As in the case of sector II-A, it can be seen that by using (3.4), the averaged output voltage of the sector II-A a-leg is maintained at the same value.

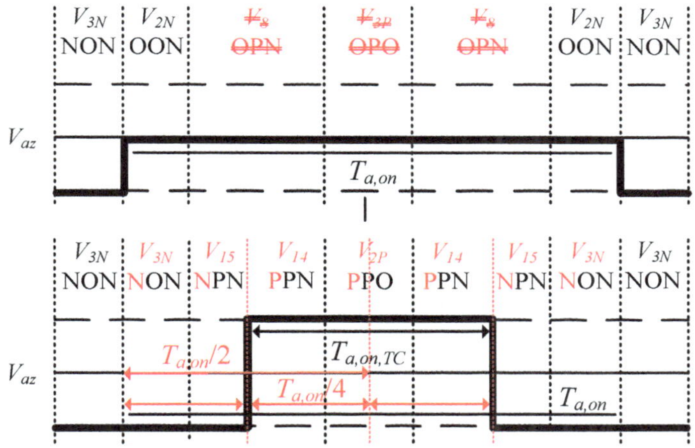

Fig. 3.23 Switching sequence of tolerance control in sector II-B (reprinted from Ref. [6], Fig. 10)

In Sector II-C and II-D, the switching on-time of the a-leg is changed as shown in Fig. 3.24 and it is expressed as:

$$T_{a,off} = T_s - T_{a,on} \tag{3.5}$$

$$T_{a,on,TC} = \frac{1}{2}T_{a,off} + T_{a.on} = \frac{1}{2}(T_{a,on} + T_s) \tag{3.6}$$

The changing vector of OPTL-TC depending on the sector is summarized in Table 3.6 based on the position of an open switch fault. Like 3LSO-TC, this tolerance control can be applied in the entire modulation index range. As mentioned above, the input voltage of the one phase that contains an open-switch fault has two levels and other phases maintain the three-level input voltage. Therefore, the current ripple of only one phase increases. Nevertheless, the neutral-point balancing control for a switching period is not performed in sectors II, III, IV, and VI, which is similar to 3LSO-TC. The N-type small vectors or the P-type small vectors are used for two-thirds of a fundamental period of the input current. Therefore, the neutral-point voltage has a large ripple that has a fundamental frequency component.

- Simulation and experimental results of OP2LS-TC.

Table 3.5 shows the parameters of simulation and experiment for the M3LS-TC.

The simulation parameters are as follows: the rated power is 1.5 kW, the grid line-to-line voltage is 127 V_{rms}, the grid frequency is 60 Hz, the L-filter value is 1 mH, the DC-link voltage is 200 V, the DC-link capacitor is 1100 μF, the load is 33.3 Ω, the switching frequency is 10 kHz, and the control period is 100 us. They are shown in Table 3.7.

OP2LS-TC makes the a-phase pole voltage into two levels in all of the sectors. The b-phase and c-phase have three-level pole voltages as shown in Fig. 3.25.

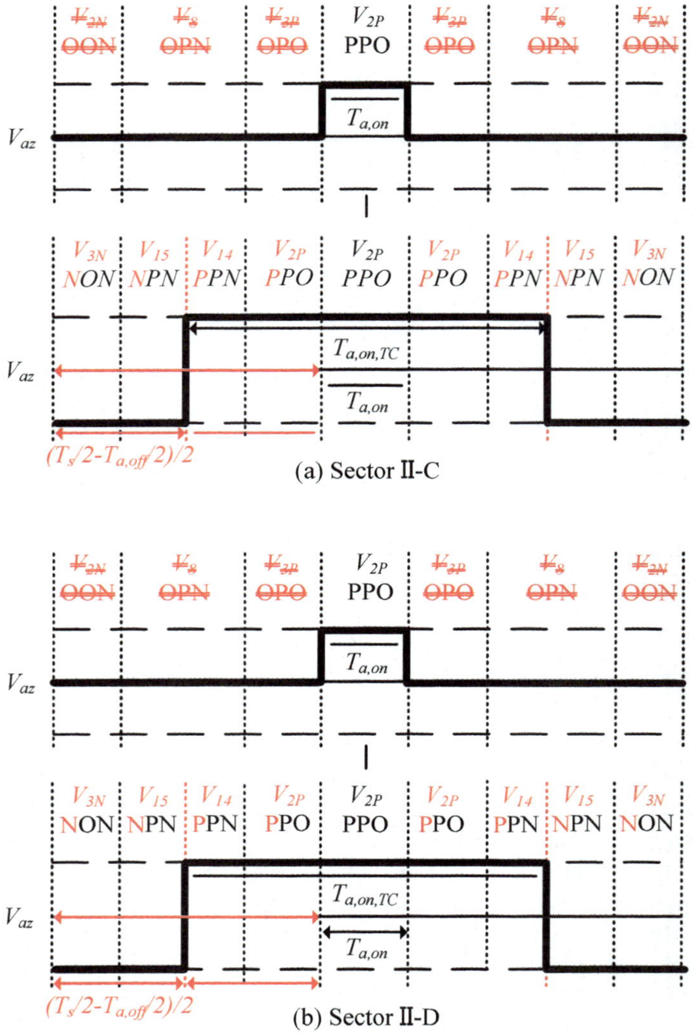

(a) Sector II-C

(b) Sector II-D

Fig. 3.24 Switching sequence of tolerance control (reprinted from Ref. [6], Fig. 11)

Table 3.5 Simulation and experimental parameters

DC-link voltage	150 V	Line-to-line voltage	100 Vrms
DC-link capacitance	1100 μF	Grid frequency	60 Hz
Switching frequency	10 kHz	L-filter	1 mH
Control period	100 μs	Load	33.3 Ω

Table 3.6 OP2LS-TC depending on the position of an open-switch fault

Position of open-switch fault	$V_{x,ref}$ $(x = a, b, c) < 0$	$V_{x,ref}$ $(x = a, b, c) > 0$
S_{a2} or S_{a3}	$T_{a,on,TC} = \frac{1}{2}T_{a,on}$	$T_{a,on,TC} = \frac{1}{2}(T_{a,off} + T_s)$
S_{b2} or S_{b3}	$T_{b,on,TC} = \frac{1}{2}T_{b,on}$	$T_{b,on,TC} = \frac{1}{2}(T_{b,off} + T_s)$
S_{c2} or S_{c3}	$T_{c,on,TC} = \frac{1}{2}T_{c,on}$	$T_{c,on,TC} = \frac{1}{2}(T_{c,off} + T_s)$

Table 3.7 Simulation and experimental parameters

DC-link voltage	200 V	Line-to-line voltage	127 V_{rms}
DC-link capacitance	1100 μF	Grid frequency	60 Hz
Switching frequency	10 kHz	L-filter	1 mH
Control period	100 μs	Load	33.3 Ω

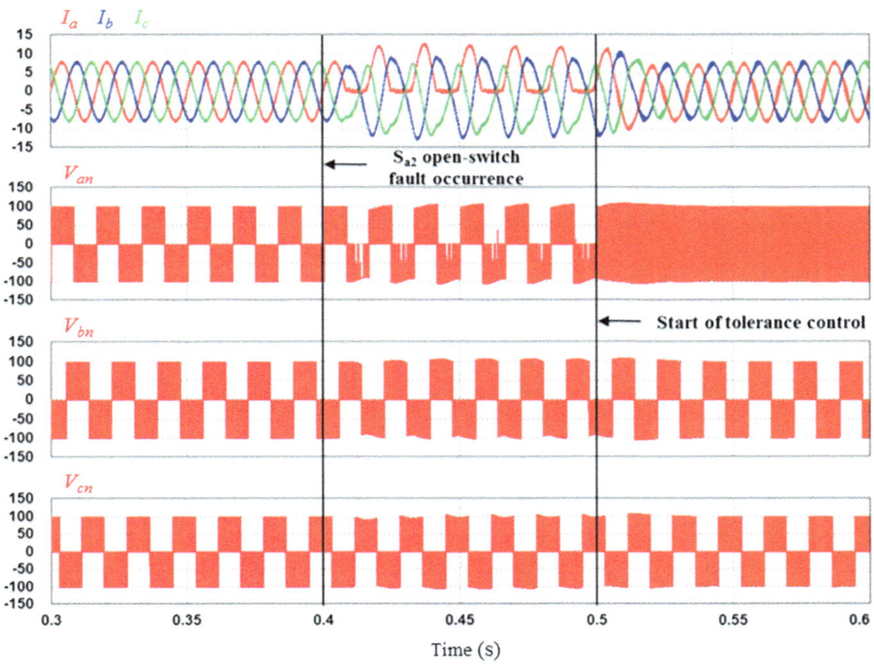

Fig. 3.25 Simulation results of OP2LS-TC when an S_{a2} open fault occurs (reprinted from Ref. [6], Fig. 13)

In Fig. 3.26, the pole voltage V_{az}, which contains an open-switch fault, operates as 2-LSM by OP2LS-TC. The DC-link voltage ripple and the neutral-point voltage are 1 and 2 V. Even if the neutral voltage equilibrium is considered in the PWM method of the three-level converter, the imbalance of the neutral voltage can be caused by a number of reasons. Differences in the capacitance of the DC short

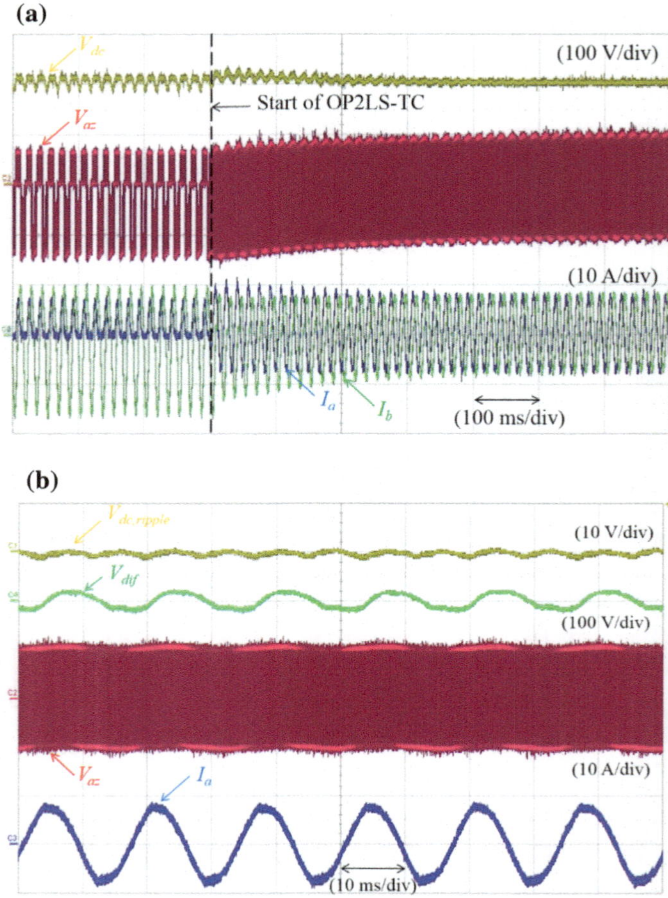

Fig. 3.26 Experimental results of OP2LS-TC when an S_{a2} open fault occurs (reprinted from Ref. [6], Figs. 20 and 21)

capacitor generated in the manufacturing process, the characteristics of each switching device, and a mismatch of the switching on-off points all contribute to the neutral point voltage imbalance.

D. Three-Level Switching Oriented Tolerance Control (3LSO-TC) [9].

3LSO-TC maintains 3-LSM in sectors as much as possible when an open-switch fault occurs. In 3LSO-TC, as shown in Fig. 3.27, the infeasible vectors, OPO, OPP, OOP, and OOO, are changed to NON, NOO, NNO, and NNN vectors in sectors III and IV when an S_{a2} open-switch fault occurs. This method was explained in the M3LS tolerance control, which adds an offset value to the switch on-times ($T_{a,on}$, $T_{b,on}$, and $T_{c,on}$) of the three-phase reference voltages. The changed vectors in

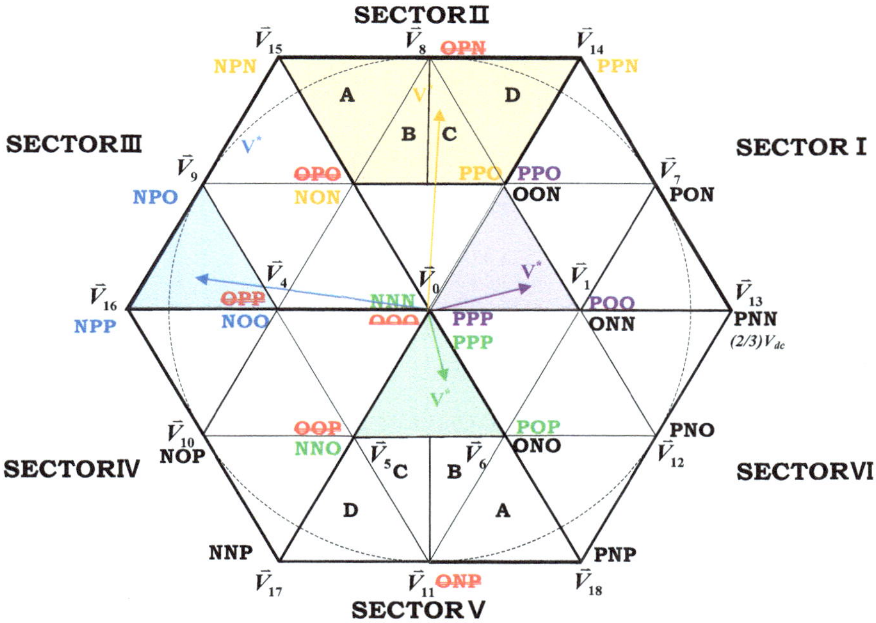

Fig. 3.27 3LSO-TC for an S_{a2} open-switch fault (reprinted from Ref. [6], Fig. 8)

Table 3.8 The range for applying the two-level switching method in 3LSO-TC, M3LS-TC and OP2LS-TC

	M3LS-TC	OP2LS-TC	3LSO-TC
The range of the two-level switching method	Two of the sectors in three legs	All of the sectors in a leg containing an open-switch fault	Two of the sectors in a leg containing an open-switch fault

sectors III and IV, are N-type small vectors NON, NOO, and NNO. Therefore, in order to consider neutral-point voltage balancing, the P-type small vectors should be chosen in sectors I and VI, as explained in the M3LS tolerance control.

Table 3.8 shows the range for applying a two-level switching method depending on the tolerant control. Among them, 3LSO-TC uses the two-level switching method in the smallest range. Table 3.9 shows the summary of 3LSO-TC.

- Simulation and Experimental results of 3LSO-TC.

Table 3.10 shows the parameters of simulation and experiment for 3LSO-TC.

The a-phase pole voltage has two levels in sectors II and V. In the rest of the sector, the a-phase pole voltage is clamped at $V_{dc}/2$ and $-V_{dc}/2$ when 3LSO-TC is applied. The b-phase and c-phase pole voltages maintain three levels as shown in Fig. 3.28.

Table 3.9 3LSO-TC depending on the position of an open-switch fault

	Position of open-switch fault		
	S_{a2} or S_{a3}	S_{b2} or S_{b3}	S_{c2} or S_{c3}
Sector I	OON → PPO ONN → POO	-A,-B: $T_{b,on,TC} = \frac{T_{b,on}}{2}$ -C,-D: $T_{b,on,TC} = \frac{T_{b,off}+T_s}{2}$	OON → PPO ONN → POO
Sector II	-A,-B: $T_{a,on,TC} = \frac{T_{a,on}}{2}$ -C,-D: $T_{a,on,TC} = \frac{T_{a,off}+T_s}{2}$	OON → PPO OPO → NON	OON → PPO OPO → NON
Sector III	OPO → NON OPP → NOO	OPO → NON OPP → NOO	-A,-B: $T_{c,on,TC} = \frac{T_{c,on}}{2}$ -C,-D: $T_{c,on,TC} = \frac{T_{c,off}+T_s}{2}$
Sector IV	OPP → NOO OOP → NOO	-A, -B: $T_{b,on,TC} = \frac{T_{b,off}+T_s}{2}$ -C, -D: $T_{b,on,TC} = \frac{T_{b,on}}{2}$	OPP → NOO OOP → NNO
Sector V	-A,-B: $T_{a,on,TC} = \frac{T_{a,off}+T_s}{2}$ -C,-D: $T_{a,on,TC} = \frac{T_{a,on}}{2}$	OOP → NNO ONO → POP	OOP → NNO ONO → POP
Sector VI	ONN → POO ONO → POP	ONN → POO ONO → POP	-A,-B: $T_{c,on,TC} = \frac{T_{c,off}+T_s}{2}$ -C,-D: $T_{c,on,TC} = \frac{T_{c,on}}{2}$

Table 3.10 Simulation and experimental parameters

DC-link voltage	200 V	Line-to-line voltage	127 Vrms
DC-link capacitance	1100 μF	Grid frequency	60 Hz
Switching frequency	10 kHz	L-filter	1 mH
Control period	100 μs	Load	33.3 Ω

Figure 3.29 shows the performance of the proposed 3LSO-TC when an S_{a2} open-switch fault occurs in the experiment. Due to the S_{a2} open-switch fault, the positive a-phase current (I_a) becomes zero and the DC-link voltage has a large ripple. After 3LSO-TC is applied, the a-phase and b-phase currents (I_a and I_b) are restored as sinusoidal waveforms and the DC-link voltage ripple decreases as shown in Fig. 3.29a. As can be seen from the experimental results, the V_{ab} of 3LSO-TC, which is different from the conventional V_{ab} in two sectors, is the same as that in the simulation. Moreover, the a-phase pole voltage V_{az}, which contains an open-switch fault, has two levels and is maintained at $V_{dc}/2$ or $-V_{dc}/2$ in four sectors. The b-phase pole voltage V_{bz} is not changed when compared to that obtained during normal operation.

E. Comparison of tolerant controls [9].

In this section, the characteristics of the tolerance controls introduced in section IV, are analyzed and compared through the simulation results. For analysis of the

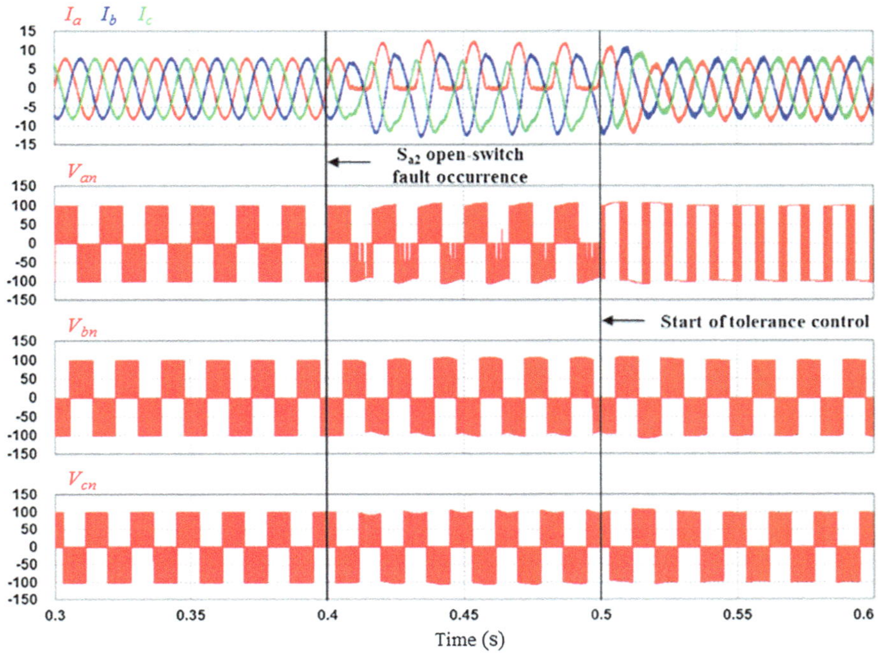

Fig. 3.28 Simulation results of 3LSO-TC when an S_{a2} open fault occurs (reprinted from Ref. [6], Fig. 13)

input current ripple, DC-link voltage ripple, and switch losses, the simulation parameters are as follows: the rated power is 1.5 kW, the grid line-to-line voltage is 127 V_{rms}, the grid frequency is 60 Hz, the L-filter value is 1 mH, the DC-link voltage is 200 V, the DC-link capacitor is 1100 μF, the load is 33.3 Ω, the switching frequency is 10 kHz, and the control period is 100 μs. The parameters of an IGBT module 4MBI300VG-120R-50 from Fuji Electric were used in the simulation.

- Input current ripple.

Figures 3.12, 3.25 and 3.28 show the pole voltage and current waveforms of the tolerance controls when an S_{a2} open-switch occurs. Three of the tolerance controls restore the distorted input currents after 0.5 s. However, each of the tolerance controls has a different pole voltage and current ripple. R2LS-TC makes the pole voltage of the three-phase as two levels ($V_{dc}/2$ and $-V_{dc}/2$) as shown in Fig. 3.12. The a-phase pole voltage has two levels in sectors II and V, In the rest of the sectors, the a-phase pole voltage is clamped at $V_{dc}/2$ and $-V_{dc}/2$ when 3LSO-TC is applied. The b-phase and c-phase pole voltages maintain three levels as shown in

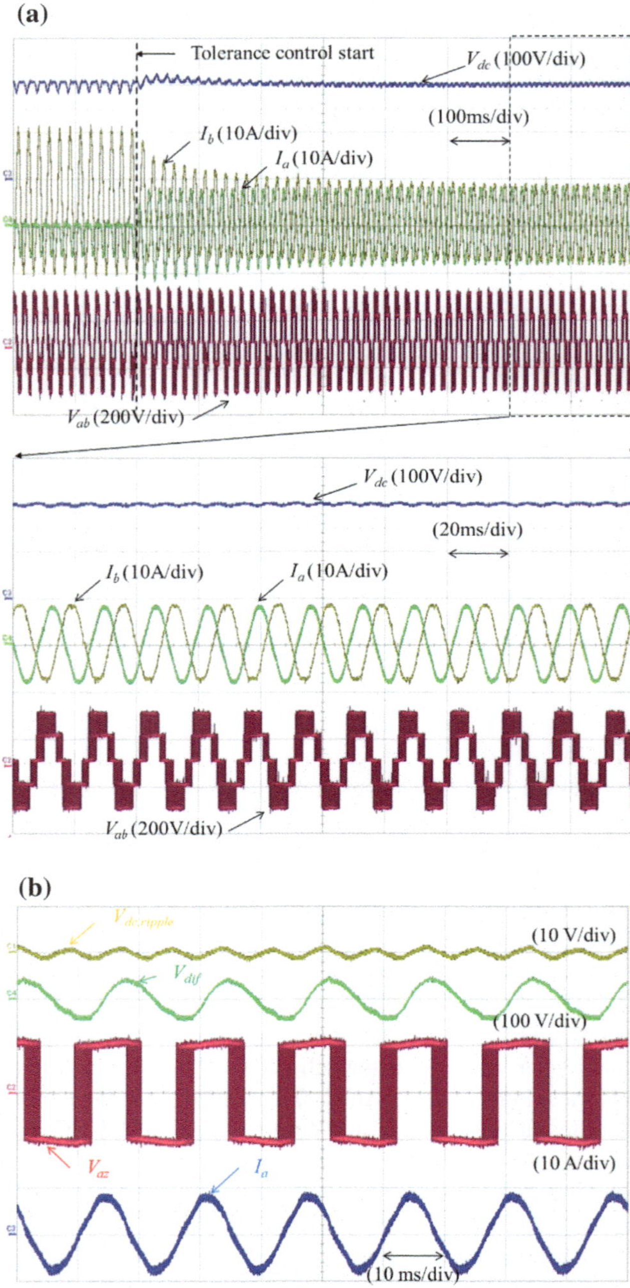

Fig. 3.29 Experimental results of 3LSO-TC when an S_{a2} open fault occurs (reprinted from Ref. [6], Figs. 19 and 21)

Fig. 3.25. OP2LS-TC makes the a-phase pole voltage as two levels in all of the sectors. The b-phase and c-phases have three-level pole voltages as shown in Fig. 3.28.

Figure 3.30 shows expanded current waveforms of the tolerance controls. R2LS-TC increases the ripple of the three-phase currents in sectors II–V as shown in Fig. 3.30b owing to 2-LSM. 3LSO-TC has a current ripple in sector II–V that is smaller than that of R2LS-TC. However, the current ripple in sectors I and IV increases compared with that of R2LS-TC as shown in Fig. 3.30c. Although OP2LS-TC only makes the a-phase pole voltage as two levels, it exhibits good performance with respect to the current ripple as shown in Fig. 3.30d. Table 3.11 shows the THD results of the three-phase currents depending on the tolerance controls. Consequently, OP2LS-TC is advantageous in terms of the current THD.

- DC-link voltage ripple.

Figure 3.31 shows the DC-link voltage and neutral-point voltage (top-voltage and bottom voltage) when each tolerance control is applied at a unity power factor. Among the three tolerance controls, R2LS-TC has the smallest ripple in the neutral-point voltage. The neutral-point voltages of 3LSO-TC and OP2LS-TC contain larger ripples. Therefore, the DC-link voltage also has ripple because it is affected by the ripple of the neutral-point voltage.

Table 3.12 lists the magnitudes of the DC-link voltage ripple and neutral-point voltage ripple ($V_{dc,ripple}$, $V_{neu,ripple}$), when the power factor is unity depending on various power levels (Load = 27, 33, and 66 Ω). It can be seen that R2LS-TC is advantageous in terms of the ripples in the DC-link voltage and neutral-point voltage.

- Loss.

The switch losses are divided into switching loss and conduction loss. Therefore, a loss analysis of the three tolerance controls is conducted by considering both of the loss components.

To analyze the switch losses, the parameters of a 4MBI300VG-120R-50 from Fuji are used. The parameters of D_{x1} and D_{x4} are used to calculate the diode losses of D_{x2} and D_{x3} that are not included in the 4MBI300VG-120R-50. The simulation is performed, with the included junction temperature set as 25 °C. Figures 3.32 and 3.33 show the efficiency analysis results.

This chapter discusses the characteristics of the tolerant controls. To evaluate these controls, the following three factors were considered: the current THD, DC-link voltage ripple, and efficiency. These factors are summarized in Table 3.13. From the analyzed results, the following conclusions can be obtained.

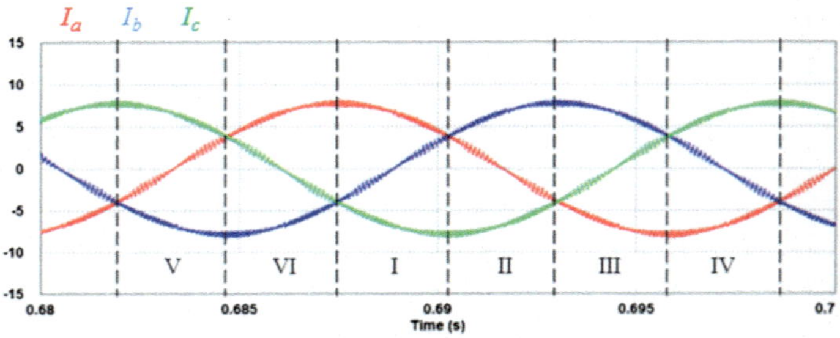

(a) Normal

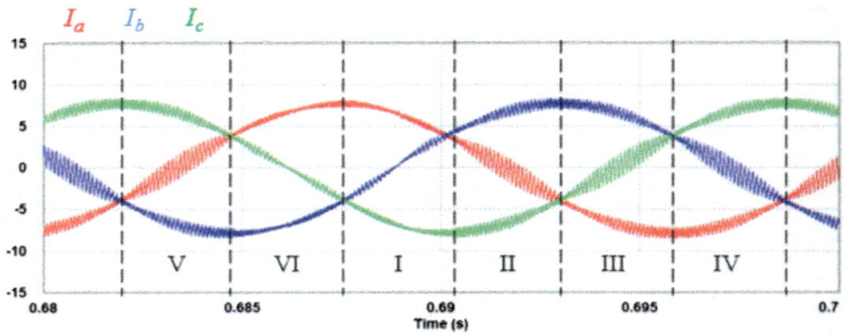

(b) R2LS-TC

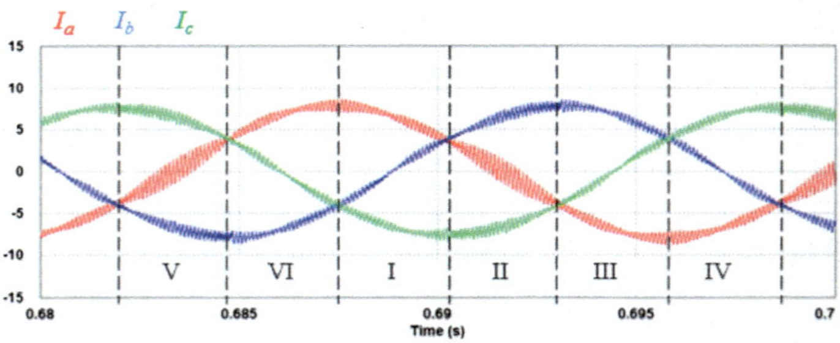

(c) 3LSO-TC

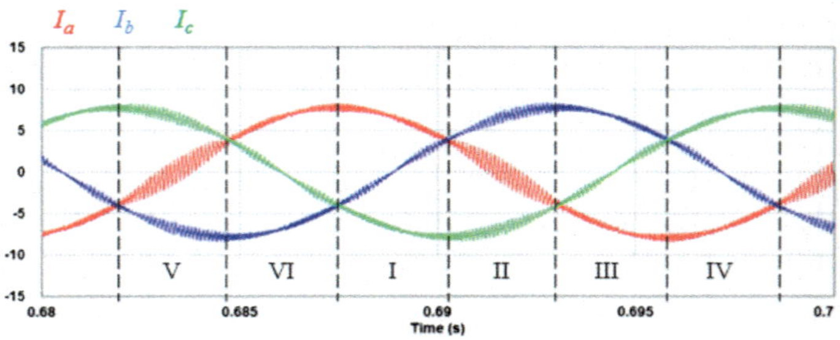

(d) OP2LS-TC

Fig. 3.30 Expanded three-phase currents of tolerance controls for an S_{a2} open-switch fault (reprinted from Ref. [6], Fig. 14)

Table 3.11 Current THD comparison of the tolerant controls

Normal			R2LS-TC		
a	b	c	a	b	c
3.5%	3.5%	3.5%	7.5%	6.7%	6.7%
3LSO-TC			OP2LS-TC		
a	b	c	a	b	c
8.2%	5.1%	5.1%	7.0%	4.6%	4.6%

Table 3.12 DC-link voltage ripple comparison of the tolerant controls

	Normal			R2LS-TC		
Power (%)	100	80	40	100	80	40
DC-link voltage ripple (V)	0.04	0.04	0.03	0.05	0.05	0.04
Neutral-point voltage ripple (V)	0.3	0.25	0.1	0.4	0.3	0.2
	3LSO-TC			OP2LS-TC		
Power (%)	100	80	40	100	80	40
DC-link voltage ripple (V)	0.28	0.22	0.1	0.07	0.06	0.05
Neutral-point voltage ripple (V)	2	1.4	0.71	0.9	0.8	0.38

Table 3.13 Performance comparison of R2LS-TC, 3LSO-TC, and OP2LS-TC

	R2LS-TC	3LSO-TC	OP2LS-TC
Current THD	Large incensement	incensement	Small incensement
Voltage ripple	No change	Large incensement	Incensement
Loss	Large incensement	Small incensement	Incensement

R2LS-TC: This control guarantees a lower DC-link voltage ripple compared to the other tolerance controls (3LSO-TC and OP2LS-TC). In sectors that operate as the convectional three-level switching method, the neutral-point voltage balancing control can be achieved. However, this method results in a high current THD and a low efficiency.

3LSO-TC: This control guarantees a higher efficiency. Still the current THD and DC-link voltage ripple increase. Additionally, the changed small vectors from the infeasible vectors due to open-switch faults are not used for the neutral-point voltage balancing control. However, the changed vectors for neutral-point voltage balancing control in opposite sectors can be used as the neutral-point voltage balancing control (the small vectors, which are the opposite type, in the opposite sector, guarantee the neutral-point balance). These vectors for neutral-point voltage

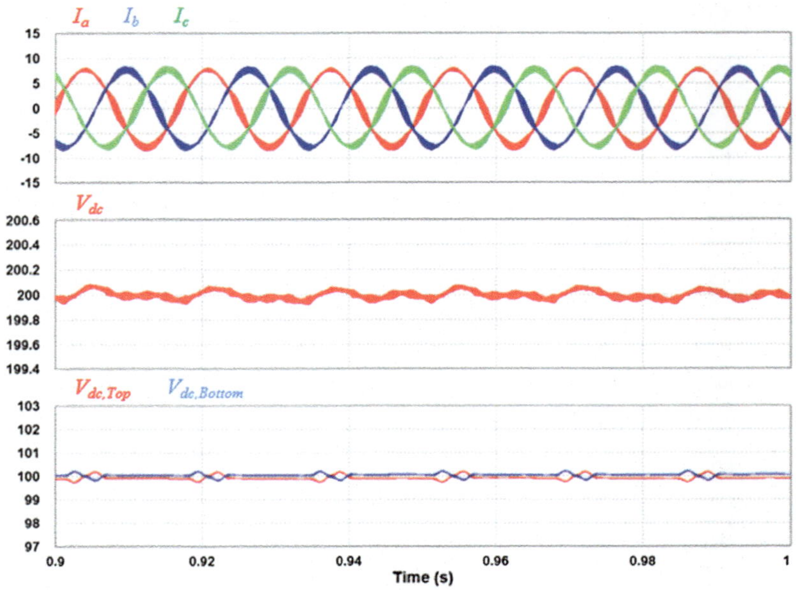

(a) R2LS-TC

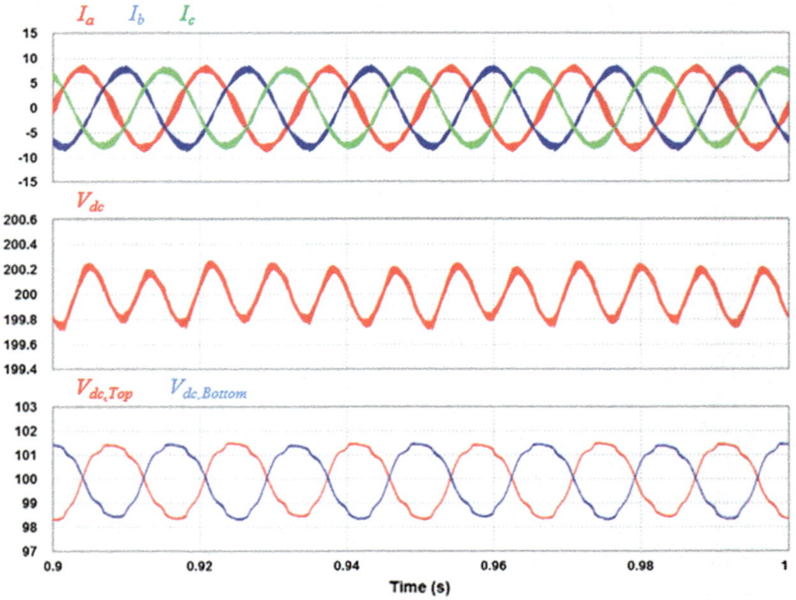

(b) 3LSO-TC

Fig. 3.31 DC-link voltages of tolerance control for an S_{a2} open-switch fault at a unity power factor (reprinted from Ref. [6], Fig. 15)

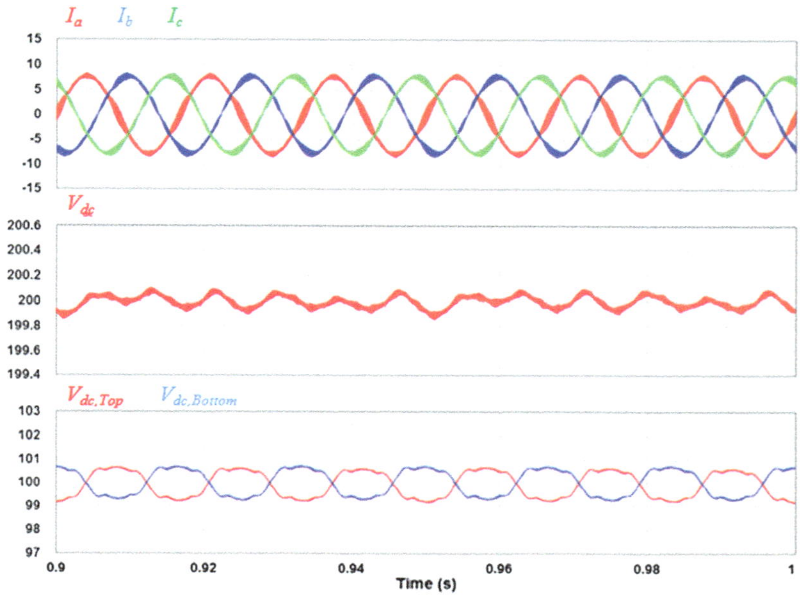

(c) OP2LS-TC

Fig. 3.31 (continued)

balancing control are decided depending on the position of an open-switch fault. Therefore, the neutral-point voltage balancing control in 3LSO-TC has a limitation since it can counteract the unidirectional neutral-point voltage unbalancing.

OP2LS-TC: This control guarantees a lower current THD. However, the efficiency decreases and the DC-link voltage ripple increases. Like 3LSO-TC, the neutral-point voltage balancing control can counteract the unidirectional neutral-point voltage unbalancing.

3.2.3 *Tolerant Control for S_{x1} and S_{x4} Open-Circuit Fault*

This chapter focuses on open-circuit faults of the outer switches in three-level topologies. In general, tolerant controls for the T-type topology cannot be applied as tolerant control for the NPC topology because of their structural differences. The effects caused by an open-circuit fault of the outer switches in both the T-type and NPC topologies are the same. Therefore, tolerant control can be applied to open-circuit faults of the outer switches in both T-type and NPC topologies.

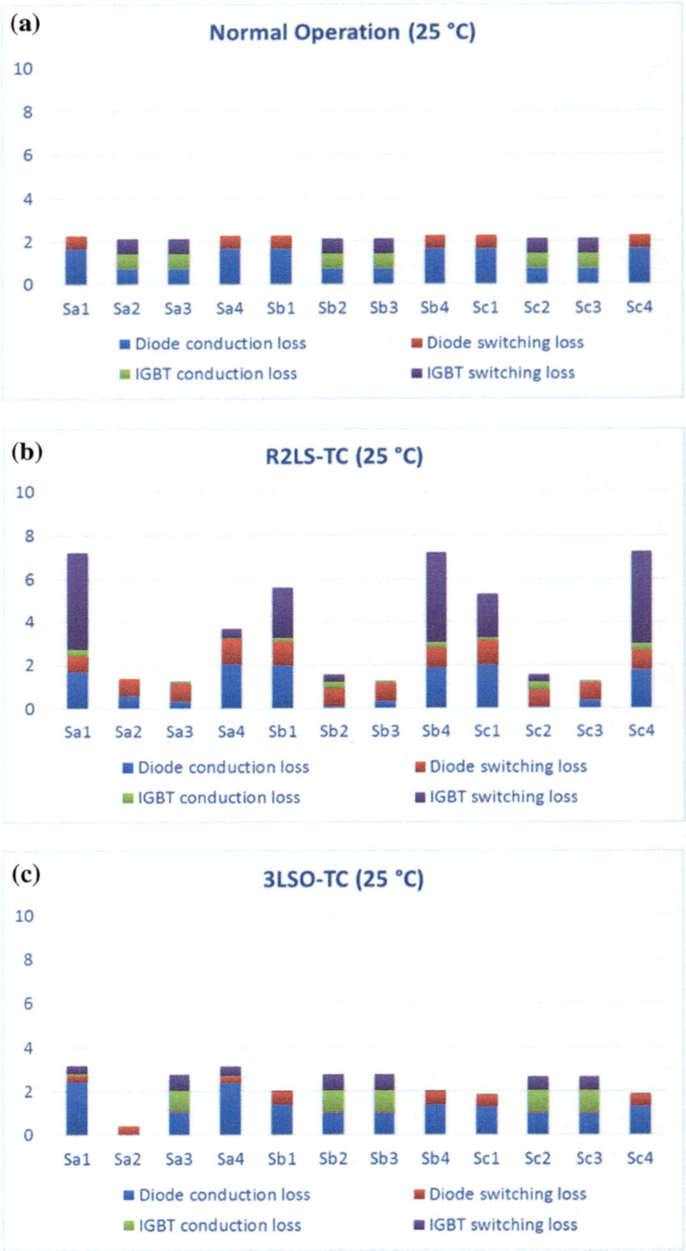

Fig. 3.32 Loss analysis results of the tolerance controls for an S_{a2} open-switch fault (reprinted from Ref. [6], Fig. 16)

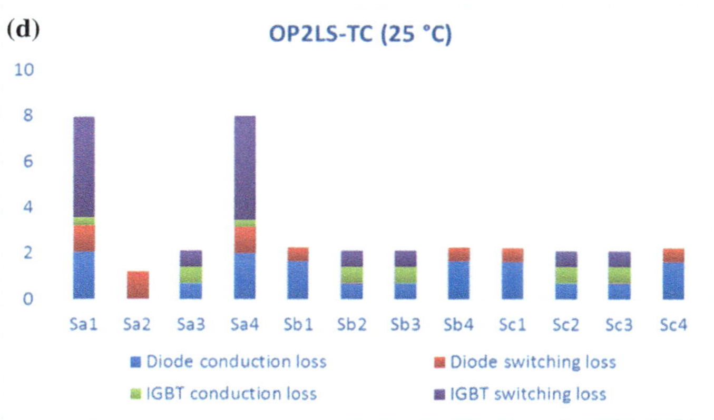

Fig. 3.32 (continued)

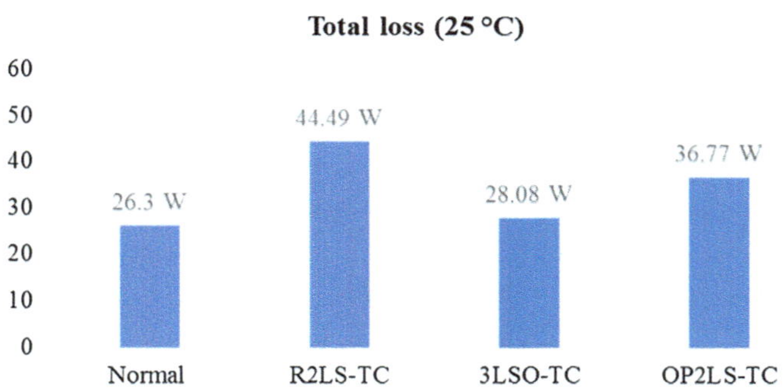

Fig. 3.33 Total loss comparison of the tolerance controls for an S_{a2} open-switch fault (reprinted from Ref. [6], Fig. 17)

However, the operating mode must be considered to develop the tolerant control because the input currents of a three-level rectifier do not flow through the outer switches (S_{x1} and S_{x4}) at a unity power factor (*PF*). If not, in the inverter mode, open-circuit faults of the outer switches become critical.

S_{x1} and S_{x4} open-circuit faults cause current distortion when a three-level topology operates in the inverter mode. In this mode, an S_{x1} open-circuit fault makes the P-switching state infeasible, and an S_{x4} open-circuit fault makes the N-switching state infeasible. Therefore, when there is an S_{a1} open-circuit fault, the voltage vectors containing the P-switching state in the a-leg cannot be generated as shown in Fig. 3.34. These vectors are [PPN], [PNN], [PNP], [PON], [PNO], [PPO], [POO], and [POP].

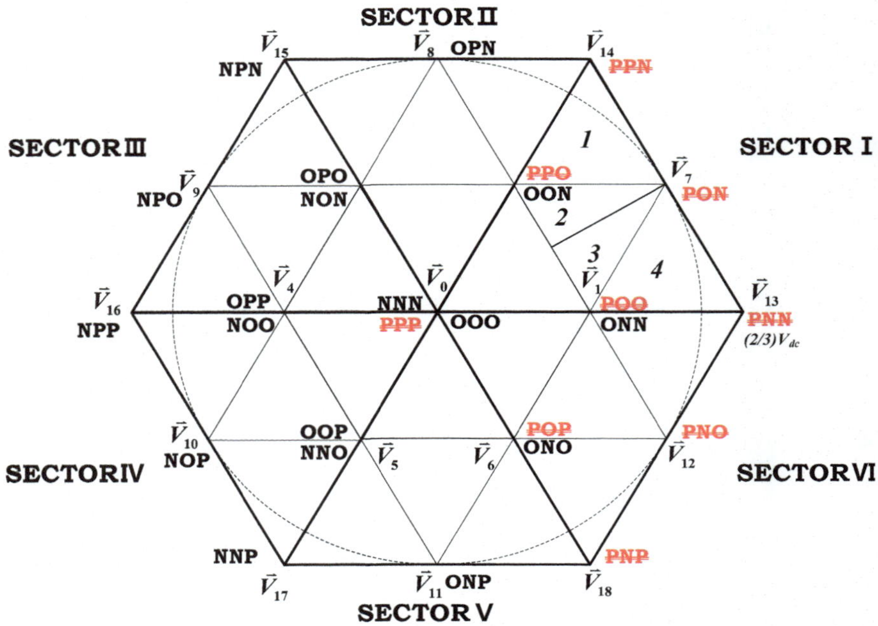

Fig. 3.34 Space vector diagram for tolerant control when an S_{a1} open-circuit fault occurs

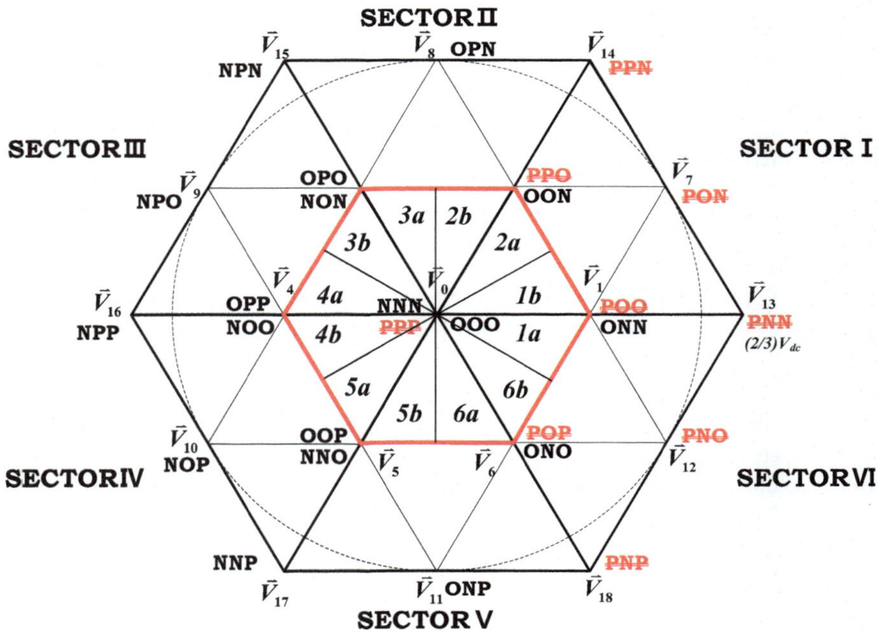

Fig. 3.35 Space vector diagram for tolerant control when an S_{a1} open-circuit fault occurs

A. Three-level inverter tolerant control with a limited output voltage range [10].

If an open-circuit fault occurs in the S_{x1} and S_{x4} switches, the output phase currents are considerably distorted and the neutral-point voltage is unbalanced as mentioned previously. This is because it is impossible to generate large-voltage and medium-voltage vectors that contain the P- or N- switching state states in the faulty leg. Hence, when an open-circuit fault occurs in the outer switches, the modulation index should be reduced to generate output without the impossible voltage vectors.

Taking an S_{a1} fault for an example, the large-voltage vectors [PNN], [PNP], and [PPN] and the medium-voltage vectors [PNO] and [PON] are impossible. To generate output without theses vectors, the modulation index should be reduced so that the reference voltage is in an inner hexagon, as shown in Fig. 3.35.

To assist in understanding of this method, the case of a fault in leg-a is considered. If an open-switch fault occurs in the switch S_{a1}, the modulation index needs to be reduced so that the voltage reference is in the inner hexagon (red region). Then add $T_{\min}$ to the three phase turn-on times (T_a, T_b, T_c) when the reference voltage is in regions 1, 2, and 6.

Figure 3.36 shows the switching sequence before $T_{\min}$ is added and after $T_{\min}$ is added when an S_{a1} open-switch fault occurs and the voltage reference is in region 1a. In this sequence, $T_{\min}$ is T_c. As shown in Fig. 3.36b, after $T_{\min}$ is added, the new switching sequence is [OON]-[ONO]-[OOO]-[ONO]-[ONN]. In this sequence, the

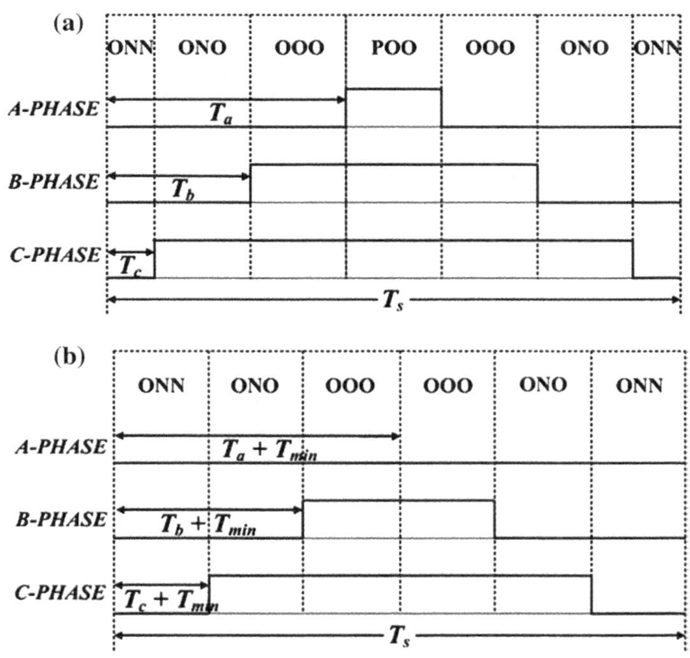

Fig. 3.36 Switching sequence when the reference voltage is in region 1a: **a** before Tmin is added and **b** after Tmin is added (reprinted from Ref. [8], Fig. 11)

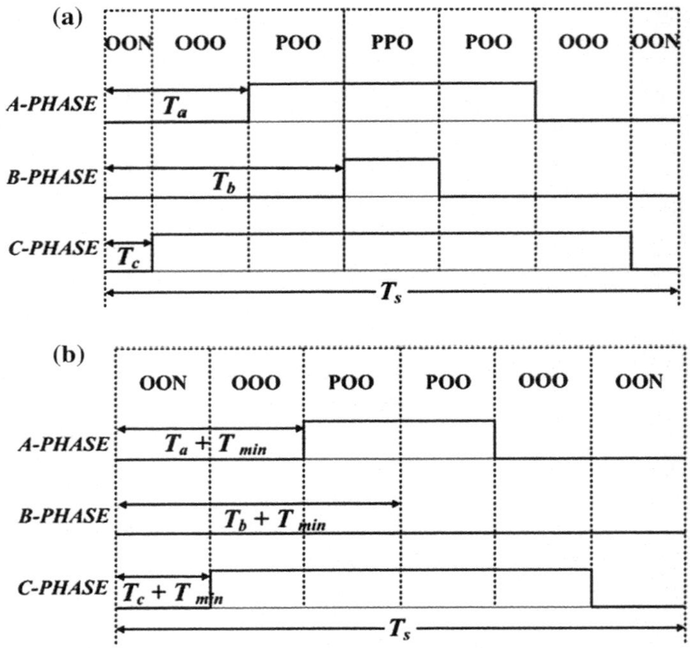

Fig. 3.37 Switching sequence when the reference voltage is in region 2a: **a** before Tmin is added and **b** after Tmin is added (reprinted from Ref. [8], Fig. 12)

dwell time of the P-type switching state [POO] becomes zero, and the dwell time of the N-type switching state [ONN] is doubled in comparison with the conventional three-level space vector modulation (SVM). This does not affect the output phase voltage because the dwell times of the other switching states are unchanged, and the small switching states [ONN] and [POO] generate the same line-to-line ac output voltage. However, in regions 2a and 6b, the P-type switching state [POO] still remains after T_{min} is added as shown in Fig. 3.37. Therefore, the P-type switching state [POO] should be replaced with the N-type switching state [ONN] as shown in Fig. 3.38a.

In regions 1, 2, and 6, the dwell time of the P-type switching state becomes zero, and the dwell time of the N-type switching state is doubled by adding T_{min} to the three phase turn on times. This causes an unbalance of the neutral-point voltage. To balance the neutral-point voltage, T_{min} should be subtracted when the reference voltage is in regions 3, 4, and 5. Furthermore, the N-type switching state [NOO] should be replaced with the P-type switching state [OPP] in regions 3b and 5a. This is because the P-type switching state [POO] is substituted with [ONN] in regions 2a and 6b. In regions 2a, 3b, 5a, and 6b, the number of switching is increased because the switching states of the small voltage vector are changed. Therefore, the switching sequence should be rearranged to reduce the number of switching. For example, in region 2a, the switching sequence is [OON]-[OOO]-[ONN]-[OOO]-[OON] as shown in Fig. 3.38a.

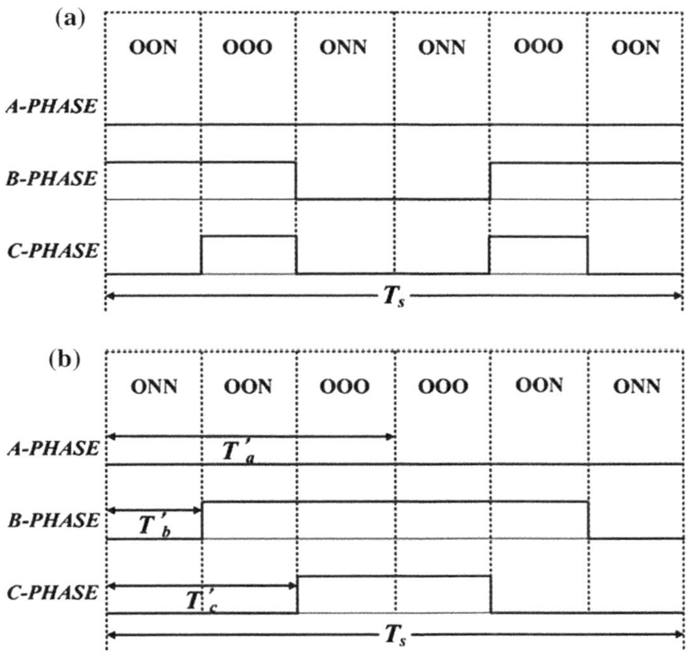

Fig. 3.38 Switching sequence when the reference voltage is in region 2a: **a** when the P-type switching state [POO] is replaced with the N-type switching state [OON] and **b** when the switching sequence is rearranged (reprinted from Ref. [8], Fig. 13)

In this switching sequence, leg c changes its switching state twice. To reduce the number of switching, the switching sequence should be rearranged as [ONN]-[OON]-[OOO]-[OON]-[ONN] as shown in Fig. 3.38b. In order to rearrange the switching sequence as shown in Fig. 3.38b, the dwell time of each switching state should be obtained. The dwell time can be expressed by the turn-on times as follows:

$$\begin{aligned} T_{dwell[OON]} &= 2(T_c + T_{\min}) = 4T_c \\ T_{dwell[OOO]} &= 2(T_a + T_c) \\ T_{dwell[POO]} &= T_{dwell[ONN]} = 2(T_b + T_a) \end{aligned} \tag{3.7}$$

Using these conditions, the three phase turn-on times (T_a, T_b, T_c) are redefined as follows:

$$\begin{aligned} T_{a,TC} &= T_s/2 \\ T_{b,TC} &= \frac{1}{2} T_{dwell[ONN]} = T_b - T_a \\ T_{c,TC} &= \frac{1}{2}\left(T_{dwell[ONN]} + T_{dwell[OON]}\right) = 2T_c + T_b - T_a \end{aligned} \tag{3.8}$$

Table 3.14 Tolerant control with limited output voltage ranges

	S_{a1}, S_{a4} open-circuit fault	S_{b1}, S_{b4} open-circuit fault	S_{c1}, S_{c4} open-circuit fault
1a	$T_{x,TC} = T_x + T_{\min}(x = a, b, c)$	$T_{a,TC} = T_a - T_b$ $T_{b,TC} = T_s/2$ $T_{c,TC} = T_s/2 + T_c - T_b$	$T_{x,TC} = T_x - T_{\min}(x = a, b, c)$
1b	$T_{x,TC} = T_x + T_{\min}(x = a, b, c)$	$T_{x,TC} = T_x - T_{\min}(x = a, b, c)$	$T_{a,TC} = T_a - T_c$ $T_{b,TC} = T_s/2 + T_b - T_c$ $T_{c,TC} = T_s/2$
2a	$T_{a,TC} = T_s/2$ $T_{b,TC} = T_b - T_a$ $T_{c,TC} = 2T_c + T_b - T_a$	$T_{x,TC} = T_x + T_{\min}(x = a, b, c)$	$T_{x,TC} = T_x - T_{\min}(x = a, b, c)$
2b	$T_{x,TC} = T_x + T_{\min}(x = a, b, c)$	$T_{a,TC} = T_a - T_b$ $T_{b,TC} = T_s/2$ $T_{c,TC} = 2T_c + T_a - T_b$	$T_{x,TC} = T_x - T_{\min}(x = a, b, c)$
3a	$T_{x,TC} = T_x - T_{\min}(x = a, b, c)$	$T_{x,TC} = T_x + T_{\min}(x = a, b, c)$	$T_{a,TC} = T_s/2 + T_a - T_c$ $T_{b,TC} = T_b - T_c$ $T_{c,TC} = T_s/2$
3b	$T_{a,TC} = T_s/2$ $T_{b,TC} = T_b - T_a$ $T_{c,TC} = T_s/2 - T_a + T_c$	$T_{x,TC} = T_x + T_{\min}(x = a, b, c)$	$T_{x,TC} = T_x - T_{\min}(x = a, b, c)$
4a	$T_{x,TC} = T_x - T_{\min}(x = a, b, c)$	$T_{a,TC} = 2T_a + T_c - T_b$ $T_{b,TC} = T_s/2$ $T_{c,TC} = T_c - T_b$	$T_{x,TC} = T_x + T_{\min}(x = a, b, c)$

(continued)

Table 3.14 (continued)

	S_{a1}, S_{a4} open-circuit fault	S_{b1}, S_{b4} open-circuit fault	S_{c1}, S_{c4} open-circuit fault
4b	$T_{x,TC} = T_x - T_{\min}(x = a, b, c)$	$T_{x,TC} = T_x + T_{\min}(x = a, b, c)$	$T_{a,TC} = 2T_s + T_b - T_c$ $T_{b,TC} = T_b - T_c$ $T_{c,TC} = T_s/2$
5a	$T_{a,TC} = T_s/2$ $T_{b,TC} = T_s/2 + T_b - T_a$ $T_{c,TC} = T_c - T_a$	$T_{x,TC} = T_x - T_{\min}(x = a, b, c)$	$T_{x,TC} = T_x + T_{\min}(x = a, b, c)$
5b	$T_{x,TC} = T_x - T_{\min}(x = a, b, c)$	$T_{a,TC} = T_s/2 + T_a - T_b$ $T_{b,TC} = T_s/2$ $T_{c,TC} = T_c - T_b$	$T_{x,TC} = T_x + T_{\min}(x = a, b, c)$
6a	$T_{x,TC} = T_x + T_{\min}(x = a, b, c)$	$T_{x,TC} = T_x - T_{\min}(x = a, b, c)$	$T_{a,TC} = T_a - T_c$ $T_{b,TC} = 2T_s + T_a - T_c$ $T_{c,TC} = T_s/2$
6b	$T_{a,TC} = T_s/2$ $T_{b,TC} = T_c + 2T_b - T_a$ $T_{c,TC} = T_c - T_a$	$T_{x,TC} = T_x - T_{\min}(x = a, b, c)$	$T_{x,TC} = T_x + T_{\min}(x = a, b, c)$

Table 3.15 Simulation and experiment parameters

System parameters			
DC-link voltage (simulation/experiment)	300 V/200 V	Load (simulation/experiment)	3mH, 15Ω/ 10mH, 10Ω
DC-link capacitor	1100 μF	Frequency of Current	60 Hz
Switching frequency	10 kHz	Control period	100 μs

Following the fault-tolerant strategy explained earlier, the redefined turn-on times for the tolerant control of a switch S_{a4} fault are the same as those of a switch S_{a1} fault. The redefined turn-on times to achieve the fault-tolerant control are presented in Table 3.14.

- Simulation and experimental results.

Simulations have been carried out to verify the validity of the algorithm. The parameters of simulation and experiment are listed in Table 3.15.

Figure 3.39 shows the simulation results of the fault tolerant strategy when an open-switch fault occurs in S_{a1}. The distortion of the phase current is eliminated after the algorithm is applied. The output line-to-line voltage is not five levels but three levels, and the magnitude of the phase current is decreased because the modulation index is reduced for the fault tolerant control. After the algorithm is applied, the output pole voltage of the A-phase is fixed to zero. This means that the switches S_{a1} and S_{a4} are not used in this method.

Figure 3.40 shows experimental results after the fault-tolerant control is applied when the switch S_{a1} is faulty. The output line-to-line voltage is not five levels but

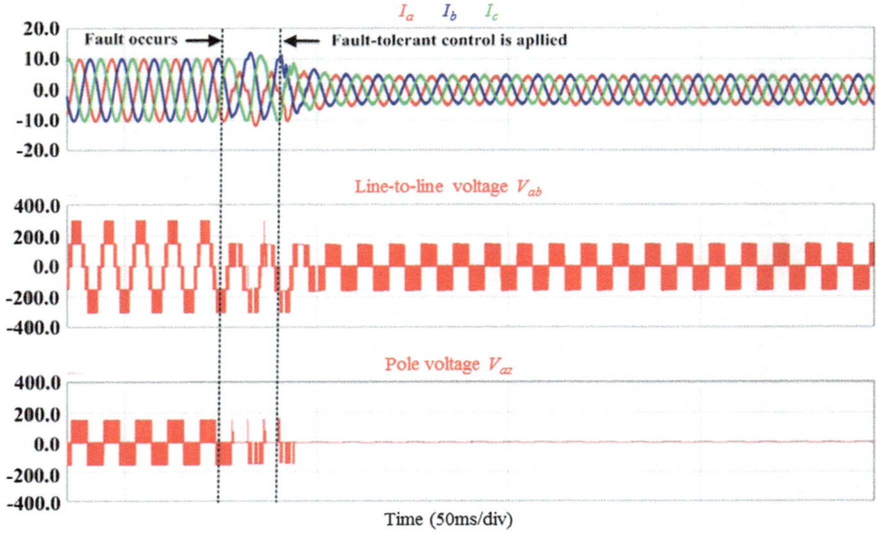

Fig. 3.39 Simulation results of fault-tolerant control when an open switch fault occurs in switch S_{a1} (reprinted from Ref. [8], Fig. 20)

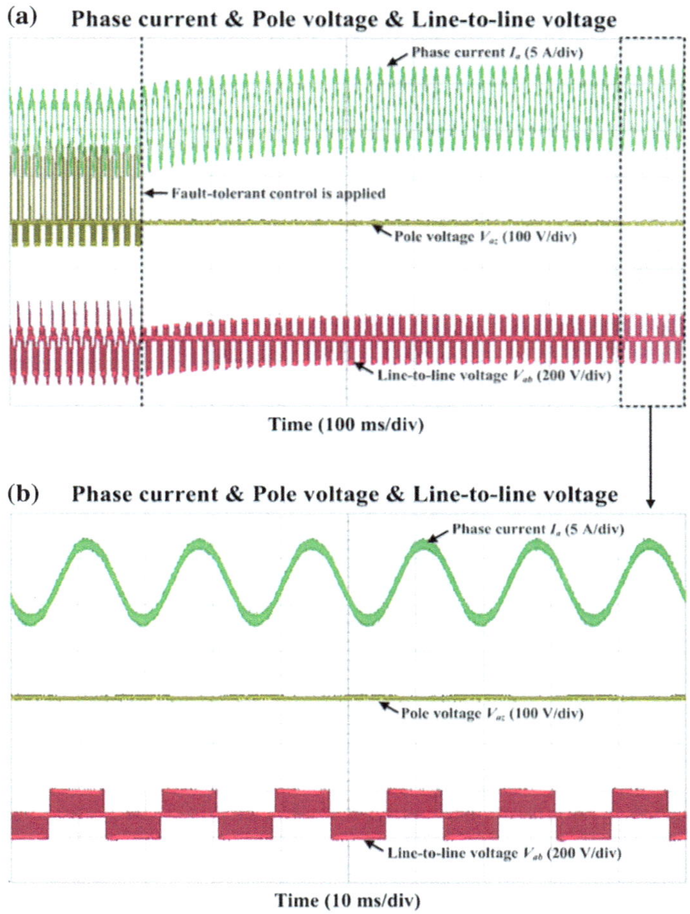

Fig. 3.40 Output of a T-type three-level inverter when tolerant control is applied under the switch S_{a1} fault condition (reprinted from Ref. [8], Fig. 32)

three levels and the magnitude of the output phase current is decreased because the modulation index is reduced. As a result, the reference voltage is in the inner hexagon of the space vector diagram. In this control, the output pole voltage is fixed to zero. This means that the switches S_{a1} and S_{a4} are not used in this method. Therefore, distortions due to the P- and N-switching states does not occur. Although the magnitude of the output phase current is decreased, the distortion of the phase current is eliminated.

B. Three-level rectifier tolerant control using reactive current injection [11].

If a three-level rectifier operates at a unity power factor, an S_{x1} or S_{x4} open circuit fault does not affect the current or cause current distortion. However, as mentioned in

Sect. 2.3, when a three-level rectifier is used in a WTS, the current distortion caused by an S_{x1} or S_{x4} open-circuit fault can be generated in following cases:

- Case 1: when the V_{EMF} is small and a large I_{rec} is required.
- Case 2: when the reactive current is injected for an IPMSG.

The current distortion is caused by the impossibility of the P-switching state in the negative current or the impossibility of the N-switching state in the positive current when an S_{x1} or S_{x4} open-switch fault occurs, respectively. The time intervals of A and C containing infeasible switching states should be eliminated to restore the current distorted by an S_{x1} or S_{x4} open-switch fault. Therefore, the proposed tolerance control moves I_{rec} to the new current $I_{rec,new}$, which is the red line, as shown in Fig. 3.41.

Moving I_{rec} is performed by injecting the d-axis current I_{de}. In a WTG system, d-axis and q-axis current controllers based on the synchronous reference frame are used to control the flux and torque of PMSGs. The q-axis current I_{qe} only indicates the torque component and is controlled to transfer power from the PMSG to the DC-link capacitor. I_{de} is normally controlled to be 0 because this indicates the flux component.

To completely remove current distortion, an exact I_{de} should be injected. An I_{de} that is more or less than the exact I_{de}, extends the time intervals of A and C containing infeasible switching states, and the current either cannot be restored or becomes more distorted. To determine the value of the exact I_{de}, the phase angle difference $\Delta\theta$ between V_{EMF} and V_{ref} should be calculated. Figure 2.23 shows a circuit model of one phase between the PMSG and the rectifier. The phase of I_{rec} corresponds with the phase of V_{EMF}, and V_{ref} is expressed with θ. I_{rec} is expressed as:

$$\frac{v_{EMF} - v_{ref}}{Z} = \frac{V_{EMF}\angle 0 - V_{ref}\angle\theta}{R_s + j2\pi f_s L_s} = I_{rec}\angle 0 \tag{3.9}$$

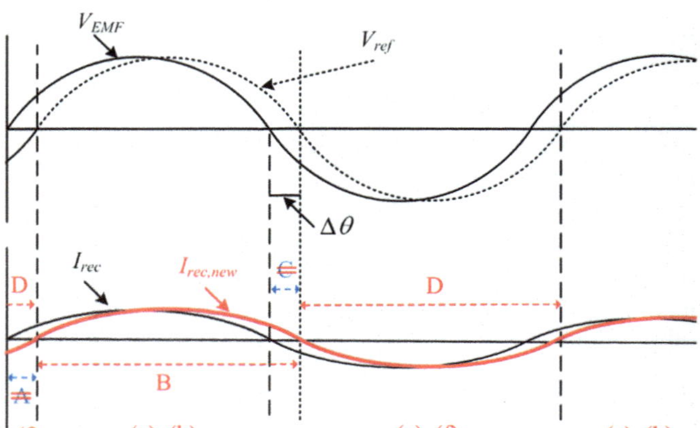

Fig. 3.41 Current change by the proposed tolerance control (reprinted from Ref. [9], Fig. 8)

where L_s and R_s are the stator inductance and resistance of the PMSG, and f_s is the angular frequency of the PMSG.

By using (1), θ is calculated as:

$$\theta = \tan^{-1}\left(\frac{-I_{rec} \times 2\pi f_s L_s}{V_{EMF} - I_{rec} R_s}\right) \tag{3.10}$$

Due to the d-axis and q-axis current controllers, I_{de} is zero, and I_{rec} is the same as I_{qe}. The injected I_{de} is calculated using the calculated θ, and this value is expressed as:

$$I_{de} = I_{qe} \tan\theta \tag{3.11}$$

A vector diagram of the tolerance control is shown in Fig. 3.42. The injected I_{de} changes I_{rec} to $I_{rec,new}$, and the phases of $I_{rec,new}$ and V_{ref} become equal. In using the proposed tolerance control, the rated current I_{rated} should be considered. Therefore, the injected I_{de} is limited by:

$$I_{de} < I_{de,\max} = \sqrt{I_{rated}^2 - I_{qe}^2} \tag{3.12}$$

The tolerance control for S_{x1} and S_{x4} open-switch faults can be applied to both NPC and T-type rectifiers. Moreover, this control guarantees normal operation when one or more open-switch faults of S_{a1}, S_{a4}, S_{b1}, S_{b4}, S_{c1}, or S_{c4} occur until the WTG system is safely stopped.

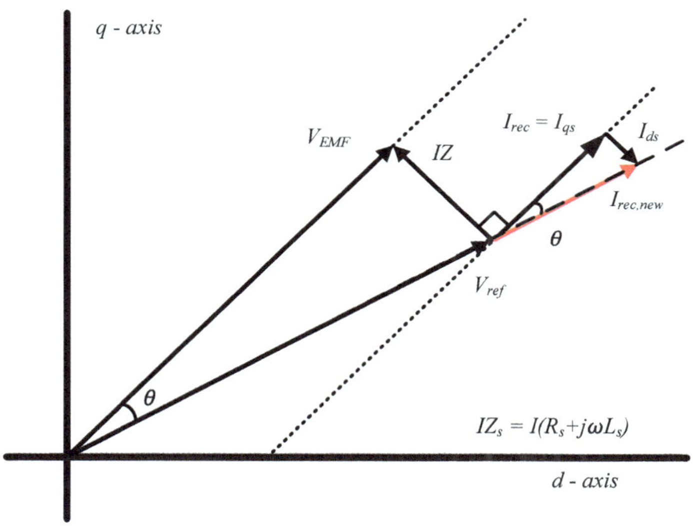

Fig. 3.42 Vector diagram of the proposed tolerance control for S_{x1} and S_{x4} open-switch faults in the stationary frame (reprinted from Ref. [9], Fig. 10)

Abruptly stopping a WTG system is difficult owing to the inertia of the PMSG and blades. The first purpose of the tolerance control is to maintain stable operation of a system without additional faults until the system is completely stopped. The proposed tolerance control satisfies the first purpose. Additionally, although the power factor of the rectifier slightly decreases, the WTG system continues on generating power.

- Simulation and experimental results of tolerant control using reactive current injection.

Simulations and experiments are performed using PSIM Tool to verify the proposed tolerance control. The parameters of the PMSG are shown in Table 3.16. The DC-link voltage V_{dc} is 600 V (V_{top} and V_{bottom} are 300 V), the total DC-link capacitance is 1100 μF, the switching frequency is 10 kHz, the control period is 100 μs, and the grid line-to-line voltage is 380 V_{rms}. I_{rec} is controlled at 10 and 20 A. The neutral-point voltage balancing control is performed on the grid-side inverter. These parameters are summarized in Table 3.16.

Figure 3.43a shows the performance of the tolerance control of a three-level rectifier at 600 rpm. In Fig. 3.34, the currents are distorted by an S_{a1} open-switch fault before the tolerance control is applied. However, the current distortion is removed by the injected I_{de}, which is about 6 A. Figure 3.43b shows the performance of the tolerance control of a three-level rectifier at 1200 rpm. Like the simulation of 600 rpm, the distorted currents are restored after the tolerance control is applied. The injected I_{de} is almost the same as that of the 600 rpm case. This value is not changed because f_s and V_{EMF} are increased at the same rate, and the voltage dip due to $I_{rec}R_s$ of the denominator is much smaller than V_{EMF}. In Fig. 3.43c, an S_{a4} open-switch fault occurs at 0.35 s. From 0.35 s, the current is distorted and V_{dc} has a ripple. The current distortion and V_{dc} ripple are removed by the tolerance control after 0.45 s. Additionally, the tolerance control guarantees a sinusoidal current for the range of the speed transition.

Figure 3.44 shows the performance of the tolerance control for S_{a1} open-switch faults of a three-level rectifier under two speed conditions (600 and 1200 rpm) of experiment. The I_{de} of the tolerance control is calculated and then injected. In this case, due to the 600 rpm of the PMSG, f_s is 30 Hz, and V_{EMF} is 64.53 V. Therefore, the calculated θ by (2) is 7.9°, and the calculated I_{de} is 1.39 A, which is identified as

Table 3.16 Simulation and experiment parameters

Generator			
Rated power	11 kW	Rated line-to-line voltage	191 V_{rms}
pole	6	Rated speed	1450 rpm
Stator resistance	0.099 Ω	d/q inductance	4.07/4.65 mH
System parameter			
DC-link voltage	600 V	Grid line-to-line voltage	380 V_{rms}
DC-link capacitance	1100 μF	Grid frequency	60 Hz
Switching frequency	10 kHz	Control period	100 μs

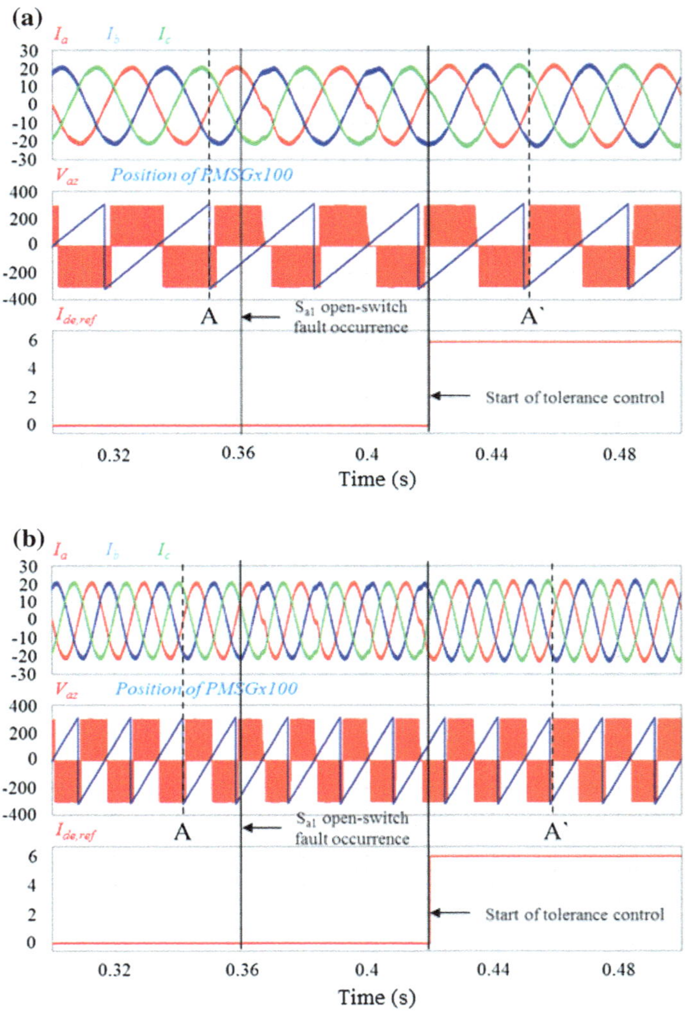

Fig. 3.43 Tolerance control for a three-level rectifier: **a** S_{a1} open-switch fault, 600 rpm, **b** S_{a1} open-switch fault, 1200 rpm, **c** S_{a4} open-switch fault, 1200 rpm (reprinted from Ref. [9], Figs. 15 and 16)

shown in Fig. 3.44a. The current distortion is eliminated after the tolerance control is applied. Figure 3.44b shows the performance of the tolerance control at 1200 rpm. The I_{de} by the tolerance control is almost the same as that at 600 rpm because f_s and V_{EMF} are increased at the same rate and the voltage dip due to $I_{rec}R_s$ of the denominator is much smaller than V_{EMF}. The current distortion is eliminated after the tolerance control is applied. The distorted and restored current THD values are measured, as shown in Table 3.17.

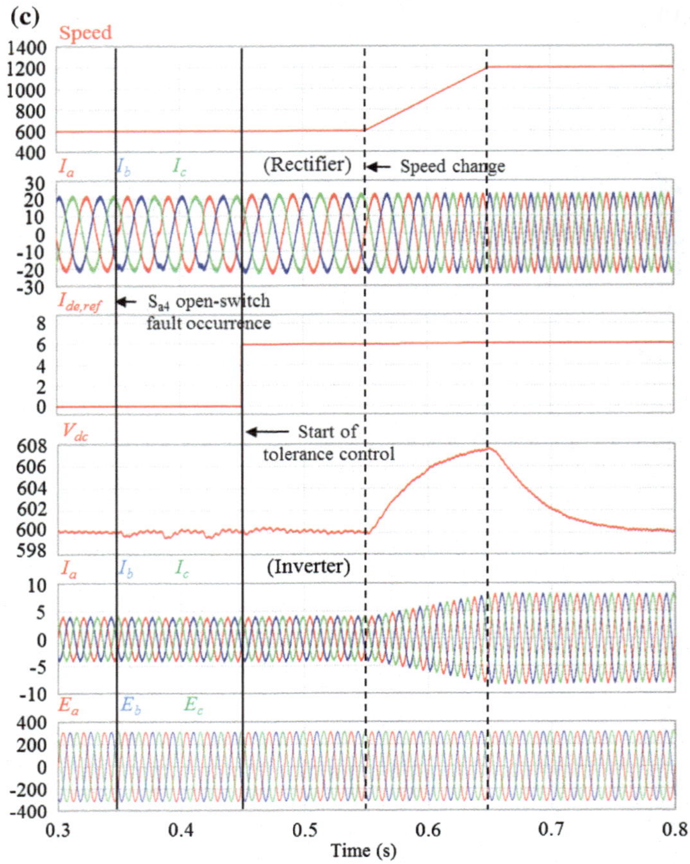

Fig. 3.43 (continued)

C. Three-level rectifier tolerant control using offset voltage injection [12].

In the above mentioned tolerant control for S_{x1} and S_{x4} open-circuit faults, the control method injects the exact reactive current required to eliminate the current distortion. However, this means that the *PF* is changed.

In tolerant control using offset voltage, a switch with an open-circuit fault is not used to generate the input voltages of the three-level rectifier by adding a compensation value to the reference voltages. The compensation value is simply calculated and the *PF* does not change in the tolerant control.

Figure 3.45 shows the input current generation process of a rectifier for cases I (when V_{EMF} is small and a large I_{rec} is required) and II (when reactive current is injected for the IPMSG).

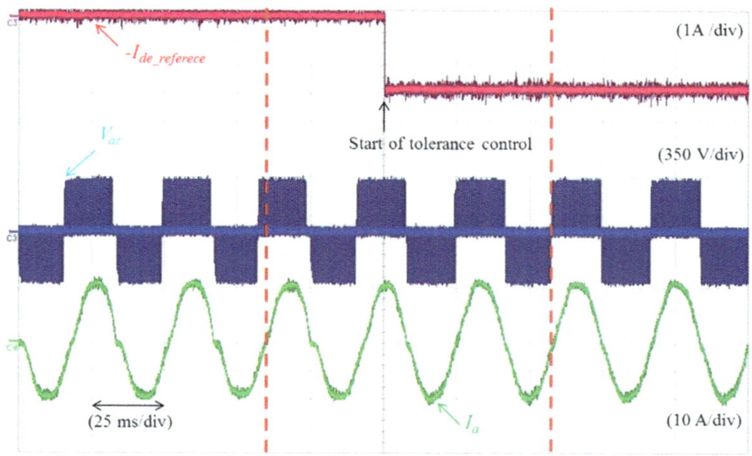

(a) S_{a1} open-switch fault

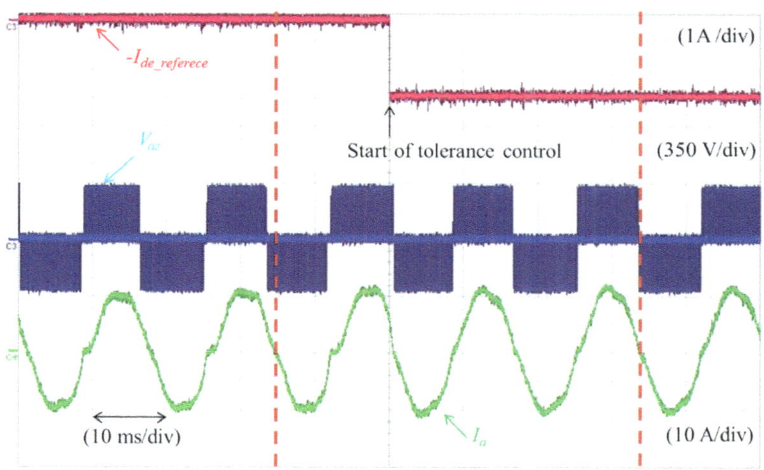

(b) S_{a4} open-switch fault

Fig. 3.44 Experimental result of the proposed tolerance control for Sx1 and Sx4 open-switch faults: **a** 600 rpm, **b** 1200 rpm (reprinted from Ref. [9], Fig. 21)

Table 3.17 Current THD values

	Normal condition	S_{a1} open fault	Appling tolerant control
Current THD of a leg (%)	11	14.5	11.1

There are two phase differences as shown in Fig. 3.45: the phase difference (θ_Z) between V_{EMF} and V_{rec} explained in [22], and the phase difference (θ_{PF}) between I_{ref} and V_{EMF} caused by the *PF*. In Fig. 3.45, part B (or part D) consists of θ_Z and

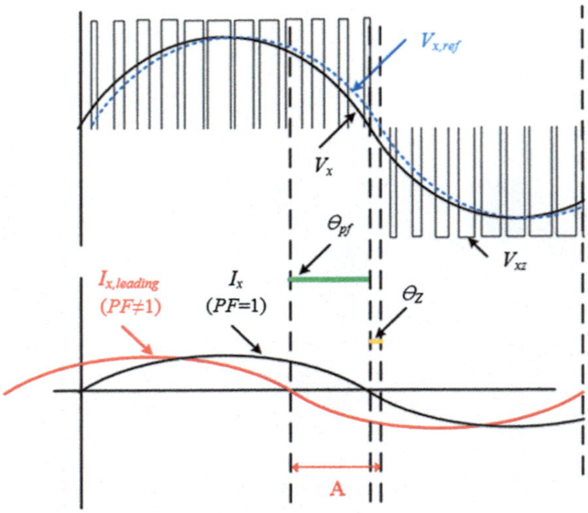

Fig. 3.45 Rectifier operation with an IPMSG at any PF (reprinted from Ref. [10], Fig. 4)

ϕ_{PF}, and their lengths increase. This means that current can be more distorted by an open-circuit fault of the outer switches compared to when θ_Z alone is considered. Case I can be ignored because θ_Z is determined based on the operating conditions of the rectifier and PMSG. However, because θ_{PF} is determined by the *PF*, Case II should be considered when an IPMSG is employed.

The current distortion caused by an open-circuit fault of the outer switches is shown in Fig. 3.46 for various *PF*s. Owing to an infeasible open-circuit fault switch, the current becomes zero during the range consisting of θ_Z and θ_{PF}.

θ_Z and θ_{PF} of A in Fig. 3.45 can be calculated by (3.9) and (3.10). If the d–q control theorem is used, θ_{PF} can be calculated as:

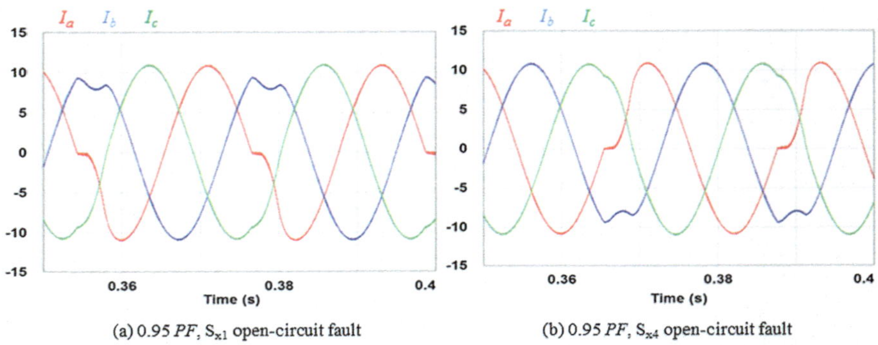

Fig. 3.46 Current distortion depending on an open-circuit fault and the PF (reprinted from Ref. [11], Fig. 5)

$$\theta_{PF} = \cos^{-1}\left(\sqrt{\frac{I_{qe}^2}{I_{qe}^2 + I_{de}^2}}\right) \tag{3.13}$$

where I_{de} indicates the d-axis current related to the flux, and I_{qe} indicates the q-axis current related to the torque. In addition, these values are in the d–q synchronous rotating frame. θ_Z and θ_{PF}, which are calculated from (3.10) and (3.13), are located near the zero-crossing point of V_{EMF} as shown in Fig. 3.47. Therefore, the compensation position for adding V_{comp} is defined on the basis of V_{EMF}'s angle (θ_{EMF}). Figure 3.48 shows three-phase V_{EMF} s and θ_{EMF}. θ_{EMF} is acquired from an encoder or a position sensor. Six zero-crossing points are expressed for every 60°, which are matched to each open-circuit fault as shown in Fig. 3.48.

Consequently, the value of θ_{EMF} representing each zero-crossing point is a criterion for adding V_{comp}. For example, when an S_{a1} open-circuit fault occurs, V_{comp} should be added for the compensation range from (0° − θ_{PF}) to (0° + θ_Z), which is based on 0°.

In the proposed tolerant control, the reference voltage of a phase containing an S_{x1} open-circuit fault is changed to zero as shown in Fig. 3.47. As a result, only the O-switching state is used in part B. To make the reference voltage zero, $|V_{comp}|$ is assigned to the magnitude of the reference voltage ($V_{x,ref,offset}$) containing an open-circuit fault, and V_{comp} can be expressed as:

$$V_{comp} = -V_{a,ref,offset} \tag{3.14}$$

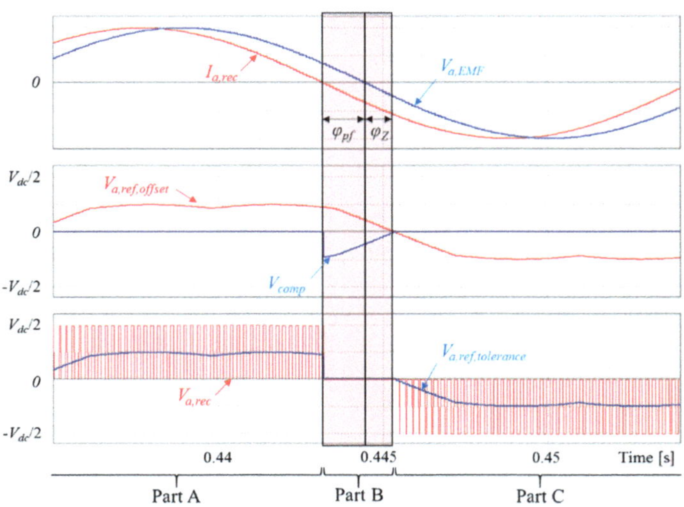

Fig. 3.47 Change of the reference voltages in the proposed tolerant control for an S_{a1} open-circuit fault (0.95 PF) (reprinted from Ref. [12], Fig. 6)

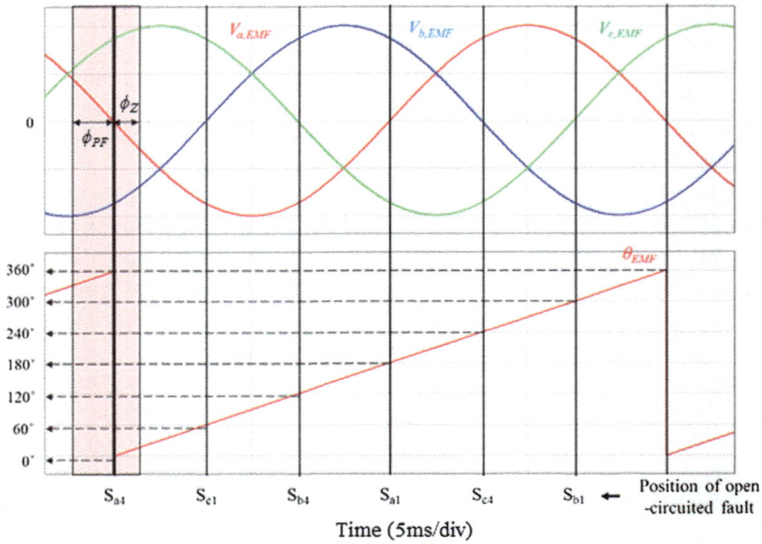

Fig. 3.48 Compensation position on the basis of an VEMF's angle (θEMF) (reprinted from Ref. [12], Fig. 7)

The proposed tolerant control is implemented by adding V_{comp} to the reference voltages ($V_{x,ref,offset}$, $x = a, b, c$). The new reference voltages ($V_{x,ref,tolerance}$, $x = a, b, c$) of the proposed tolerant control are expressed as:

$$\begin{aligned} V_{a,ref,tolerance} &= V_{a,ref,offset} + V_{comp} \\ V_{b,ref,tolerance} &= V_{b,ref,offset} + V_{comp} \\ V_{c,ref,tolerance} &= V_{b,ref,offset} + V_{comp} \end{aligned} \tag{3.15}$$

- Considering neutral-point voltage balance.

Compensation voltage, which is one of the offset voltages, can cause neutral-point voltage unbalance because V_{comp} calculated from Table 3.17 is a one-sided voltage. Therefore, two DC-link capacitors have different values depending on the polarity of V_{comp} generated for an open-circuit fault. Neutral-point voltage unbalance increases both the voltage stress on the switch and the current THD. Tolerant control for open-circuit faults of the outer switches has to incorporate a solution for the neutral-point voltage unbalance problem. Therefore, as mentioned earlier, V_{comp} is added to the corresponding compensation position based on the position of the open-circuit fault. Then V_{comp} is also added to the diametrically opposite compensation position to balance the neutral-point voltage. Figure 3.49 shows the concept of tolerant control considering the neutral-point voltage balance when an

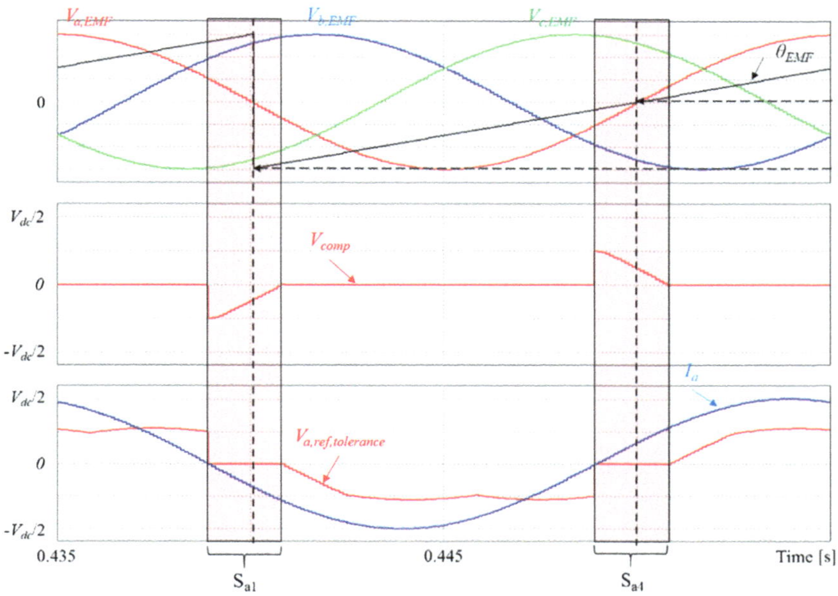

Fig. 3.49 Tolerant control considering the neutral-point voltage balance under an S_{a1} open-circuit fault (reprinted from Ref. [12], Fig. 8)

S_{a1} open-circuit fault occurs. In Fig. 3.49, V_{comp} is added to the compensation range $[(0° - \theta_{PF}) \sim (0° + \theta_Z)]$, which corresponds to the position of the S_{a1} open-circuit fault. In addition, V_{comp} is also added to the diametrically opposite compensation range $[(180° - \theta_{PF}) \sim (180° + \theta_Z)]$, which is the range for the S_{a4} open-circuit fault. The two V_{comp}s added to two positions have opposite polarities, which results in a balanced neutral-point voltage.

The final principles of tolerant control with neutral-point voltage balance are summarized in Table 3.18.

- Limitation of this tolerant control.

$V_{x,ref,tolerance}$ cannot exceed a limitation voltage (V_{limit}), which is restricted by the DC-link voltage (V_{dc}). Therefore, V_{comp} is limited as follows:

$$V_{comp} < V_{\lim it} - V_{ref,\max} \tag{3.16}$$

where V_{limit} is $V_{dc}/2$.

On the basis of (3.16), the applicable operation range of the tolerant control is determined depending on M_a and the PF. Figure 3.50 shows $V_{x,ref,tolerance}$ and V_{comp} of the tolerant control depending on the PF when M_a is 0.5. In Fig. 3.50, V_{comp}

Table 3.18 Principle of the proposed tolerant control depending on the position of an open-circuit fault

Position of open-circuit fault	V_{comp}	Compensation range
S_{a1}, (S_{a4})	$V_{comp} = -V_{a,ref,offset}$	$0° - \theta_Z - 0° + \theta_{PF}$
S_{c4}, (S_{c1})	$V_{comp} = -V_{c,ref,offset}$	$60° - \theta_Z - 60° + \theta_{PF}$
S_{b1}, (S_{b4})	$V_{comp} = -V_{b,ref,offset}$	$120° - \theta_Z - 120° + \theta_{PF}$
S_{a4}, (S_{a1})	$V_{comp} = -V_{a,ref,offset}$	$120° - \theta_Z - 120° + \theta_{PF}$
S_{c1}, (S_{c4})	$V_{comp} = -V_{c,ref,offset}$	$120° - \theta_Z - 120° + \theta_{PF}$
S_{b4}, (S_{b1})	$V_{comp} = -V_{b,ref,offset}$	$120° - \theta_Z - 120° + \theta_{PF}$

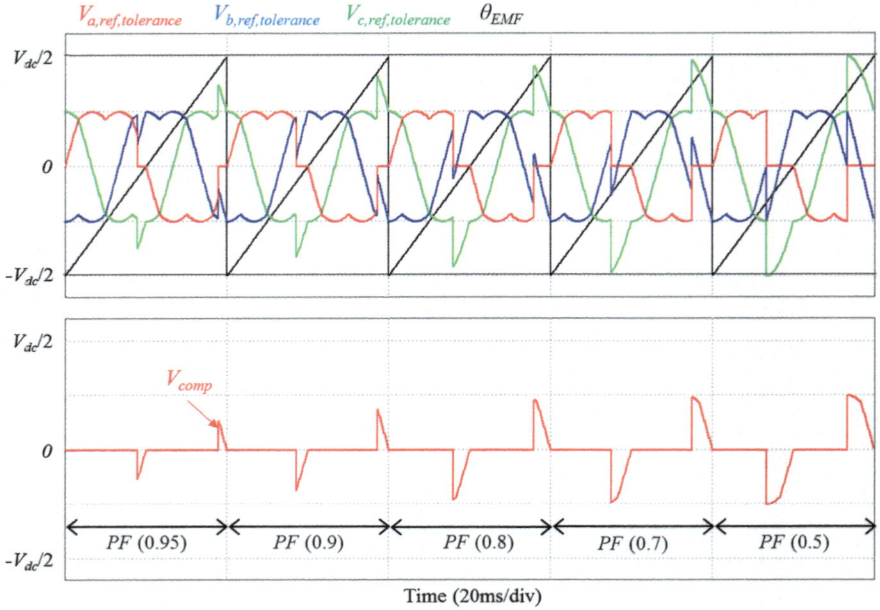

Fig. 3.50 Vx,ref,tolerance (x = a, b, c) and Vcomp depending on the PF and Ma (reprinted from Ref. [12], Fig. 9)

leads to $V_{a,ref,tolerance}$ with a zero value in the corresponding compensation range. Moreover, the peak value of $V_{c,ref,tolerance}$ increases due to V_{comp}. When the *PF* decreases, this peak value increases. However, it does not exceed V_{limit}. Consequently, when M_a is smaller than 0.5, V_{comp} can be added regardless of the *PF* because $V_{c,ref,tolerance}$ cannot exceed V_{limit}. When M_a is larger than 0.5, the

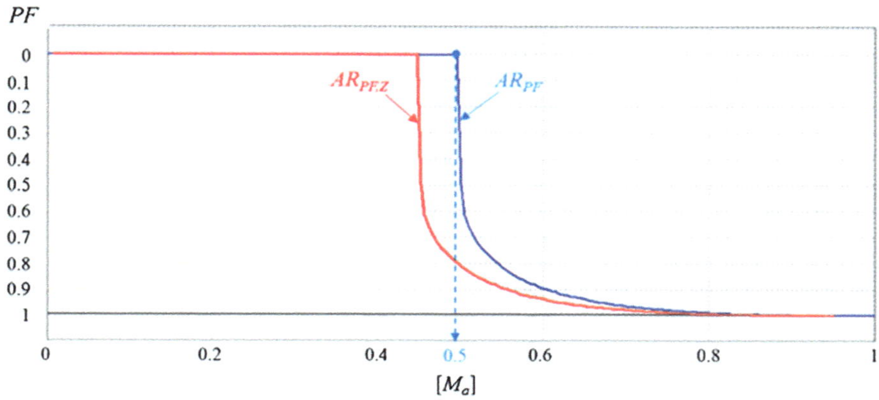

Fig. 3.51 Applicable PF range of tolerant control depending on Ma (reprinted from Ref. [12], Fig. 10)

applicable *PF* range is determined by M_a. This is because a low M_a provides a large margin for V_{comp}. However, a large V_{comp} cannot be acceptable for a high M_a.

Figure 3.51 shows the applicable *PF* range for various values of M_a. The shaded part of Fig. 3.51 represents the applicable operation range. The tolerant control is feasible over the entire factor range when M_a is smaller than 0.5. By making M_a larger than 0.5, the applicable operation range decreases. The applicable operation range shown in AR_{PF}'s is largest when only the *PF* related to θ_{PF} is taken into account. However, a large θ_Z means that a large portion of the compensation range is reserved for θ_Z, and that the rest can be used to compensate θ_{PF}. Therefore, θ_Z caused by the impedance of a PMSG reduces the applicable *PF* range, as shown in $AR_{PF,Z}$ with $\theta_Z = 10°$. θ_Z is determined by the operation conditions and parameters of the PMSG shown in (3.10).

The tolerant control has a limitation in terms of its operation range that depends on the *PF* and M_a. However, considering that wind turbine systems do not always operate at the rated wind speed (high M_a) and that the operating *PF* of the rectifier of an IPMSG is not too low, the tolerant control can clearly be effective.

- Simulation and experimental results of tolerant control using offset voltage.

Simulations and experiments are conducted to identify the performance of the tolerant control for open-circuit faults of the outer switches. The low-power prototype hardware setup consists of a power converter and an M–G set. An induction motor operates as the blades of a wind turbine systems. Therefore, a gearbox is not used in the experiment. In addition, 3L-NPC IGBT modules (Semikron, SK75MLI066T) are used in the experiment. The simulation and experimental setup

Table 3.19 Simulation and experimental parameters

Generator			
Rated power	11 kW	Rated line-to-line voltage	380 V_{rms}
Pole	6	Rated speed	1750 rpm
Stator resistance	0.349 Ω	d/q inductance	15.6/4.1 mH
System parameter			
DC-link voltage	600 V	Grid line-to-line voltage	380 V_{rms}
DC-link capacitance	1100 μF	Grid frequency	60 Hz
Switching frequency	10 kHz	Control period	100 μs

parameters are as follows: the switching frequency is 10 kHz, the control period is 100 μs, the DC-link capacitor is 1100 μF, and the DC-link voltage is 600 V. The IPMSG parameters used in the simulation and experiment are shown in Table 3.19.

Figure 3.52 shows simulation results of the tolerant control when an S_{a1} open-circuit fault occurs. The speed of the PMSG is 900 rpm, M_a is 0.5, and the *PF* of the rectifier is 0.95. Due to the S_{a1} open-circuit fault, the negative current is distorted as shown in Fig. 3.52a. After the tolerant control is applied, the reference voltages are changed by V_{comp} in the corresponding ranges [$(0° - \theta_{PF}) \sim (0° + \theta_Z)$, $(180° - \theta_{PF}) \sim (180° + \theta_Z)$], which are defined in Table 3.18. As a result, the a-phase pole voltage (V_{an}) is clamped to 0 in these ranges as shown in Fig. 3.52b, and the current distortion is eliminated completely. In addition, the two DC-link capacitor voltages are balanced.

Figure 3.53 shows the performance of the tolerant control when an S_{a1} open-circuit fault occurs in the experiment. The speed of the PMSG is 900 rpm, M_a is 0.5, and the *PF* is 0.95. Like the simulation results, the S_{a1} open-circuit fault causes current distortion, as shown in Fig. 3.53a. After the tolerant control is applied, the reference voltages are changed by V_{comp}. V_{comp} is added to the corresponding compensation ranges as shown in Fig. 3.53b. The a-phase pole voltage (V_{an}) is zero for these ranges. Therefore, the current path related to the S_{a1} open-circuit fault is not used, and current distortion is eliminated.

Figure 3.54 shows the compensation range of the tolerant control for different *PF*s. When the *PF* decreases, the compensation range is extended. Closer inspection reveals that the *PF* is only related to θ_{PF}. Therefore, θ_Z is almost unchanged in Fig. 3.54a, b.

The tolerant control can respond to open-circuit faults of two switches. In Fig. 3.55, S_{a1} and S_{b1} open-circuit faults occur, and the tolerant control for each fault is applied in order. First, the tolerant control for the S_{a1} open-circuit fault is applied and the distorted a-phase current (I_a) is corrected. Second, the b-phase current (I_b) distortion caused by the S_{b1} open-circuit fault is also eliminated.

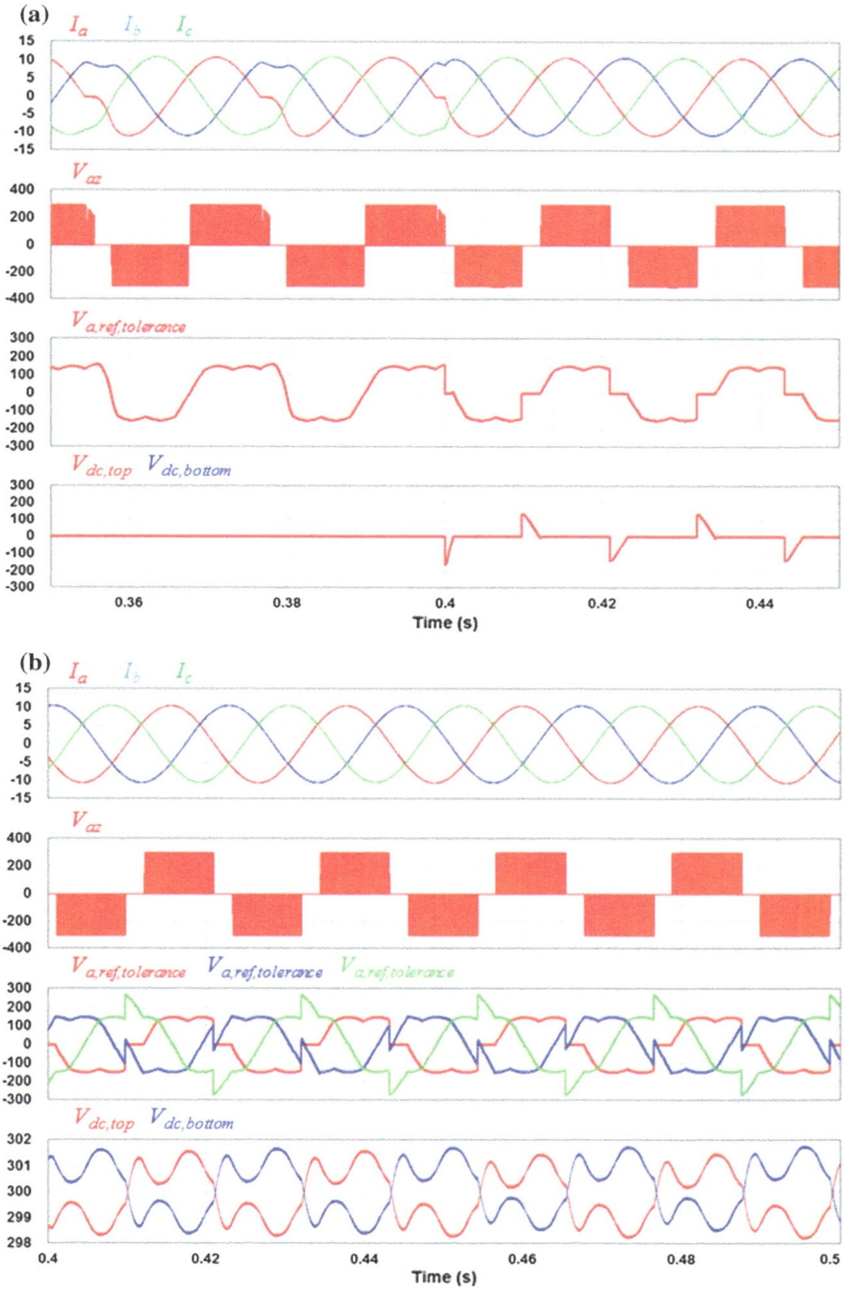

Fig. 3.52 Simulation results with the proposed tolerant control under an S_{a1} open-circuit fault (600 rpm)

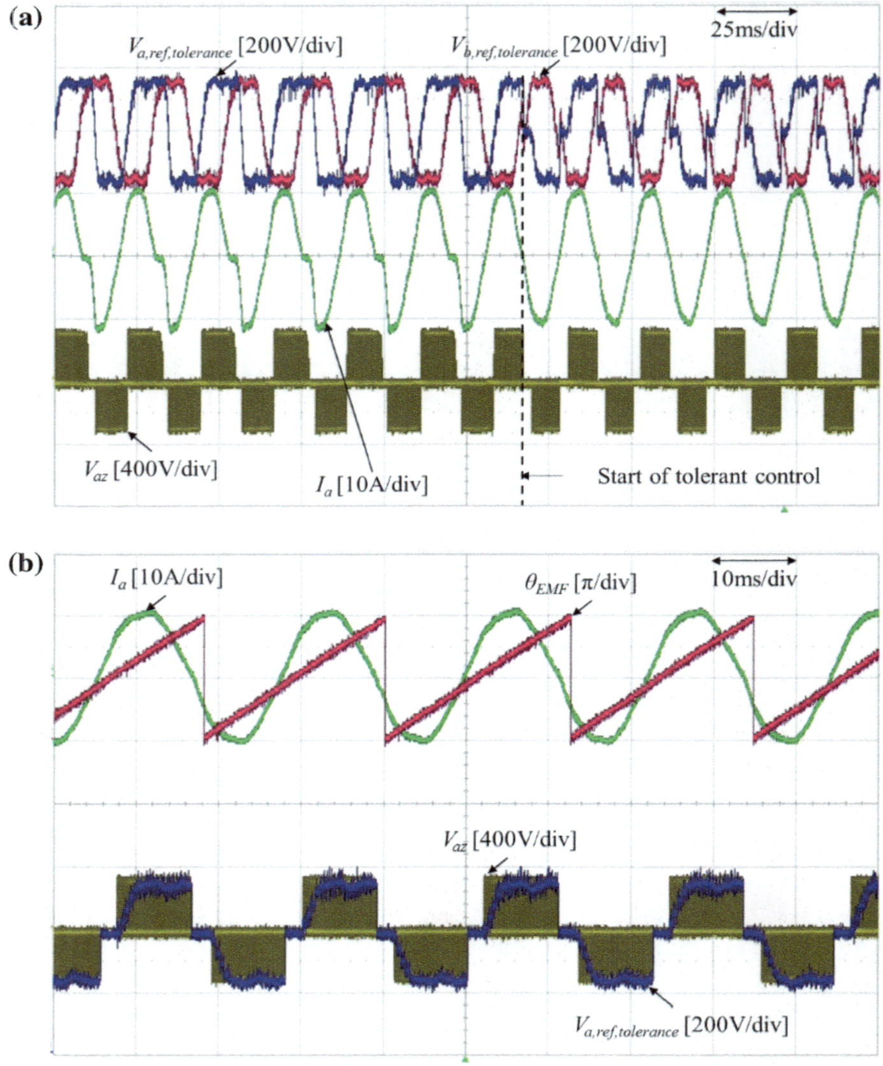

Fig. 3.53 Experimental results with the proposed tolerant control under an S_{a1} open-circuit fault (900 rpm) (reprinted from Ref. [12], Fig. 15)

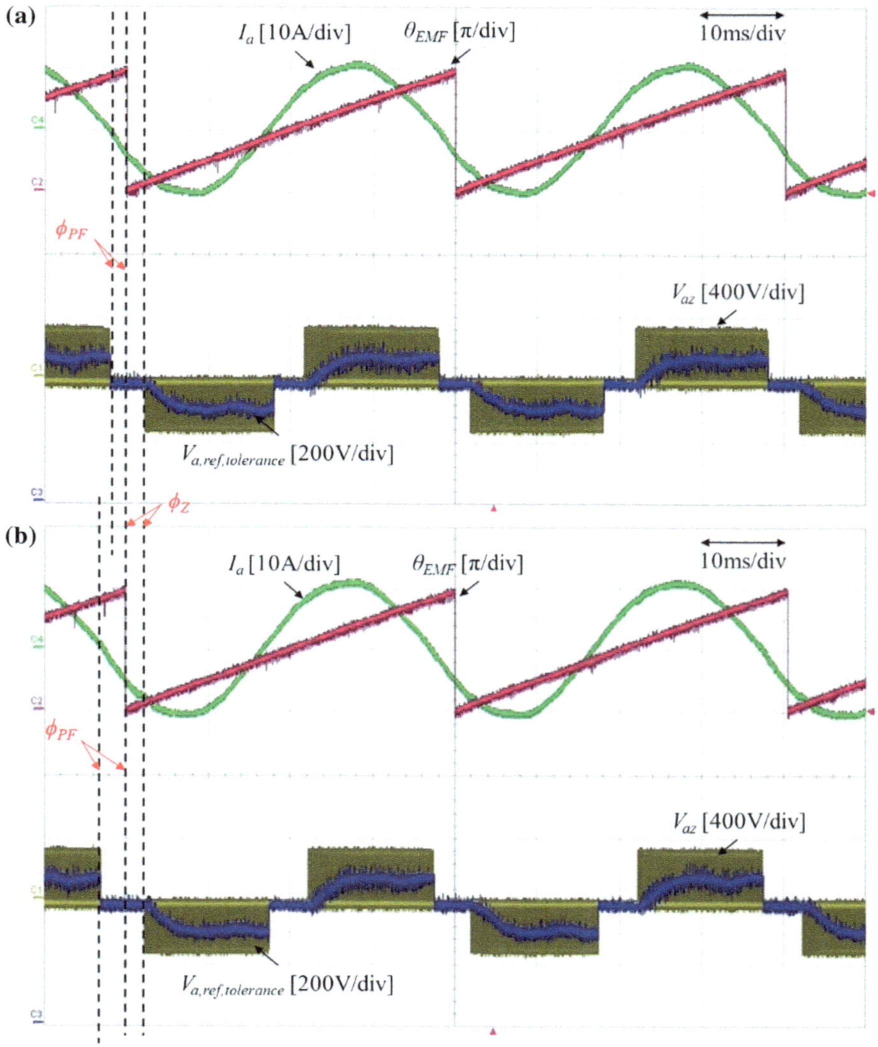

Fig. 3.54 Experimental results with the proposed tolerant control at different PFs (450 rpm, Ma = 0.25): **a** 0.95 PF, **b** 0.9 PF (reprinted from Ref. [12], Fig. 16)

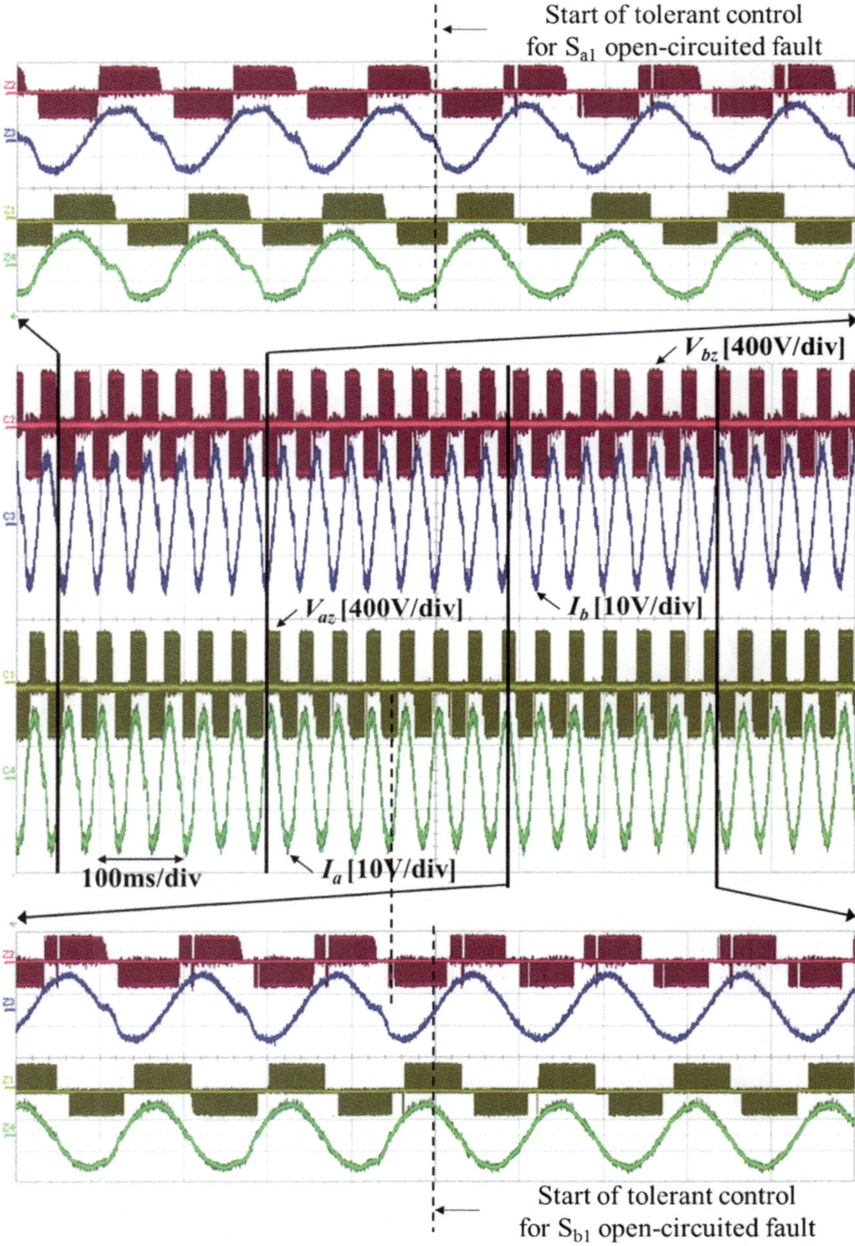

Fig. 3.55 Experimental results with the tolerant control under S_{a1} and S_{b1} open-circuit faults (450 rpm, Ma = 0.25, 0.95 PF) (reprinted from Ref. [12], Fig. 17)

References

1. Ceballos, S., J. Pou, E. Robles, I. Gabiola, J. Zaragoza, J.L. Villate, and D. Boroyevich. 2000. Three-level converter topologies with switch breakdown fault-tolerance capability. *IEEE Transaction on Industrial Electronics* 55 (3): 982–995.
2. Park, G.T., T.J. Kim, D.W. Kang, and D.S. Hyun. 2004. Control method of npc inverter for continuous operation under one phase fault condition. In *Proceeding of IAS* (pp. 2188–2193).
3. Silva, E.R., W.S. Lima, A.S. Oliveira, C.B. Jacobina, and H. Razik. 2006. Detection and compensation of switch faults in a three level inverter. In *Proceeding of PESC* (pp. 1–7).
4. Ceballos, S., J. Pou, E. Robles, J. Zaragoza, and J.L. Martín. 2010. Performance evaluation of fault-tolerant neutral-point-clamped converters. *IEEE Transaction on Industrial Electronics* 57 (8): 2709–2718.
5. Ceballos, S., J. Pou, and J. Zaragoza. 2011. Fault-tolerant neutral-point-clamped converter solutions based on including a fourth resonant leg. *IEEE Transaction on Industrial Electronics* 58 (6): 2293–2303.
6. Lee, J.S., K.B. Lee, and F. Blaabjerg. 2015. Open-Switch Fault Detection Method of a Back-to-Back Converter using NPC Topology for Wind Turbine Systems. *IEEE Transactions on Industry Applications* 51 (1): 325–335.
7. Lee, J.S., and K.B. Lee. 2014. Tolerance Control for Inner Open-Switch Faults of a T-type Three-Level Rectifier. *Journal of Power Electronics* 14 (6): 1157–1165.
8. Lee, J.S., and K.B. Lee. 2014. An Open-Switch Fault Detection Method and Tolerance Controls Based on SVM in a Grid-Connected T-Type Rectifier with Unity Power Factor. *IEEE Transactions on Industrial Electronics* 61 (12): 7092–7104.
9. Lee, J.S., U.M. Choi, and K.B. Lee. 2015. Comparison of Tolerance Controls for Open-Switch Fault in a Grid-Connected T-Type Rectifier. *IEEE Transactions on Power Electronics* 30 (10): 5810–5820.
10. Choi, U.M., K.B. Lee, and F. Blaabjerg. 2014. Diagnosis and tolerant strategy of an open-switch fault for T-type three-level inverter systems. *IEEE Transactions on Industry Applications* 50 (1): 495–508.
11. Lee, J.S., and K.B. Lee. 2015. Open-switch fault tolerance control for a three-level NPC/T-type rectifier in wind turbine systems. *IEEE Transactions on Industrial Electronics* 62 (2): 1012–1021.
12. Lee, J.S., and K.B. Lee. 2016. Open-circuit fault tolerant control for outer switches of three-level rectifiers in wind turbine systems. *IEEE Transactions on Power Electronics* 31 (5): 3806–3815.

Chapter 4
Neutral-Point Voltage Reduction

A three-level converter has two DC-links and it makes output voltage with three levels. If the two DC-link voltages are not the same, the output voltage is distorted. To prevent this problem, the PWM methods in three-level converters generate output voltage while guaranteeing that the two DC-link voltages are the same. However, due to hardware problems, a voltage difference between the two DC-link voltages can be generated. We call it as the DC component of neutral-point voltage. To eliminate it, lots of methods have been proposed and existing methods compensate the voltage difference of two DC-link voltages by measuring them [1–4].

In addition to the DC component of neutral-point voltage, an AC component, which influences the lifetime of DC-link capacitors, is generated. However, it is not generated due to hardware problems [5, 6]. The AC component is a unique characteristic of PWM methods. Different AC components are produces depending on the PWM method. A large AC component of neutral-point voltage leads to a lifetime reduction of DC-link capacitors and it requires a large value for DC-link capacitors.

In this chapter, only the AC component of neutral-point voltage is taken into account. A PWM method for reducing the AC component of neutral-point voltage and for increasing the lifetime of DC-link capacitors will be introduced.

4.1 Neutral-Point Voltage of Three-Level Converters

The main reason for the neutral-point voltage in a three-level converter is shown in this section. As mentioned in Chap. 1, space vector PWM (SVPWM) makes output voltage by combining the voltage vectors nearest to the reference voltage.

There are six areas in sector I as shown in Fig. 4.1 and Table 4.1 shows the voltage vectors in order depending on the area containing the reference voltage.

In sector I-1a, the order of the voltage vectors is [ONN]-[OON]-[OOO]-[POO]-[OOO]-[OON]-[ONN]. Among them, the [ONN] and [OON] N-type small vectors and the [POO] P-type small vector effect the variation of the neutral-point voltage.

K.-B. Lee and J.-S. Lee, *Reliability Improvement Technology for Power Converters*, Power Systems, DOI 10.1007/978-981-10-4992-7_4

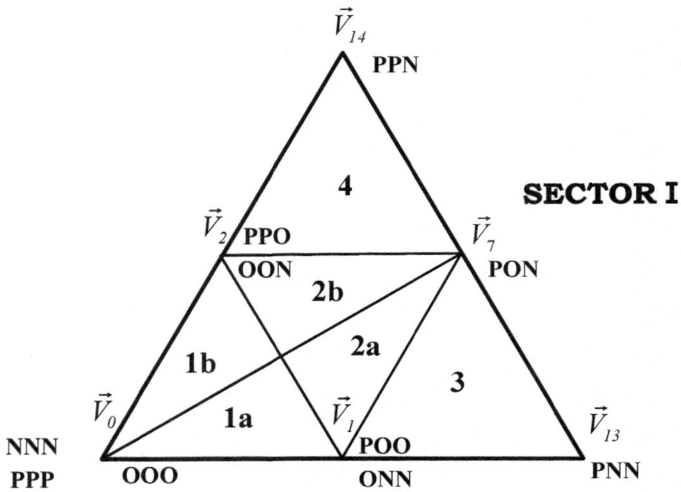

Fig. 4.1 Space vectors in sector I

Table 4.1 Voltage vectors depending on the area of sector I

Sector I						
Order	1a	1b	2a	2b	3	4
1	V_{1N}[ONN]	V_{2N}[OON]	V_{1N}[ONN]	V_{2N}[OON]	V_{1N}[ONN]	V_{2N}[OON]
2	V_{2N}[OON]	V_0[OOO]	V_{2N}[OON]	V_7[PON]	V_{13}[PNN]	V_7[PON]
3	V_0[OOO]	V_{1P}[POO]	V_7[PON]	V_{1P}[POO]	V_7[PON]	V_{14}[PPN]
4	V_{1P}[POO]	V_{2P} [PPO]	V_{1P}[POO]	V_{2P}[PPO]	V_{1P}[POO]	V_{2P}[PPO]
5	V_0[OOO]	V_{1P}[POO]	V_7[PON]	V_{1P}[POO]	V_7[PON]	V_{14}[PPN]
6	V_{2N}[OON]	V_0[OOO]	V_{2N}[OON]	V_7[PON]	V_{13}[PNN]	V_7[PON]
7	V_{1N}[ONN]	V_{2N}[OON]	V_{1N}[ONN]	V_{2N}[OON]	V_{1N}[ONN]	V_{2N}[OON]

The [ONN] and [POO] vectors have the same output voltage and they output the same dwell time. However, the [OON] N-type small vector causes variation of the neutral-point voltage and it makes the neutral-point voltage unbalanced in a switching period. In sector I-3, the [PON] medium vector leads to variation of the neutral-point voltage.

The neutral-point voltage variation in a switching period produces an AC voltage ripple with a constant frequency component in the neutral-point voltage. The frequency is depended on the PWM method used in the three-level converter. The SVPWM method and the carrier based PWM method [7, 8] using the offset voltage mentioned in Chap. 1 generate an AC voltage ripple with a frequency that is three times the fundamental frequency of the output voltage as shown in Fig. 4.2.

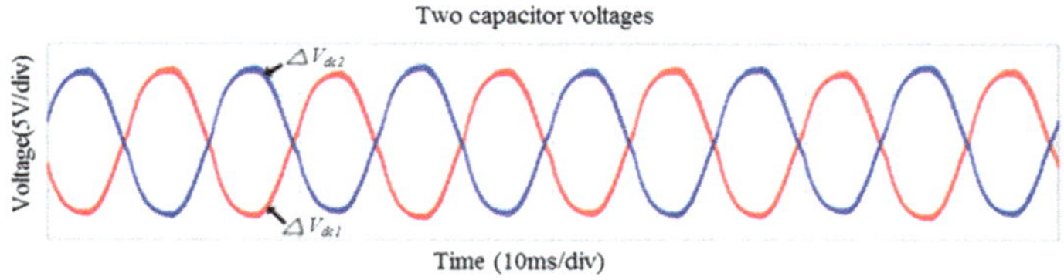

Fig. 4.2 Neutral-point voltage with 150 Hz (fundamental frequency: 50 Hz) (reprinted from [9], Fig. 4)

- Neutral-point current calculation

In a three-level converter, the neutral-point voltage ripple can be calculated by using the neutral-point current. In this section, the neutral-point current of the carrier based PWM method with the offset voltage mentioned in Chap. 1 will be calculated.

The reference voltages for the a, b, and c phases are expressed as:

$$\begin{aligned} V_{a,ref} &= V_{mag}\cos(2\pi f_f t) \\ V_{b,ref} &= V_{mag}\cos(2\pi f_f t - 2\pi/3) \\ V_{c,ref} &= V_{mag}\cos(2\pi f_f t + 2\pi/3) \end{aligned} \quad (4.1)$$

where V_{mag} is the magnitude of the reference voltages, and f_f is the fundamental frequency.

The offset value (V_{offset}) is added to each reference voltage to expand M_a. V_{offset} can be calculated using the maximum, medium, and minimum values. These three values are expressed as:

$$\begin{aligned} V_{ref,max} &= \max(V_{a,ref}, V_{b,ref}, V_{c,ref}) \\ V_{ref,mid} &= \mathrm{mid}(V_{a,ref}, V_{b,ref}, V_{c,ref}) \\ V_{ref,min} &= \min(V_{a,ref}, V_{b,ref}, V_{c,ref}) \end{aligned} \quad (4.2)$$

V_{offset} and the changed reference voltages ($V_{a,ref,offset}$, $V_{b,ref,offset}$, $V_{c,ref,offset}$) are shown in Fig. 4.3, and are expressed as:

$$V_{offset} = -\frac{(V_{ref,max} + V_{ref,min})}{2} \quad (4.3)$$

$$\begin{aligned} V_{a,ref,offset} &= V_{a,ref} + V_{offset} \\ V_{b,ref,offset} &= V_{b,ref} + V_{offset} \\ V_{c,ref,offset} &= V_{c,ref} + V_{offset} \end{aligned} \quad (4.4)$$

$V_{a,ref,offset}$, $V_{b,ref,offset}$, and $V_{c,ref,offset}$ can be represented as $V_{ref,offset,max}$, $V_{ref,offset,mid}$, and $V_{ref,offset,min}$ depending on the value, which is similar to (4.2). $V_{ref,offset,max}$,

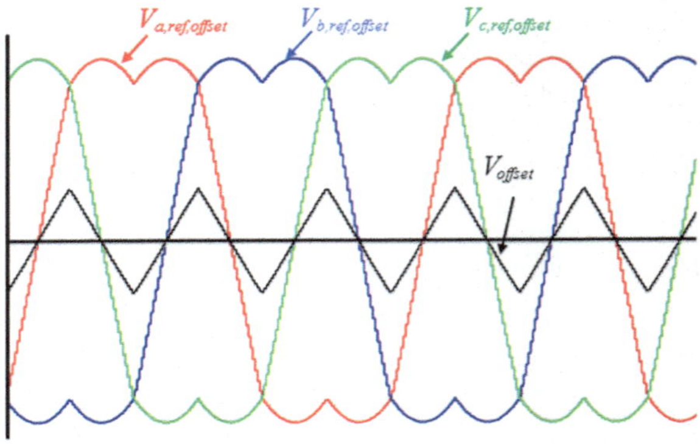

Fig. 4.3 Reference voltages (reprinted from [10], Fig. 3)

$V_{ref,offset,mid}$, and $V_{ref,offset,min}$ are compared with two carrier signals as shown in Fig. 4.4. The relationship among the reference voltages, the reference duty signals for each of the phases, and the inverter output voltages ($V_{out,offset,min}$, $V_{out,offset,mid}$, $V_{out,offset,max}$) is shown. The reference duty signals ($d_{ref,offset,max}$, $d_{ref,offset,mid}$, $d_{ref,offset,min}$) for each of the phases are represented as:

$$\begin{aligned} d_{ref,offset,max} &= 2 \times V_{ref,offset,max} / V_{dc} \\ d_{ref,offset,mid} &= 2 \times V_{ref,offset,mid} / V_{dc} \\ d_{ref,offset,min} &= 2 \times V_{ref,offset,min} / V_{dc} \end{aligned} \tag{4.5}$$

$d_{ref,offset,max}$ always causes $V_{out,offset,max}$ to be connected to point P of the DC-link during $d_{ref,offset,min} \times T_s$, and $d_{ref,offset,min}$ always causes $V_{out,offset,min}$ to be connected to point N of the DC-link during ($[1 - d_{ref,offset,min}] \times T_s$). However, the $V_{out,offset,mid}$ produced by $d_{ref,offset,mid}$ can be connected to point P or N of the DC-link depending on the sign of $d_{ref,offset,mid}$. If the dc-link voltage (V_{dc}), the three phase currents, and the reference duty signals do not change for a switching period (T_s) and the sum of the three phase currents is zero, the average neutral-point current (I_{NP}) for T_s, which is expressed in [11], can be expressed as:

$$I_{NP} = -\left(\left|d_{ref,offset,max}\right| \cdot I_{V_{ref,offset,max}} + \left|d_{ref,offset,mid}\right| \cdot I_{V_{ref,offset,mid}} + \left|d_{ref,offset,min}\right| \cdot I_{V_{ref,offset,min}}\right) \tag{4.6}$$

where I_{max}, I_{mid}, and I_{min} are the currents of the maximum, medium, and minimum reference voltages.

I_{NP} is calculated by the reference duty signals and the three phase currents. I_{NP} affects the neutral-point AC ripple voltage, and the frequency of I_{NP} is the same as that of the neutral-point AC ripple voltage. Figure 4.5 shows the two voltages and

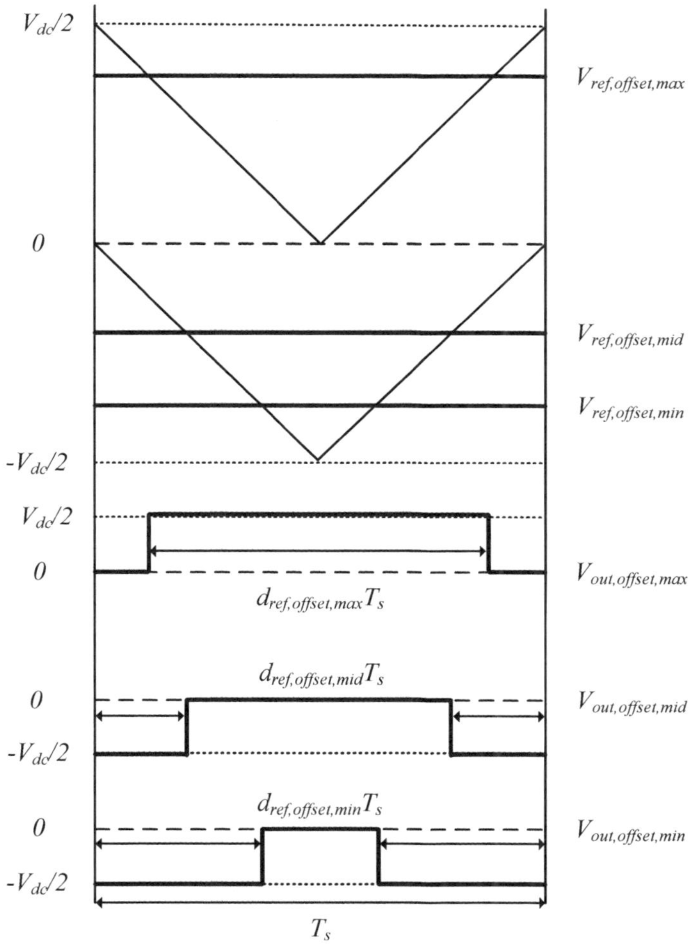

Fig. 4.4 Relationship between the reference voltages and the duty signals (reprinted from [10], Fig. 4)

I_{NP} calculated from (4.6) when the previous mentioned CB-PWM method is applied with a unity PF. The frequency of I_{NP} is three times the fundamental frequency of the output current. The top-side voltage (V_{top}) is decreased and the bottom-side voltage (V_{bottom}) is increased when I_{NP} is positive. A negative I_{NP} causes V_{top} to be increased and V_{bottom} to be decreased. The magnitude of the neutral-point AC ripple voltage ($V_{NP,mag,ripple}$) is determined by the magnitude of the output current, the DC-link capacitor value, and the PF. A small value of the DC-link capacitor, a large output current, and a low-PF increase $V_{NP,mag,ripple}$.

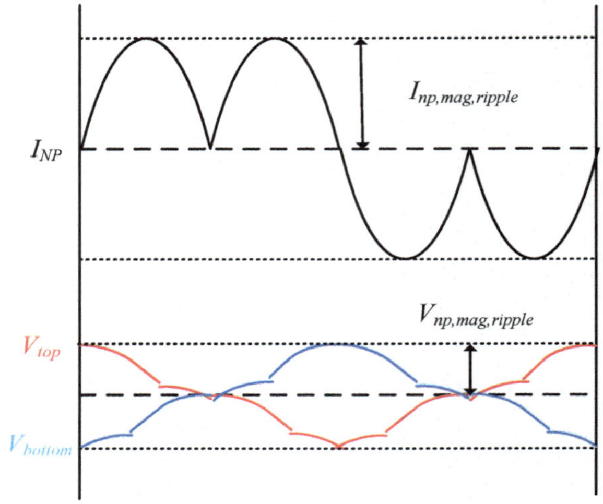

Fig. 4.5 Change in the neutral-point voltage due to the average neutral-point current (I_{NP}) (reprinted from [10], Fig. 5)

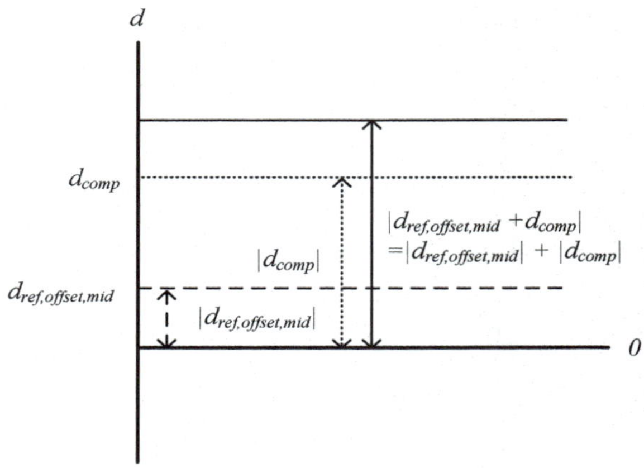

Fig. 4.6 Case 1: $d_{comp} > 0$ and $d_{ref,offset,mid} > 0$ (reprinted from [10], Fig. 6)

4.2 Neutral-Point AC Ripple Voltage Reduction Methods

The main reason for neutral-point voltage ripple was introduced and the neutral-point voltage of the conventional PWM method was calculated in the above section. Neutral-point AC ripple voltage reduction methods are introduced in this section.

4.2.1 Dwell Time Compensation for Neutral-Point Voltage Reduction in the Carrier Based PWM Method [10]

I_{NP} can be zero by adding the optimal compensation value (d_{comp}) to the reference duty signals. This means that $V_{NP,mag,ripple}$ can be reduced. When the optimal compensation value (d_{comp}) is added to the reference duty signals, (4.6) is represented as:

$$I_{NP,casex} = -\left(\left|d_{ref,offset,max} + d_{comp}\right| \cdot I_{V_{ref,offset,max}} + \left|d_{ref,offset,mid} + d_{comp}\right| \cdot I_{V_{ref,offset,mid}} + \left|d_{ref,offset,min} + d_{comp}\right| \cdot I_{V_{ref,offset,min}}\right) \tag{4.7}$$

where $I_{NP,casex}$ represents the average neutral-point current affected by d_{comp}.

A. Optimal compensation value (d_{comp}) calculation

In (4.7), $d_{ref,offset,max}$ is always positive and $d_{ref,offset,min}$ is always negative. However, $d_{ref,offset,mid}$ and d_{comp} can be positive or negative. When considering that $|d_{comp}|$ should be smaller than $|d_{ref,offset,min}|$ (or $|d_{ref,offset,max}|$) because M_a must not exceeded 1, there are four cases.

- Case 1: $d_{comp} > 0$, $d_{ref,offset,mid} > 0$

In this case, the sign of $d_{ref,offset,mid}$ is not changed as shown in Fig. 4.6. Equation (4.7) can be represented as:

$$\begin{aligned} I_{NP,case1} = & -\left(\left|d_{ref,offset,max}\right| \cdot I_{V_{ref,offset,max}} + \left|d_{ref,offset,mid}\right| \cdot I_{V_{ref,offset,mid}} + \left|d_{ref,offset,min}\right| \cdot I_{V_{ref,offset,min}}\right. \\ & - \left(\left|d_{comp}\right| \cdot I_{V_{ref,offset,max}} - \left|d_{comp}\right| \cdot I_{V_{ref,offset,min}} + \left|d_{comp}\right| \cdot I_{V_{ref,offset,mid}}\right) \end{aligned} \tag{4.8}$$

By using I_{NP}, which is the initial value before adding d_{comp} to the reference duty signals, (4.8) is represented as:

$$I_{NP,case1} = I_{NP} - \left(\left|d_{comp}\right| \cdot I_{V_{ref,offset,max}} - \left|d_{comp}\right| \cdot I_{V_{ref,offset,min}} + \left|d_{comp}\right| \cdot I_{V_{ref,offset,mid}}\right) \tag{4.9}$$

To make $I_{NP,\text{case1}}$ equal to zero, d_{comp} can be calculated as:

$$\left|d_{comp}\right| = \frac{I_{NP}}{I_{V_{ref,offset,max}} - I_{V_{ref,offset,min}} + I_{V_{ref,offset,mid}}}, \; I_{NP,case1} = 0 \tag{4.10}$$

Then, because ($I_{max} - I_{min} + I_{mid}$) > 0, (4.10) is finally expressed as

$$d_{comp} = \frac{I_{NP}}{I_{V_{ref,offset,max}} - I_{V_{ref,offset,min}} + I_{V_{ref,offset,mid}}} \quad \text{for } I_{NP} > 0 \tag{4.11}$$

- Case 2: $d_{comp} > 0$, $d_{ref,offset,mid} < 0$

Depending on the magnitude of d_{comp} and $d_{ref,offset,mid}$, Case 2 can be divided into two cases (Cases 2-1 and 2-2) as shown in Fig. 4.7. For Case 2–1, in which $|d_{comp}|$ is larger than $|d_{ref,offset,mid}|$, (4.7) can be represented as:

Fig. 4.7 Case 2: $d_{comp} > 0$ and $d_{ref,offset,mid} < 0$ (reprinted from [10], Fig. 7)

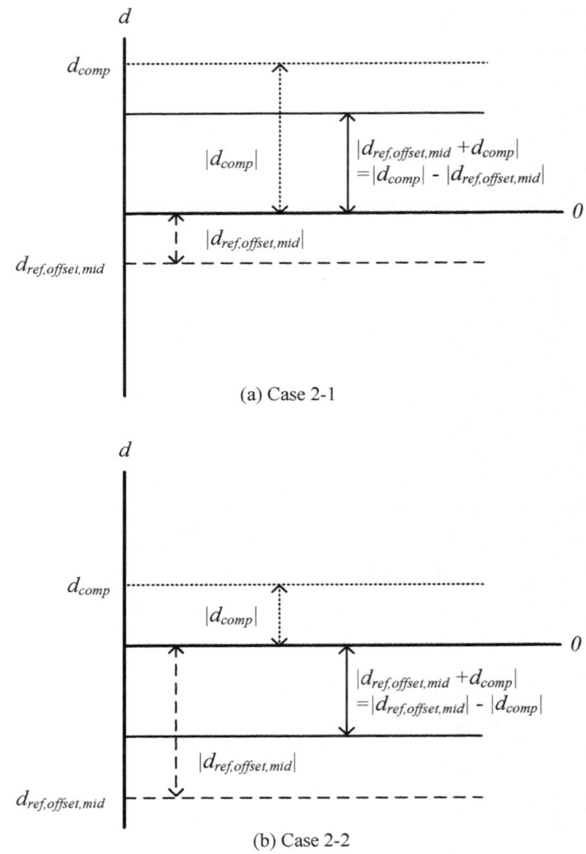

$$\begin{aligned} I_{NP,case2-1} = &-\left(\left|d_{ref,offset,max}\right| \cdot I_{V_{ref,offset,max}} + \left|d_{ref,offset,mid}\right| \cdot I_{V_{ref,offset,mid}} + \left|d_{ref,offset,min}\right| \cdot I_{V_{ref,offset,min}}\right) \\ &- \left(\left|d_{comp}\right| \cdot I_{V_{ref,offset,max}} - \left|d_{comp}\right| \cdot I_{V_{ref,offset,min}} + \left(\left|d_{comp}\right| - 2\left|d_{ref,offset,mid}\right|\right) \cdot I_{V_{ref,offset,mid}}\right) \end{aligned} \tag{4.12}$$

To make $I_{NP,\text{case2-1}}$ equal to zero, d_{comp} can be calculated as:

$$\left|d_{comp}\right| = \frac{I_{NP} + 2\left|d_{ref,offset,mid}\right| \cdot I_{V_{ref,offset,mid}}}{I_{V_{ref,offset,max}} - I_{V_{ref,offset,min}} + I_{V_{ref,offset,mid}}}, \; I_{NP,case2-1} = 0 \tag{4.13}$$

Then, because $(I_{max} - I_{min} + I_{mid}) > 0$, (4.13) can be finally expressed as:

$$\begin{aligned} &d_{comp} = \frac{I_{NP} + 2\left|d_{ref,offset,mid}\right| \cdot I_{V_{ref,offset,mid}}}{I_{V_{ref,offset,max}} - I_{V_{ref,offset,min}} + I_{V_{ref,offset,mid}}} \\ &\text{for } I_{NP} + 2\left|d_{ref,offset,mid}\right| \cdot I_{V_{ref,offset,mid}} > 0 \end{aligned} \tag{4.14}$$

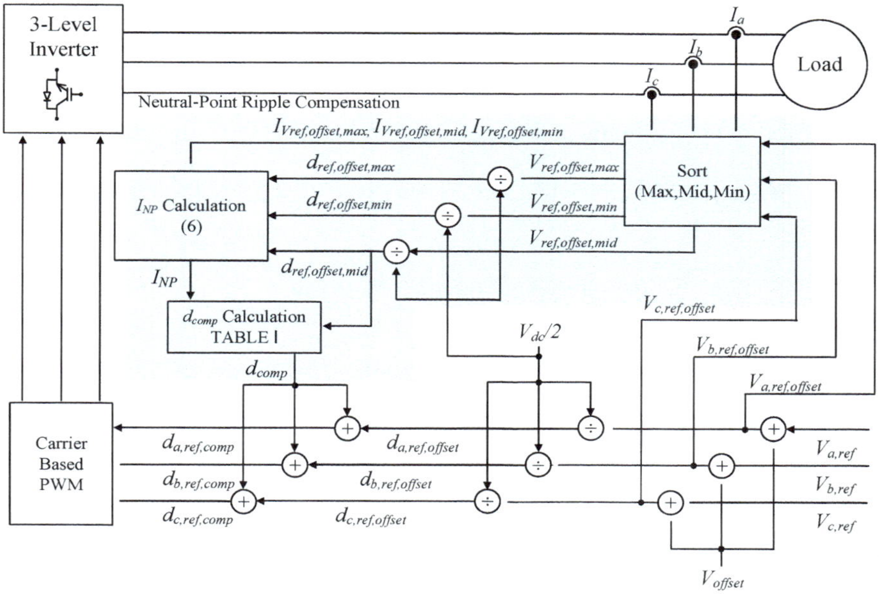

Fig. 4.8 Block diagram for neutral-point AC ripple voltage reduction (reprinted from [10], Fig. 10)

For Case 2–2, in which $|d_{comp}|$ is smaller than $|d_{ref,offset,mid}|$, (4.7) can be represented as:

$$\begin{aligned} I_{NP,case2-2} = & -\left(\left|d_{ref,offset,max}\right| \cdot I_{V_{ref,offset,max}} + \left|d_{ref,offset,mid}\right| \cdot I_{V_{ref,offset,mid}} + \left|d_{ref,offset,min}\right| \cdot I_{V_{ref,offset,min}}\right) \\ & - \left(\left|d_{comp}\right| \cdot I_{V_{ref,offset,max}} - \left|d_{comp}\right| \cdot I_{V_{ref,offset,min}} - \left|d_{comp}\right| \cdot I_{V_{ref,offset,mid}}\right) \end{aligned} \quad (4.15)$$

Like Case 1, to make $I_{NP,\text{ Case2-2}}$ equal to zero, (4.15) can be expressed as:

$$d_{comp} = \frac{I_{NP}}{I_{V_{ref,offset,max}} - I_{V_{ref,offset,min}} - I_{V_{ref,offset,mid}}} \quad \text{for } I_{NP} > 0 \quad (4.16)$$

Additionally, d_{comp} of Case 3 ($d_{comp} < 0$ and $d_{ref,offset,mid} < 0$) and Case 4 ($d_{comp} < 0$ and $d_{\text{ref,offset,mid}} > 0$) can be calculated in a manner that is similar to Case 1 and Case 2.

- Case 3: $d_{comp} < 0$, $d_{ref,offset,mid} > 0$

d_{comp} of Case 3-1, in which $|d_{comp}|$ is larger than $|d_{ref,offset,mid}|$, is represented as:

$$\begin{aligned} d_{comp} &= \frac{I_{NP} + 2\left|d_{ref,offset,mid}\right| \cdot I_{V_{ref,offset,mid}}}{I_{V_{ref,offset,max}} - I_{V_{ref,offset,min}} - I_{V_{ref,offset,mid}}} \\ \text{for } & I_{NP} + 2\left|d_{ref,offset,mid}\right| \cdot I_{V_{ref,offset,mid}} < 0 \end{aligned} \tag{4.17}$$

d_{comp} of Case 3-2, in which $|d_{comp}|$ is smaller than $|d_{ref,offset,mid}|$, is represented as:

$$d_{comp} = \frac{I_{NP}}{I_{V_{ref,offset,max}} - I_{V_{ref,offset,min}} + I_{V_{ref,offset,mid}}} \quad \text{for } I_{NP} < 0 \tag{4.18}$$

- Case 4: $d_{comp} < 0$, $d_{ref,offset,mid} < 0$

d_{comp} of Case 4 is represented as:

$$d_{comp} = \frac{I_{NP}}{I_{V_{ref,offset,max}} - I_{V_{ref,offset,min}} - I_{V_{ref,offset,mid}}}, \quad \text{for } I_{NP} < 0 \tag{4.19}$$

The requirements of (4.14) and (4.17) are not obvious. Therefore, this should be verified mathematically. The magnitudes of $|d_{ref,offset,max}|$ and $|d_{ref,offset,min}|$ are the same as those shown in Fig. 4.3. Thus, (4.7) can be represented as:

$$\begin{aligned} I_{NP} &= -\left(K \cdot I_{V_{ref,offset,max}} + \left|d_{ref,offset,mid}\right| \cdot I_{V_{ref,offset,mid}} + K \cdot I_{V_{ref,offset,min}}\right) \\ K &= \left|d_{ref,offset,max}\right| = \left|d_{ref,offset,min}\right| \end{aligned} \tag{4.20}$$

Equation (4.20) is simplified as:

$$\begin{aligned} I_{NP} &= -\left(K \cdot \left(I_{V_{ref,offset,max}} + I_{V_{ref,offset,min}}\right) + \left|d_{ref,offset,mid}\right| \cdot I_{V_{ref,offset,mid}}\right) \\ &= \left(\left(K - \left|d_{ref,offset,mid}\right|\right) \cdot I_{V_{ref,offset,mid}}\right) \end{aligned} \tag{4.21}$$

From (4.21), the sign of I_{NP} is always the same as the sign of I_{mid} because (K − $|d_{ref,offset,mid}|$) of (4.21) is always positive. Consequently, the requirements of (4.14) and (4.17) are:

$$d_{comp} = \frac{I_{NP} + 2\left|d_{ref,offset,mid}\right| \cdot I_{V_{ref,offset,mid}}}{I_{V_{ref,offset,max}} - I_{V_{ref,offset,min}} + I_{V_{ref,offset,mid}}} \quad \text{for } I_{NP} > 0 \tag{4.22}$$

$$d_{comp} = \frac{I_{NP} + 2\left|d_{ref,offset,mid}\right| \cdot I_{V_{ref,offset,mid}}}{I_{V_{ref,offset,max}} - I_{V_{ref,offset,min}} - I_{V_{ref,offset,mid}}} \quad \text{for } I_{NP} < 0 \tag{4.23}$$

From (4.11), (4.19), (4.22), and (4.23), which are mathematical results considering the conditions of d_{comp} and $d_{ref,offset,mid}$, it is known that the sign of d_{comp} is always the same as the sign of I_{NP}. Therefore, two requirements (I_{NP} and $d_{ref,offset,mid}$)

Table 4.2 Compensation value calculations

Condtion	$d_{ref,offset,mid} > 0$	$d_{ref,offset,mid} < 0$
$I_{NP} > 0$	Case 1	Case 2-1: $\lvert d_{comp}\rvert > \lvert d_{ref,offset,mid}\rvert$
	$d_{comp} = \frac{I_{NP}}{I_{V_{ref,offset,max}} - I_{V_{ref,offset,min}} + I_{V_{ref,offset,mid}}}$	$d_{comp} = \frac{I_{NP}}{I_{V_{ref,offset,max}} - I_{V_{ref,offset,min}} - I_{V_{ref,offset,mid}}}$
		Case 2-2: $\lvert d_{comp}\rvert < \lvert d_{ref,offset,mid}\rvert$ (If $\lvert d_{comp}\rvert > \lvert d_{ref,offset,mid}\rvert$ is not true)
		$d_{comp} = \frac{I_{NP} + 2\lvert d_{ref,offset,mid}\rvert \cdot I_{V_{ref,offset,mid}}}{I_{V_{ref,offset,max}} - I_{V_{ref,offset,min}} + I_{V_{ref,offset,mid}}}$
$I_{NP} < 0$	Case 3-1: $\lvert d_{comp}\rvert > \lvert d_{ref,offset,mid}\rvert$	Case 4
	$d_{comp} = \frac{I_{NP}}{I_{V_{ref,offset,max}} - I_{V_{ref,offset,min}} + I_{V_{ref,offset,mid}}}$	$d_{comp} = \frac{I_{NP}}{I_{V_{ref,offset,max}} - I_{V_{ref,offset,min}} - I_{V_{ref,offset,mid}}}$
	Case 3-2: $\lvert d_{comp}\rvert < \lvert d_{ref,offset,mid}\rvert$ (If $\lvert d_{comp}\rvert > \lvert d_{ref,offset,mid}\rvert$ is not true)	
	$d_{comp} = \frac{I_{NP} + 2\lvert d_{ref,offset,mid}\rvert \cdot I_{V_{ref,offset,mid}}}{I_{V_{ref,offset,max}} - I_{V_{ref,offset,min}} - I_{V_{ref,offset,mid}}}$	

are provided to calculate d_{comp}. In summary, d_{comp} is calculated depending on the values of $d_{ref,offset,mid}$ and the original I_{NP}. In Case 2 and Case 3, first, Case x – 1 (x = 2, 3) is calculated. Then Case x – 2 (x = 2, 3) is calculated if d_{comp} calculated from Case x–1 does not satisfy its requirement ($\lvert d_{comp}\rvert > \lvert d_{ref,offset,mid}\rvert$). The d_{comp} calculation method for neutral-point AC ripple voltage reduction is shown in Table 4.2, and the compensated reference duty signals are expressed as:

$$d_{x,ref,comp} = d_{x,ref,offset} + d_{comp}(x = a, b, c) \tag{4.24}$$

Although the phase of the currents is not the same as that of the reference duty signals in a nonunity *PF*, all of the equations in this chapter can be applied for the case of a nonunity *PF* because these equations only consider the signs of $d_{ref,offset,min}$, $d_{ref,offset,mid}$, and $d_{ref,offset,max}$.

B. The applicable range of the proposed neutral-point AC ripple voltage reduction method

d_{comp} is calculated from Table 4.1. The value of $d_{x,ref,comp}$ cannot exceed 1. Therefore, the margin (d_{margin}) for adding d_{comp} is expressed as:

$$d_{margin} = \lvert 1 - d_{ref,offset,max}\rvert \tag{4.25}$$

Table 4.3 Simulation and experiment parameters

DC-link voltage	200 V	Output frequency	60 Hz
DC-link capacitor	165 μF	Output current	10 A_{peak}
Control period	100 μs	Switching frequency	10 kHz

d_{comp} should satisfy the following condition:

$$|d_{comp}| < d_{margin} \tag{4.26}$$

Additionally, the PF and M_a affect the applicable range of the proposed neutral-point voltage AC ripple reduction method. Figure 4.9 shows the magnitude of the neutral-point AC ripple voltage depending on the PF and M_a. This result is the same as that in some papers which show that a low PF causes the neutral-point AC ripple voltage to increase [3, 5, 12].

Figure 4.10 shows the required d_{comp} for the zero neutral-point AC ripple voltage when the CB-PWM method is applied. This is provided from simulation results obtained using the simulation parameters in section IV. A large M_a or a low PF require a large d_{comp} to make the neutral-point AC ripple voltage zero. In addition, Fig. 4.10 illustrates the applicable range of the proposed method. For example, when M_a is 0.7, d_{margin} is 0.3 and the applicable range means that the proposed method can eliminate the neutral-point AC ripple voltage completely when the PF is from 0.6 to 1. For a low PF, large d_{comp} values are needed for neutral-point AC ripple voltage reduction. In addition, because a high M_a provides a small d_{margin} to add to d_{comp}, the neutral-point AC ripple voltage cannot be eliminated completely in the high M_a situation with a low PF. Consequently, d_{comp} is restricted to d_{margin}, which varies depending on M_a. When M_a increases, d_{margin} becomes smaller. The total control diagram of this method is shown in Fig. 4.8.

C. Simulation and experimental results

Figure 4.11 shows waveforms of the CB-PWM method. Each waveform of Fig. 4.11a signifies the values in the block diagram of Fig. 4.8. The reference voltages ($V_{a,ref}$, $V_{b,ref}$, and $V_{c,ref}$) are modified by V_{offset}. The changed reference voltages ($V_{a,ref,offset}$, $V_{b,ref,offset}$, and $V_{c,ref,offset}$) are used for determining the inverter

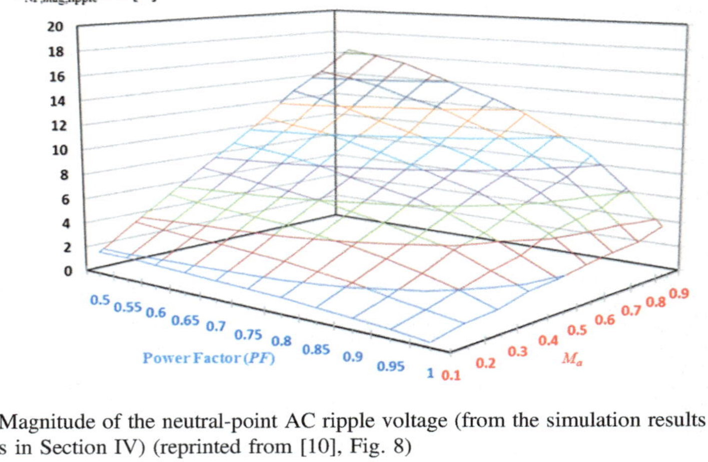

Fig. 4.9 Magnitude of the neutral-point AC ripple voltage (from the simulation results using the parameters in Section IV) (reprinted from [10], Fig. 8)

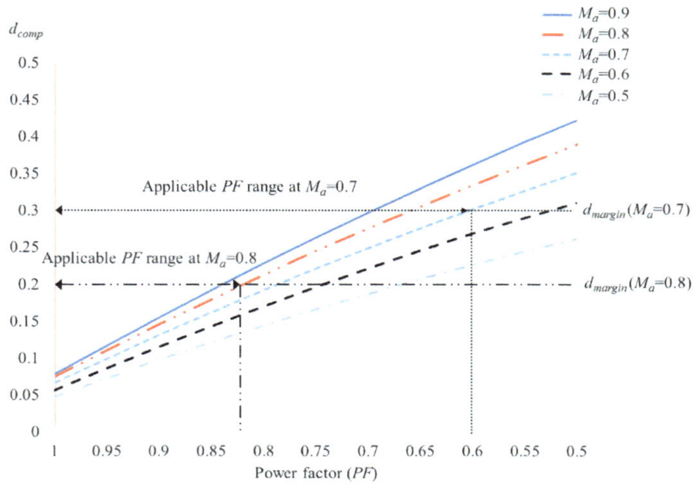

Fig. 4.10 d_{comp} for the zero neutral-point AC ripple voltage according to M_a (from the simulation results using the parameters in Table 4.3) (reprinted from [10], Fig. 9)

output voltages by comparing them with two carrier signals. In this case, the *PF* is unity because the phase of the currents corresponds with that of the reference voltages. The frequency of the neutral-point AC ripple voltage, which is shown in Fig. 4.11b, is three times as large as the fundamental frequency of the current. The I_{NP} and the variables for the I_{NP} calculation are shown in Fig. 4.11b.

Figures 4.12 and 4.13 show the performance of the proposed neutral-point AC ripple voltage reduction method according to the *PF* and M_a. The neutral-point AC ripple voltage decreases after the proposed method is applied at 4.05 s. In Fig. 4.12a, $V_{NP,mag,ripple}$ decreases from 1.7 to 0.1 V. Because $d_{x,ref,comps}$ (x = a, b, c) cannot exceed 1, the proposed method does not completely eliminate the AC ripple voltage (see Fig. 4.12b). Instead it is decreased from 5.8 to 2 V. When the *PF* decreases, $V_{NP,mag,ripple}$ increases, and a large magnitude of d_{comp} is needed to reduce it as shown in Fig. 4.12b, c.

The x-dot in Fig. 4.13a–c means that M_a exceeds 1 by adding d_{comp} to $d_{x,ref}$ (x = a, b, c). Therefore, the neutral-point AC ripple voltage is not completely eliminated. As M_a decreases, d_{margin} becomes large and the range where the compensation can be achieved completely is expanded. When M_a is below 0.6, this method guarantees complete reduction of the neutral-point AC ripple voltage, when the inverters are operated at a *PF* from 0.5 to 1 as shown in Fig. 4.13d.

Figure 4.14 shows the experimental results of this method. The magnitude of the neutral-point AC ripple voltage increases when the *PF* decreases. According to the *PF* and M_a, a suitable d_{comp} is calculated for the neutral-point AC ripple voltage reduction. Then, $d_{x,ref,comp}$ (x = a, b, c,) is changed by adding d_{comp}. The changed $d_{a,ref,comp}$ causes the neutral-point voltage to be reduced. However, the a-phase current is not changed. In Fig. 4.14a, c, the neutral-point AC ripple voltage is

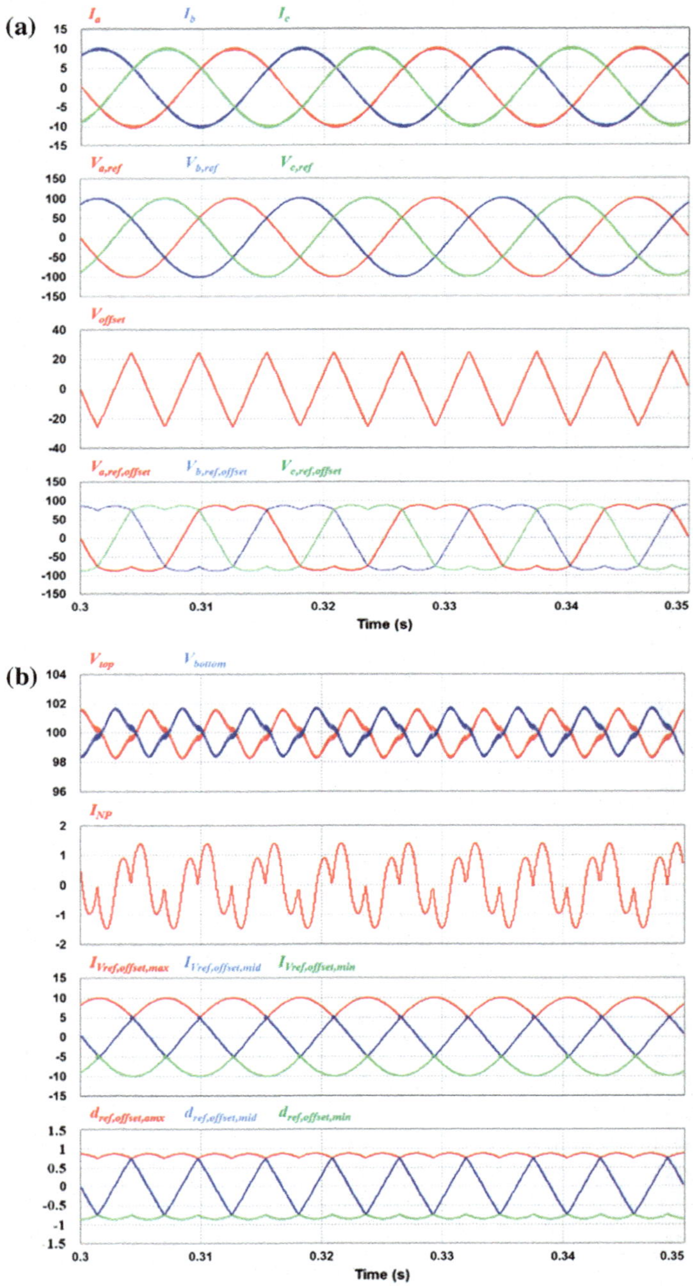

Fig. 4.11 Waveforms: **a** conventional CB-PWM method, **b** I_{NP} calculation (reprinted from [10], Fig. 11)

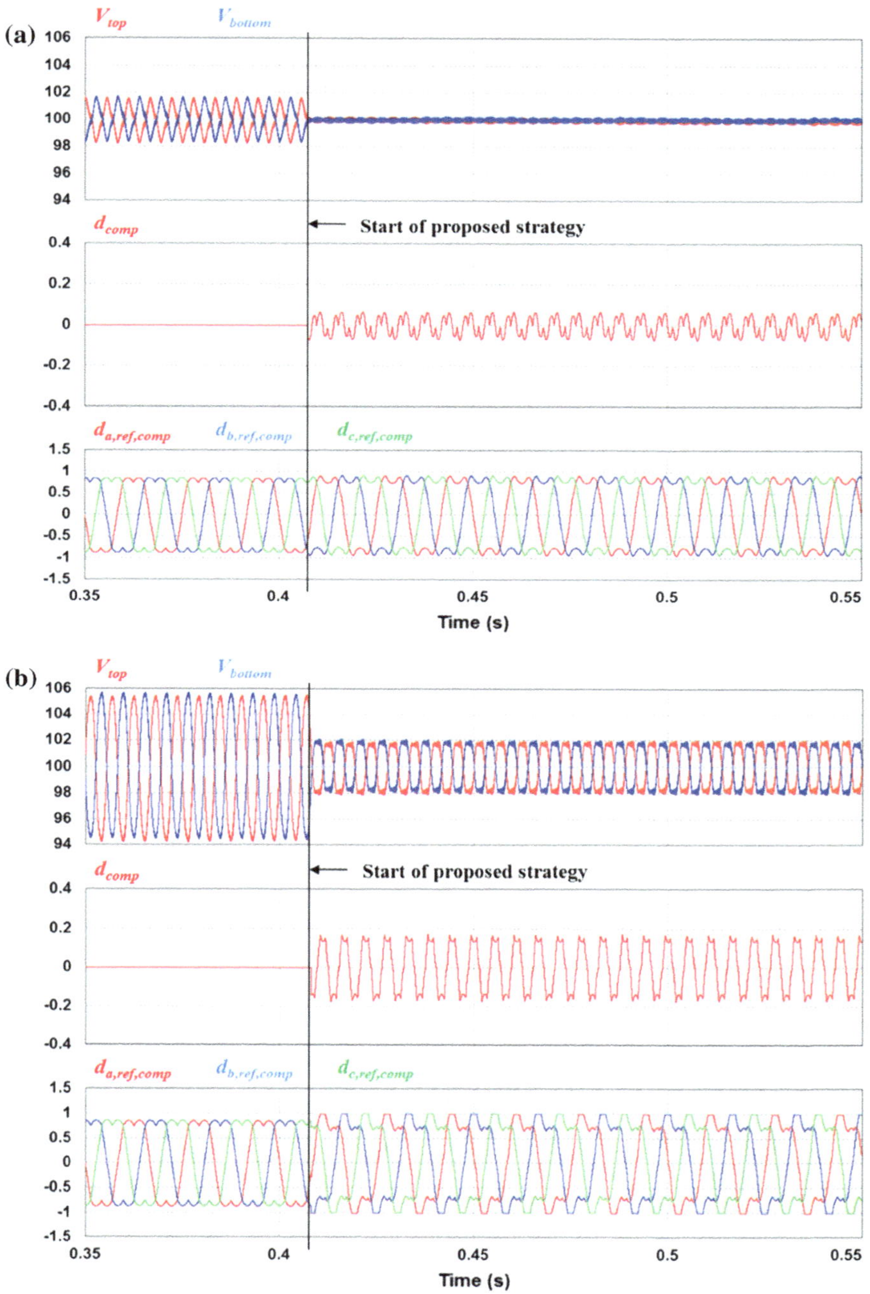

Fig. 4.12 Waveforms of the CB-PWM method with neutral-point AC ripple voltage reduction: **a** $M_a = 0.87$, $PF = 1$. **b** $M_a = 0.87$, $PF = 0.8$ (reprinted from [10], Fig. 12)

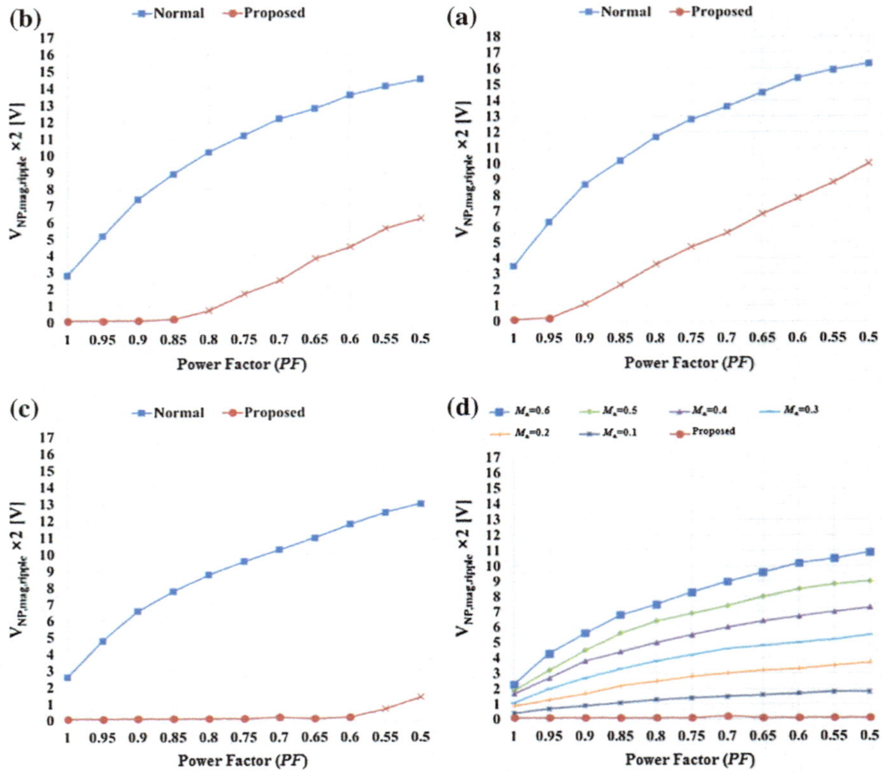

Fig. 4.13 Performance of the neutral-point AC ripple voltage reduction method according to M_a and *PF*: **a** M_a = 0.87, **b** M_a = 0.8, **c** M_a = 0.7, **d** $\mathrm{M_a}$ = 0.6–0.1 (reprinted from [10], Fig. 13)

almost eliminated and $d_{a,ref,comp}$ has a value from −1 to 1. However, the neutral-point AC ripple voltage of Fig. 4.14b, where M_a is 0.87 and the *PF* is 0.8, is not completely eliminated. This is because $|d_{a,ref,comp}|$ does not have to exceed 1 and is limited by 1 and −1 as shown in the dotted circles of Fig. 4.14b.

4.2.2 Dwell Time Compensation for Neutral-Point Voltage Reduction in SVPWM Method [9]

Table 4.4 shows the space voltage vector, switching states, and neutral currents of three-level inverters. As previous mentioned, the voltage vectors, except for the large-voltage and zero-voltage vectors, affect the neutral-point voltage oscillations. I_{NP} is defined depending on the switching states of the small and medium vectors.

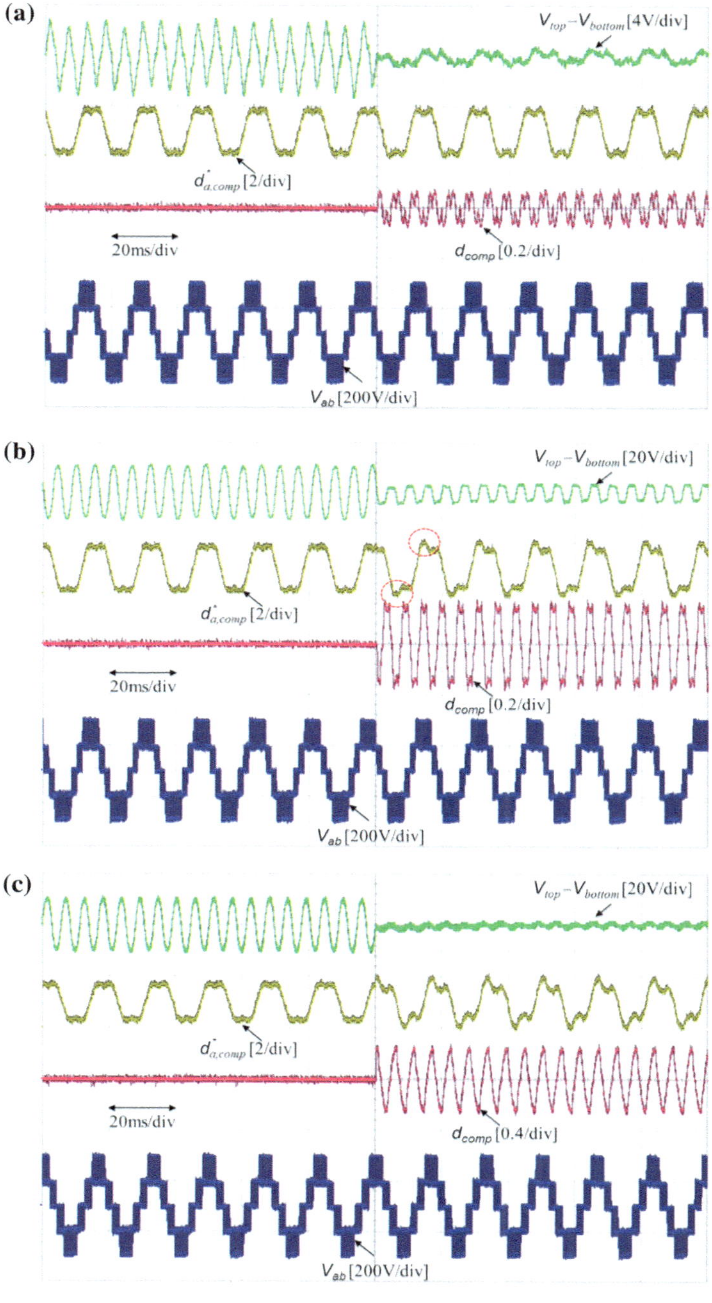

Fig. 4.14 Experimental results of the CB-PWM method with this neutral-point AC ripple voltage reduction according to M_a and PF: **a** $M_a = 0.87$, $PF = 1$, **b** $M_a = 0.87$, $PF = 0.8$, **c** $M_a = 0.7$, $PF = 0.75$ (reprinted from [10], Fig. 18)

Table 4.4 Switching states, voltage vectors, and neutral current

Vector	Magnitude	Switching state		Neutral current	
Zero vector	0	[PPP], [OOO], [NNN]		–	
Small vector	$\frac{1}{3}V_{dc}$	P-type	N-type		
		[POO]	[ONN]	$I_b + I_c$	I_a
		[PPO]	[OON]	I_c	$I_a + I_b$
		[OPO]	[NON]	$I_a + I_c$	I_b
		[OPP]	[NOO]	I_a	$I_b + I_c$
		[POP]	[ONO]	I_b	$I_a + I_c$
		[OOP]	[NNO]	$I_a + I_b$	I_c
Medium vector	$\frac{\sqrt{3}}{3}V_{dc}$	[PON]		I_b	
		[OPN]		I_a	
		[NPO]		I_c	
		[NOP]		I_b	
		[ONP]		I_a	
		[PNO]		I_c	
Large vector	$\frac{2}{3}V_{dc}$	[PNN]		–	
		[PPN]		–	
		[NPN]		–	
		[NPP]		–	
		[NNP]		–	
		[PNP]		–	

A. Time-offset injection

The small-voltage and medium-voltage vectors oscillate the two capacitor voltages. To minimize the oscillation, I_{NP} should be zero during the switching sequence. This can be obtained by adding a proper time offset (T_{off}) to the three-phase turn-on times (T_a, T_b, T_c).

The method to obtain a proper time offset (T_{off}) is expressed in this paragraph. Sector I in Fig. 4.15 will be considered as an example to explain the method to reduce the oscillation of the neutral-point voltage. The algorithm is explained under the assumption that the inverter is in the normal operating (inverting) mode. A block diagram of the method is shown in Fig. 4.16.

- Region 1

If the reference voltage V_{ref} is in region 1 of sector I, it is made by the four switching states [ONN], [PNN], [PON], and [POO] using symmetrical SVM. Using these switching states, the switching sequence is formed as [ONN]–[PNN]–[PON]–[POO]–[PON]–[PNN]–[ONN], and the dwell time of the N-type small switching

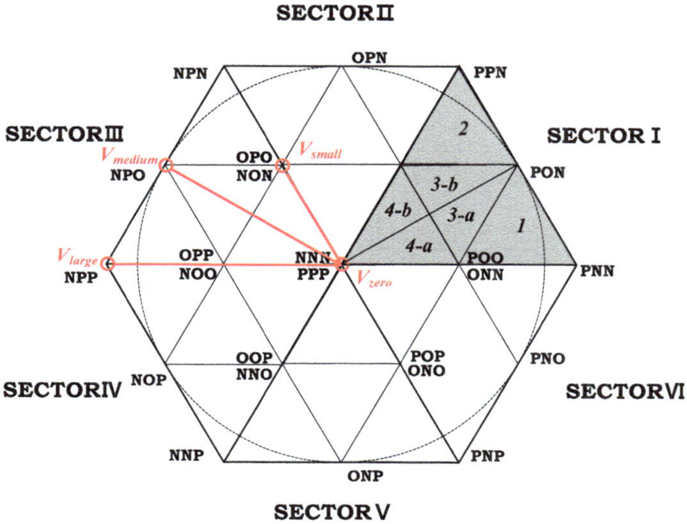

Fig. 4.15 Space vector diagram of a three-level inverter (reprinted from Ref. [9], Fig. 2)

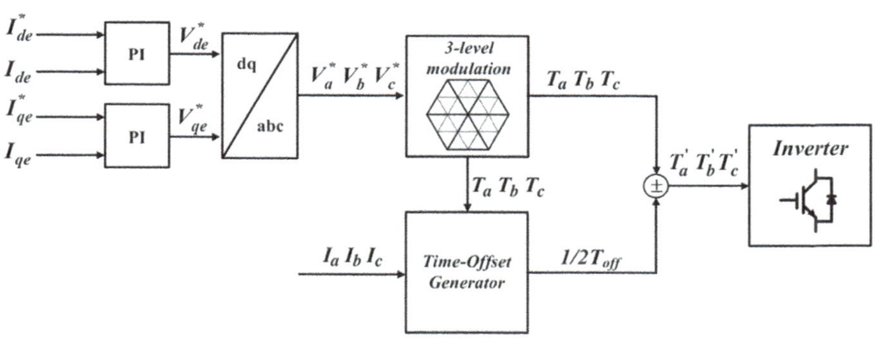

Fig. 4.16 Block diagram of the neutral-point voltage oscillation reduction method (reprinted from [9], Fig. 5)

state [ONN] and the P-type small switching state [POO] is equally divided as shown in Fig. 4.17a. In this sequence, the medium switching state [PON] makes I_{NP} not zero during the switching sequence. The neutral-point voltage is increased or decreased depending on the current direction of phase B.

To make I_{NP} to zero during the switching sequence, a time offset (T_{off}) should be added. The dwell times of the medium and small switching states should be obtained by calculating a proper T_{off}. This can be easily calculated by the turn-on times. In this region, the dwell time of [PON], [ONN], and [POO] can be expressed as:

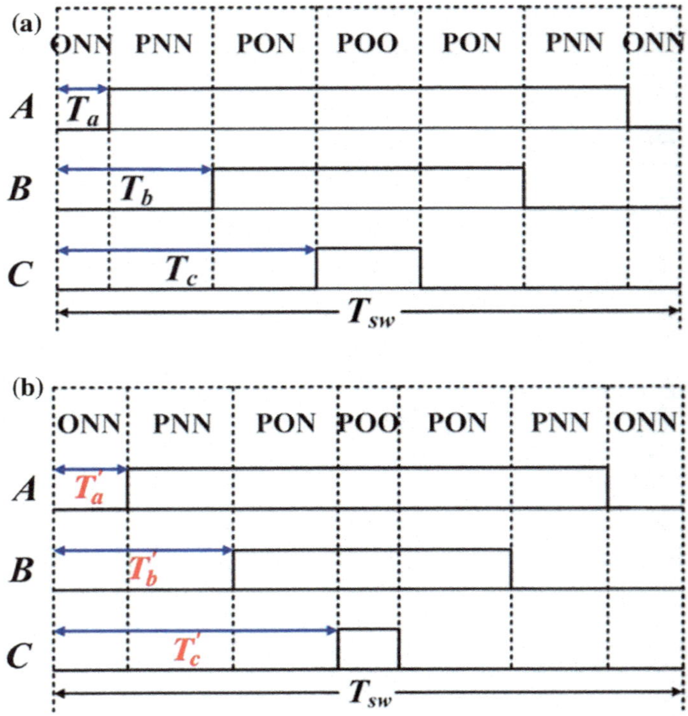

Fig. 4.17 **a** Switching sequence in region 1 of sector I, **b** conventional SVM after T_{off} is added (reprinted from [9], Fig. 6)

$$\begin{aligned} T_{dwell[PON]} &= 2(T_c - T_b) \\ T_{dwell[ONN]} &= 2T_a \\ T_{dwell[POO]} &= 2(T_{sw}/2 - T_c) = 2T_a \end{aligned} \tag{4.27}$$

The magnitude of the neutral-point voltage change ΔV by [PON] can be expressed as:

$$\Delta V = \frac{1}{\mathrm{C}} \int_{t=0}^{t=T_{dwell[PON]}} I_n(t)dt \tag{4.28}$$

where I_n is the neutral current, C is the size of the capacitor, and $T_{dwell[PON]}$ is the dwell time of the medium vector [PON]. Assuming that there is no change in the current during a short period, it can be rewritten as:

$$\Delta V = \frac{1}{C} \int_{t=0}^{t=2(T_c - T_b)} I_n(t)dt = \frac{1}{C} \cdot 2(T_c - T_b) \cdot I_b \tag{4.29}$$

This magnitude should be the same with a change of the neutral-point voltage using T_{off} as follows:

$$\begin{aligned} |\Delta V| &= \left| \frac{1}{C} \cdot 2(T_c - T_b) \cdot I_b \right| \\ &= \left| \frac{1}{C} \cdot 2(T_a + T_{off}) \cdot I_a + \frac{1}{C} \cdot (T_{sw} - 2T_c - T_{off}) \cdot (-I_a) \right| \end{aligned} \tag{4.30}$$

$$\left| \frac{1}{C} \cdot 2(T_c - T_b) \cdot I_b \right| = \left| \frac{1}{C} \cdot 2T_{off} \cdot I_a \right| \tag{4.31}$$

$$\left| T_{off} \right| = \frac{|I_b|}{|I_a|} (T_c - T_b) \tag{4.32}$$

If [PON] increases the neutral-point voltage, T_{off} should be added so that the dwell times of the P-type and N-type small switching states are changed to decrease the neutral-point voltage. This means that the dwell time of P-type small switching states decreases and the dwell time of N-type small switching states increases. The time offset is added simply by redefining the three-phase turn-on times (T_a, T_b, T_c). In this region, [PON] increases the neutral-point voltage with P-type small switching states because I_b is negative. Therefore, the dwell time of P-type small switching states should be reduced by T_{off}, and the dwell time of N-type small switching states should be increased by T_{off}. The turn-on times (T_a, T_b, T_c) are redefined as:

$$\begin{aligned} T_a' &= T_a + \frac{1}{2} \frac{|I_b|}{|I_a|} (T_c - T_b) \\ T_b' &= T_b + \frac{1}{2} \frac{|I_b|}{|I_a|} (T_c - T_b) \\ T_c' &= T_c + \frac{1}{2} \frac{|I_b|}{|I_a|} (T_c - T_b) \end{aligned} \tag{4.33}$$

where the switching sequences of the redefined turn-on times are shown in Fig. 4.17b. In this sequence, the dwell time of [ONN] and [POO] can be changed as:

$$\begin{aligned} T_{dwell[ONN]} &= 2T_a + \frac{|I_b|}{|I_a|} (T_c - T_b) \\ T_{dwell[POO]} &= 2T_a - \frac{|I_b|}{|I_a|} (T_c - T_b) \end{aligned} \tag{4.34}$$

In the opposite case, if I_b has a positive value, T_{off} should be added so that the dwell times of the P-type and N-type switching states are changed to increase the neutral-point voltage. $T_{dwell[ONN]}$ should be decreased by T_{off}, and $T_{dwell[POO]}$ should be increased by T_{off}.

- Region 2

The principle to calculate a precise T_{off} is the same as the above case. In this sequence, the dwell times of [PON], [OON], and [PPO] should be considered and can be expressed as

$$\begin{aligned} T_{dwell[PON]} &= 2(T_b - T_a) \\ T_{dwell[OON]} &= 2T_a \\ T_{dwell[PPO]} &= 2(T_{sw}/2 - T_c) = 2T_a \end{aligned} \tag{4.35}$$

where the magnitude of the neutral-point voltage change ΔV by [PON] is:

$$\begin{aligned} |\Delta V| &= \left| \frac{1}{C} \cdot 2(T_b - T_a) \cdot I_b \right| \\ &= \left| \frac{1}{C} \cdot (2T_a - T_{off}) \cdot (-I_c) + \frac{1}{C}(T_{sw} - 2T_c + T_{off}) \cdot I_c \right| \end{aligned} \tag{4.36}$$

and T_{off} is calculated as:

$$\left| \frac{1}{C} \cdot 2(T_b - T_a) \cdot I_b \right| = \left| \frac{1}{C} \cdot 2T_{off} \cdot I_c \right| \tag{4.37}$$

Then:

$$|T_{off}| = \frac{|I_b|}{|I_c|}(T_b - T_a) \tag{4.38}$$

In this region, [PON] decreases the neutral-point voltage because the direction of I_b is positive. Therefore, $T_{dwell[OON]}$ should be decreased by T_{off}, and $T_{dwell[PPO]}$ should be increased by T_{off}. The turn-on times (T_a, T_b, T_c) are redefined as:

$$\begin{aligned} T_a' &= T_a - \frac{1}{2}\frac{|I_b|}{|I_c|}(T_b - T_a) \\ T_b' &= T_b - \frac{1}{2}\frac{|I_b|}{|I_c|}(T_b - T_a) \\ T_c' &= T_c - \frac{1}{2}\frac{|I_b|}{|I_c|}(T_b - T_a) \end{aligned} \tag{4.39}$$

- Region 3-a

Unlike the above cases, in this region, the switching sequence is [ONN]–[OON]–[PON]–[POO]–[PON]–[OON]–[ONN]. There is one more switching state of a small voltage vector. Therefore, two voltage vectors should be considered to calculate T_{off}. The dwell times are defined as:

$$\begin{aligned} T_{dwell[PON]} &= 2(T_c - T_a) \\ T_{dwell[OON]} &= 2(T_a - T_b) \\ T_{dwell[ONN]} &= 2T_b \\ T_{dwell[POO]} &= 2(T_{sw}/2 - T_c) = 2T_b \end{aligned} \tag{4.40}$$

The oscillation by [OON] and the time offset $T_{off[OON]}$ are:

$$|\Delta V| = \left|\frac{1}{C}\cdot 2(T_a - T_b)\cdot I_c\right| = \left|\frac{1}{C}\cdot 2T_{off[OON]}\cdot I_a\right| \tag{4.41}$$

$$\left|T_{off[PON]}\right| = \frac{|I_c|}{|I_a|}(T_a - T_b) \tag{4.42}$$

The oscillation by [PON] and the time offset $T_{off[PON]}$ are:

$$|\Delta V| = \left|\frac{1}{C}\cdot 2(T_c - T_a)\cdot I_b\right| = \left|\frac{1}{C}\cdot 2T_{off[PON]}\cdot I_a\right| \tag{4.43}$$

$$\left|T_{off[PON]}\right| = \frac{|I_b|}{|I_a|}(T_c - T_a) \tag{4.44}$$

Since [OON] decreases the neutral-point voltage and [PON] increases the neutral-point voltage, the dwell times of $T_{dwell[ONN]}$ and $T_{dwell[POO]}$ are changed as:

$$\begin{aligned} T_{dwell[ONN]} &= 2T_a - T_{off[ONN]} + T_{off[PON]} \\ T_{dwell[POO]} &= 2T_a + T_{off[OON]} - T_{off[PON]} \end{aligned} \tag{4.45}$$

The turn-on times (T_a, T_b, T_c) are redefined as:

$$\begin{aligned} T_a' &= T_a - \frac{1}{2}\left(T_{off[OON]} - T_{off[PON]}\right) \\ T_b' &= T_b - \frac{1}{2}\left(T_{off[OON]} - T_{off[PON]}\right) \\ T_c' &= T_c - \frac{1}{2}\left(T_{off[OON]} - T_{off[PON]}\right) \end{aligned} \tag{4.46}$$

- Region 3-b

In this region, the switching sequence is [ONN]–[PON]–[POO]–[PPO]–[POO]–[PON]–[ONN]. The two switching states [PON] and [POO] are also considered in this region. The time offsets by [PON] and [POO] are calculated as given in (4.48) and (4.50). Thus:

$$|\Delta V| = \left|\frac{1}{C} \cdot 2(T_c - T_a) \cdot I_b\right| = \left|\frac{1}{C} \cdot 2T_{off[PON]} \cdot I_c\right| \tag{4.47}$$

$$\left|T_{off[PON]}\right| = \frac{|I_b|}{|I_c|}(T_c - T_a) \tag{4.48}$$

$$|\Delta V| = \left|\frac{1}{C} \cdot 2(T_b - T_c) \cdot I_a\right| = \left|\frac{1}{C} \cdot 2T_{off[POO]} \cdot I_c\right| \tag{4.49}$$

$$\left|T_{off[POO]}\right| = \frac{|I_a|}{|I_c|}(T_b - T_c) \tag{4.50}$$

- Region 4-a

In this region, an additional N-type small switching state [OON] makes I_{NP} not zero in the sequence. The sequence is [ONN]–[OON]–[OOO]–[POO]–[OOO]–[OON]–[ONN]. The dwell times of switching states [ONN], [OON], and [POO] are defined as:

$$\begin{aligned} T_{dwell[OON]} &= 2T_b \\ T_{dwell[OON]} &= 2(T_c - T_b) \\ T_{dwell[POO]} &= 2(T_{sw}/2 - T_a) = 2T_b \end{aligned} \tag{4.51}$$

The offset T_{off} for the zero I_{NP} is obtained by:

$$|\Delta V| = \left|\frac{1}{C} \cdot 2(T_c - T_b) \cdot I_c\right| = \left|\frac{1}{C} \cdot 2T_{off[OON]} \cdot I_a\right| \tag{4.52}$$

$$\left|T_{off[POO]}\right| = \frac{|I_c|}{|I_a|}(T_c - T_b) \tag{4.53}$$

The turn-on times are redefined as:

$$\begin{aligned} T_a' &= T_a - \frac{|I_c|}{|I_a|}(T_c - T_b) \\ T_b' &= T_b - \frac{|I_c|}{|I_a|}(T_c - T_b) \\ T_c' &= T_c - \frac{|I_c|}{|I_a|}(T_c - T_b) \end{aligned} \tag{4.54}$$

- Region 4-b

The switching sequence in this region is [OON]–[OOO]–[POO]–[PPO]–[POO]–[OOO]–[OON]. In this sequence, [POO] is considered to find the time offset T_{off}. The dwell time of [POO] is $T_b - T_a$. Therefore, T_{off} is calculated as:

$$|\Delta V| = \left|\frac{1}{C} \cdot 2(T_b - T_a) \cdot I_c\right| = \left|\frac{1}{C} \cdot 2T_{off[POO]} \cdot I_c\right| \tag{4.55}$$

$$|T_{off}| = \frac{|I_a|}{|I_c|}(T_b - T_a) \tag{4.56}$$

The $T_{dwell[OON]}$ of the N-type switching state should be increased by T_{off}, and the $T_{dwell[PPO]}$ of the P-type switching state should be decreased by T_{off}. The turn-on times (T_a, T_b, T_c) are redefined as:

$$\begin{aligned} T_a' &= T_a + \frac{1}{2}\frac{|I_a|}{|I_c|}(T_b - T_a) \\ T_b' &= T_b + \frac{1}{2}\frac{|I_a|}{|I_c|}(T_b - T_a) \\ T_c' &= T_c + \frac{1}{2}\frac{|I_a|}{|I_c|}(T_b - T_a) \end{aligned} \tag{4.57}$$

A proper T_{off} can be obtained in the other sectors using the same principles of calculation and compensation.

Table 4.5 shows the time offset T_{off} and the redefined turn-on times when the reference voltage V_{ref} is in sector I. If the magnitude of T_{off} is larger than T_{min}, distortions appear in the output voltages. Therefore, the magnitude of T_{off} should be limited by T_{min}. T_{min} refers to the minimum turn-on time among T_a, T_b, and T_c in each sequence. The proper time offset and the redefined turn-on times for other sectors are arranged in Tables 4.6, 4.7, 4.8, 4.9 and 4.10.

This time-offset injection method only changes the distribution of the small vector's dwell time between the N-type and P-type switching states depending on the conditions to reduce neutral-point voltage oscillations. The N-type and P-type

Table 4.5 Time offset and redefined turn-ON times for sector I

R	$\lvert T_{off}\rvert$	T_x $(x = a, b, c)$
1	$\lvert T_{off}\rvert = \frac{\lvert I_b\rvert}{\lvert I_a\rvert}(T_c - T_b)$	$T_x' = T_x + \frac{1}{2}\lvert T_{off}\rvert$
2	$\lvert T_{off}\rvert = \frac{\lvert I_b\rvert}{\lvert I_c\rvert}(T_b - T_a)$	$T_x' = T_x - \frac{1}{2}\lvert T_{off}\rvert$
3-a	$\lvert T_{off[OON]}\rvert = \frac{\lvert I_c\rvert}{\lvert I_a\rvert}(T_a - T_b)$ $\lvert T_{off[PON]}\rvert = \frac{\lvert I_b\rvert}{\lvert I_a\rvert}(T_c - T_a)$	$T_x' = T_x - \frac{1}{2}(\lvert T_{off[OON]}\rvert - \lvert T_{off[PON]}\rvert)$
3-b	$\lvert T_{off[PON]}\rvert = \frac{\lvert I_b\rvert}{\lvert I_c\rvert}(T_c - T_a)$ $\lvert T_{off[POO]}\rvert = \frac{\lvert I_a\rvert}{\lvert I_c\rvert}(T_b - T_c)$	$T_x' = T_x + \frac{1}{2}(\lvert T_{off[PON]}\rvert - \lvert T_{off[POO]}\rvert)$
4-a	$\lvert T_{off}\rvert = \frac{\lvert I_c\rvert}{\lvert I_a\rvert}(T_c - T_b)$	$T_x' = T_x - \frac{1}{2}\lvert T_{off}\rvert$
4-b	$\lvert T_{off}\rvert = \frac{\lvert I_a\rvert}{\lvert I_c\rvert}(T_b - T_a)$	$T_x' = T_x + \frac{1}{2}\lvert T_{off}\rvert$

switching states produce the same line-to-line voltages and only affect the neutral-point voltage. The total dwell time of the small-voltage vector is not changed, and the dwell times of the other voltage vectors are not changed. Therefore, the outputs are not distorted by this method.

B. Sequence change

As previously mentioned, if the time offset is properly injected, the neutral-point voltage oscillations are considerably decreased. However, there is a limitation to injecting time offset to minimize the effect of time-offset injection on the outputs. If the power factor is lower, the neutral current by the medium voltage vector is increased. In this case, the required time offset is bigger than the minimum turn-on times. This means that it is impossible to make the average of the neutral-point voltage zero. Therefore, the magnitude of the neutral-point still remains. The ability to reduce neutral-point voltage oscillations can be improved by changing the switching sequences in regions 3 and 4.

For example, in the case of sector I, if the reference voltage is in region 3-a of sector I, the switching sequence is formed as [ONN]–[OON]–[PON]–[POO]–[PON]–[OON]–[ONN]. This sequence is called sequence 1 in this chapter. In this sequence, [PON] acts similar to a P-type switching state of small voltage vectors. In the case of region 3-b of sector I, the switching sequence is formed as [OON]–[PON]–[POO]–[PPO]–[POO]–[PON]–[ONN], and [PON] acts as an N-type switching sequence. This sequence is called sequence 2 in this chapter. However, for example, if the power factor is 0.5, [PON] in region 3-b acts as a P-type switching state. Although the maximum time offset is injected to increase the N-type switching state, the effect of [PON] is higher than the increased N-type switching state due to the time offset. Therefore, the upper capacitor voltage is decreased, and the lower capacitor voltage is increased in this period.

Table 4.6 Time offset and redefined turn-ON times for sector II

R	$\lvert T_{off}\rvert$	Oscillations	Balancing ($V_{dc1} > V_{dc2}$)	Balancing ($V_{dc1} < V_{dc2}$)
1	$\lvert T_{off}\rvert = \frac{\lvert I_a\rvert}{\lvert I_c\rvert}(T_a - T_b)$	$T_x' = T_x - \frac{1}{2}\lvert T_{off}\rvert$	$T_x' = T_x - \frac{1}{2}\lvert T_{off}\rvert$	$T_x' = T_x$
2	$\lvert T_{off}\rvert = \frac{\lvert I_a\rvert}{\lvert I_b\rvert}(T_c - T_a)$	$T_x' = T_x + \frac{1}{2}\lvert T_{off}\rvert$	$T_x' = T_x$	$T_x' = T_x + \frac{1}{2}\lvert T_{off}\rvert$
3-a	$\lvert T_{off[OPN]}\rvert = \frac{\lvert I_a\rvert}{\lvert I_c\rvert}(T_c - T_b)$ $\lvert T_{off[OPO]}\rvert = \frac{\lvert I_b\rvert}{\lvert I_c\rvert}(T_a - T_c)$	$T_x' = T_x - \frac{1}{2}(\lvert T_{off[OPN]}\rvert - \lvert T_{off[OPO]}\rvert)$	$T_x' = T_x - \frac{1}{2}\lvert T_{off[OPN]}\rvert$	$T_x' = T_x + \frac{1}{2}\lvert T_{off[OPO]}\rvert$
3-b	$\lvert T_{off[OPN]}\rvert = \frac{\lvert I_a\rvert}{\lvert I_b\rvert}(T_c - T_b)$ $\lvert T_{off[OON]}\rvert = \frac{\lvert I_c\rvert}{\lvert I_b\rvert}(T_b - T_a)$	$T_x' = T_x + \frac{1}{2}(\lvert T_{off[OPN]}\rvert - \lvert T_{off[OON]}\rvert)$	$T_x' = T_x - \frac{1}{2}\lvert T_{off[OON]}\rvert$	$T_x' = T_x + \frac{1}{2}\lvert T_{off[OPN]}\rvert$
4-a	$\lvert T_{off}\rvert = \frac{\lvert I_b\rvert}{\lvert I_c\rvert}(T_a - T_b)$	$T_x' = T_x + \frac{1}{2}\lvert T_{off}\rvert$	$T_x' = T_x$	$T_x' = T_x + \frac{1}{2}\lvert T_{off}\rvert$
4-b	$\lvert T_{off}\rvert = \frac{\lvert I_c\rvert}{\lvert I_b\rvert}(T_c - T_a)$	$T_x' = T_x - \frac{1}{2}\lvert T_{off}\rvert$	$T_x' = T_x - \frac{1}{2}\lvert T_{off}\rvert$	$T_x' = T_x$

Table 4.7 Time offset and redefined turn-ON times for sector III

R	$\lvert T_{off}\rvert$	Oscillations	Balancing ($V_{dc1} > V_{dc2}$)	Balancing ($V_{dc1} < V_{dc2}$)
1	$\lvert T_{off}\rvert = \frac{\lvert I_c\rvert}{\lvert I_b\rvert}(T_a - T_c)$	$T_x' = T_x + \frac{1}{2}\lvert T_{off}\rvert$	$T_x' = T_x$	$T_x' = T_x + \frac{1}{2}\lvert T_{off}\rvert$
2	$\lvert T_{off}\rvert = \frac{\lvert I_c\rvert}{\lvert I_a\rvert}(T_c - T_b)$	$T_x' = T_x - \frac{1}{2}\lvert T_{off}\rvert$	$T_x' = T_x - \frac{1}{2}\lvert T_{off}\rvert$	$T_x' = T_x$
3-a	$\lvert T_{off[NPO]}\rvert = \frac{\lvert I_c\rvert}{\lvert I_b\rvert}(T_a - T_b)$ $\lvert T_{off[NOO]}\rvert = \frac{\lvert I_a\rvert}{\lvert I_b\rvert}(T_b - T_c)$	$T_x' = T_x + \frac{1}{2}(\lvert T_{off[NPO]}\rvert - \lvert T_{off[NOO]}\rvert)$	$T_x' = T_x - \frac{1}{2}\lvert T_{off[NOO]}\rvert$	$T_x' = T_x + \frac{1}{2}\lvert T_{off[NPO]}\rvert$
3-b	$\lvert T_{off[NPO]}\rvert = \frac{\lvert I_c\rvert}{\lvert I_a\rvert}(T_a - T_b)$ $\lvert T_{off[OPO]}\rvert = \frac{\lvert I_b\rvert}{\lvert I_a\rvert}(T_c - T_a)$	$T_x' = T_x - \frac{1}{2}(\lvert T_{off[NPO]}\rvert - \lvert T_{off[OPO]}\rvert)$	$T_x' = T_x - \frac{1}{2}\lvert T_{off[NPO]}\rvert$	$T_x' = T_x + \frac{1}{2}\lvert T_{off[OPO]}\rvert$
4-a	$\lvert T_{off}\rvert = \frac{\lvert I_a\rvert}{\lvert I_b\rvert}(T_a - T_c)$	$T_x' = T_x - \frac{1}{2}\lvert T_{off}\rvert$	$T_x' = T_x - \frac{1}{2}\lvert T_{off}\rvert$	$T_x' = T_x$
4-b	$\lvert T_{off}\rvert = \frac{\lvert I_b\rvert}{\lvert I_a\rvert}(T_c - T_b)$	$T_x' = T_x + \frac{1}{2}\lvert T_{off}\rvert$	$T_x' = T_x$	$T_x' = T_x + \frac{1}{2}\lvert T_{off}\rvert$

Table 4.8 Time offset and redefined turn-ON times for sector IV

R	$\lvert T_{off} \rvert$	Oscillations	Balancing ($V_{dc1} > V_{dc2}$)	Balancing ($V_{dc1} < V_{dc2}$)
1	$\lvert T_{off} \rvert = \frac{\lvert I_b \rvert}{\lvert I_a \rvert}(T_b - T_c)$	$T_x' = T_x - \frac{1}{2}\lvert T_{off} \rvert$	$T_x' = T_x - \frac{1}{2}\lvert T_{off} \rvert$	$T_x' = T_x$
2	$\lvert T_{off} \rvert = \frac{\lvert I_b \rvert}{\lvert I_a \rvert}(T_a - T_b)$	$T_x' = T_x + \frac{1}{2}\lvert T_{off} \rvert$	$T_x' = T_x$	$T_x' = T_x + \frac{1}{2}\lvert T_{off} \rvert$
3-a	$\lvert T_{off[NOP]} \rvert = \frac{\lvert I_b \rvert}{\lvert I_c \rvert}(T_b - T_c)$ $\lvert T_{off[OOP]} \rvert = \frac{\lvert I_c \rvert}{\lvert I_a \rvert}(T_b - T_a)$	$T_x' = T_x - \frac{1}{2}(\lvert T_{off[NOP]} \rvert - \lvert T_{off[OOP]} \rvert)$	$T_x' = T_x - \frac{1}{2}\lvert T_{off[NOP]} \rvert$	$T_x' = T_x + \frac{1}{2}\lvert T_{off[OOP]} \rvert$
3-b	$\lvert T_{off[NOP]} \rvert = \frac{\lvert I_b \rvert}{\lvert I_c \rvert}(T_a - T_c)$ $\lvert T_{off[NOO]} \rvert = \frac{\lvert I_a \rvert}{\lvert I_c \rvert}(T_c - T_b)$	$T_x' = T_x + \frac{1}{2}(\lvert T_{off[NOP]} \rvert - \lvert T_{off[NOO]} \rvert)$	$T_x' = T_x - \frac{1}{2}\lvert T_{off[NOO]} \rvert$	$T_x' = T_x + \frac{1}{2}\lvert T_{off[NOP]} \rvert$
4-a	$\lvert T_{off} \rvert = \frac{\lvert I_c \rvert}{\lvert I_a \rvert}(T_b - T_c)$	$T_x' = T_x + \frac{1}{2}\lvert T_{off} \rvert$	$T_x' = T_x$	$T_x' = T_x + \frac{1}{2}\lvert T_{off} \rvert$
4-b	$\lvert T_{off} \rvert = \frac{\lvert I_a \rvert}{\lvert I_c \rvert}(T_a - T_b)$	$T_x' = T_x - \frac{1}{2}\lvert T_{off} \rvert$	$T_x' = T_x - \frac{1}{2}\lvert T_{off} \rvert$	$T_x' = T_x$

Table 4.9 Time offset and redefined turn-ON times for sector V

R	$\lvert T_{off}\rvert$	Oscillations	Balancing ($V_{dc1} > V_{dc2}$)	Balancing ($V_{dc1} < V_{dc2}$)
1	$\lvert T_{off}\rvert = \frac{\lvert I_a\rvert}{\lvert I_c\rvert}(T_b - T_a)$	$T_x' = T_x + \frac{1}{2}\lvert T_{off}\rvert$	$T_x' = T_x$	$T_x' = T_x + \frac{1}{2}\lvert T_{off}\rvert$
2	$\lvert T_{off}\rvert = \frac{\lvert I_a\rvert}{\lvert I_b\rvert}(T_b - T_a)$	$T_x' = T_x - \frac{1}{2}\lvert T_{off}\rvert$	$T_x' = T_x - \frac{1}{2}\lvert T_{off}\rvert$	$T_x' = T_x$
3-a	$\lvert T_{off[ONP]}\rvert = \frac{\lvert I_a\rvert}{\lvert I_b\rvert}(T_a - T_c)$ $\lvert T_{off[ONO]}\rvert = \frac{\lvert I_b\rvert}{\lvert I_c\rvert}(T_c - T_a)$	$T_x' = T_x + \frac{1}{2}(\lvert T_{off[ONP]}\rvert - \lvert T_{off[ONO]}\rvert)$	$T_x' = T_x - \frac{1}{2}\lvert T_{off[ONO]}\rvert$	$T_x' = T_x + \frac{1}{2}\lvert T_{off[ONP]}\rvert$
3-b	$\lvert T_{off[ONP]}\rvert = \frac{\lvert I_a\rvert}{\lvert I_b\rvert}(T_b - T_c)$ $\lvert T_{off[OOP]}\rvert = \frac{\lvert I_c\rvert}{\lvert I_b\rvert}(T_a - T_b)$	$T_x' = T_x - \frac{1}{2}(\lvert T_{off[ONP]}\rvert - \lvert T_{off[OOP]}\rvert)$	$T_x' = T_x - \frac{1}{2}\lvert T_{off[ONP]}\rvert$	$T_x' = T_x + \frac{1}{2}\lvert T_{off[OOP]}\rvert$
4-a	$\lvert T_{off}\rvert = \frac{\lvert I_b\rvert}{\lvert I_c\rvert}(T_b - T_a)$	$T_x' = T_x - \frac{1}{2}\lvert T_{off}\rvert$	$T_x' = T_x - \frac{1}{2}\lvert T_{off}\rvert$	$T_x' = T_x$
4-b	$\lvert T_{off}\rvert = \frac{\lvert I_c\rvert}{\lvert I_b\rvert}(T_a - T_c)$	$T_x' = T_x + \frac{1}{2}\lvert T_{off}\rvert$	$T_x' = T_x$	$T_x' = T_x + \frac{1}{2}\lvert T_{off}\rvert$

Table 4.10 Time offset and redefined turn-ON times for sector VI

R	$\lvert T_{off}\rvert$	Oscillations	Balancing ($V_{dc1} > V_{dc2}$)	Balancing ($V_{dc1} < V_{dc2}$)
1	$\lvert T_{off}\rvert = \frac{\lvert I_c\rvert}{\lvert I_b\rvert}(T_c - T_a)$	$T_x' = T_x - \frac{1}{2}\lvert T_{off}\rvert$	$T_x' = T_x - \frac{1}{2}\lvert T_{off}\rvert$	$T_x' = T_x$
2	$\lvert T_{off}\rvert = \frac{\lvert I_c\rvert}{\lvert I_a\rvert}(T_b - T_c)$	$T_x' = T_x + \frac{1}{2}\lvert T_{off}\rvert$	$T_x' = T_x$	$T_x' = T_x + \frac{1}{2}\lvert T_{off}\rvert$
3-a	$\lvert T_{off[PNO]}\rvert = \frac{\lvert I_c\rvert}{\lvert I_a\rvert}(T_b - T_a)$ $\lvert T_{off[POO]}\rvert = \frac{\lvert I_a\rvert}{\lvert I_b\rvert}(T_c - T_b)$	$T_x' = T_x - \frac{1}{2}(\lvert T_{off[PNO]}\rvert - \lvert T_{off[POO]}\rvert)$	$T_x' = T_x - \frac{1}{2}\lvert T_{off[PNO]}\rvert$	$T_x' = T_x + \frac{1}{2}\lvert T_{off[POO]}\rvert$
3-b	$\lvert T_{off[PNO]}\rvert = \frac{\lvert I_b\rvert}{\lvert I_c\rvert}(T_a - T_c)$ $\lvert T_{off[ONO]}\rvert = \frac{\lvert I_b\rvert}{\lvert I_a\rvert}(T_a - T_c)$	$T_x' = T_x + \frac{1}{2}(\lvert T_{off[PNO]}\rvert - \lvert T_{off[ONO]}\rvert)$	$T_x' = T_x - \frac{1}{2}\lvert T_{off[ONO]}\rvert$	$T_x' = T_x + \frac{1}{2}\lvert T_{off[PNO]}\rvert$
4-a	$\lvert T_{off}\rvert = \frac{\lvert I_a\rvert}{\lvert I_b\rvert}(T_c - T_a)$	$T_x' = T_x + \frac{1}{2}\lvert T_{off}\rvert$	$T_x' = T_x$	$T_x' = T_x + \frac{1}{2}\lvert T_{off}\rvert$
4-b	$\lvert T_{off}\rvert = \frac{\lvert I_b\rvert}{\lvert I_a\rvert}(T_b - T_c)$	$T_x' = T_x - \frac{1}{2}\lvert T_{off}\rvert$	$T_x' = T_x - \frac{1}{2}\lvert T_{off}\rvert$	$T_x' = T_x$

To reduce the magnitude of the neutral-point voltage oscillations, more N-type switching states are needed. The dwell time of N-type switching states can be increased by changing switching sequence 2 to switching sequence 1 in region 3-b. The redefined turn-on times for the change from sequence 2 to sequence 1 can be obtained as follows. The dwell time of each switching state can be defined as:

$$\begin{aligned} T_a' &= T_a + \frac{1}{2}\frac{|I_a|}{|I_c|}(T_b - T_a) \\ T_b' &= T_b + \frac{1}{2}\frac{|I_a|}{|I_c|}(T_b - T_a) \\ T_c' &= T_c + \frac{1}{2}\frac{|I_a|}{|I_c|}(T_b - T_a) \end{aligned} \tag{4.58}$$

The dwell time of each switching state should be equal as when sequence 2 was changed to sequence 1 in region 3-b of sector I. Therefore, the turn-on times are redefined as:

$$\begin{aligned} T_a' &= \frac{1}{2}T_{dwell[ONN]} + \frac{1}{2}T_{dwell[OON]} + T_{dwell[PPO]} = \frac{1}{2}(T_b - T_c) + 2T_a \\ T_b' &= \frac{1}{2}T_{dwell[ONN]} = \frac{1}{2}(T_b - T_c) \\ T_c' &= T_a' + \frac{1}{2}T_{dwell[PON]} = \frac{1}{2}(T_b - T_c) + 2T_a + (T_c - T_a) \end{aligned} \tag{4.59}$$

By changing the switching sequence, the dwell times of $T_{dwell[OON]}$ and $T_{dwell[ONN]}$ are increased by $2T_a$ and $(T_b - T_c)$, respectively. Conversely, the dwell times of $T_{dwell[PPO]}$ and $T_{dwell[POO]}$ is decreased to 0 and from $2(T_b - T_c)$ to $(T_b - T_c)$, respectively. This has the same effect as injecting the extra time offset. The extra time offset can be represented by the time offset as:

$$|T_{off\_ex}| = 2 \cdot T_a + (T_b - T_c) \cdot \frac{|I_a|}{|I_c|} \tag{4.60}$$

By changing the switching sequence, the required time offset in sequence 2 is reduced by:

$$\left|T_{off\_seq2}'\right| = \left|T_{off\_seq2}\right| - \left|T_{off\_ex}\right| = \left|T_{off}\right| - (2T_a + (T_b - T_c) \cdot \frac{|I_a|}{|I_c|}) \tag{4.61}$$

The required time offset in sequence 2 should be changed in terms of sequence 1. The time offset for sequence 1 can be defined as:

$$\begin{aligned} \left(\left|T_{off\_seq2}\right| - \left|T_{off\_ex}\right|\right) \cdot |I_c| &= \left|T_{off\_seq1}\right| \cdot |I_c| \\ \left|T_{off\_seq1}\right| &= \left|T_{off\_seq2}\right| - \left|T_{off\_ex}\right| \frac{|I_c|}{|I_a|} \end{aligned} \tag{4.62}$$

By changing switching sequence 2 to sequence 1 in region 3-b, the dwell times of N-type switching states are increased. This has the same effect since the magnitude of the maximum time offset that can be injected is increased.

C. Simulation and experiment results

Simulations and experiments have been carried out to verify the validity of the algorithm. The simulations are performed with a grid-connected T-type inverter under the conditions shown in Tables 4.11 and 4.12.

Figure 4.18a shows the outputs when the modulation index M_a is 0.83 and a conventional SVM is applied. The capacitor voltages oscillate, and the output phase current is 43 A_{peak}. The output line-to-line voltage has five levels at a high modulation index. Figure 4.18b shows simulated outputs when the neutral-point voltage ripple reduction method is applied. After the method is applied, the neutral-point voltage oscillation is considerably reduced.

The outputs of the conventional SVM when the modulation index is 0.415 are shown in Fig. 4.19a. The output line-to-line voltages have three levels because the modulation index is below 0.5. After the neutral-point voltage ripple reduction method is applied, the neutral-point voltage oscillations are decreased as shown in Fig. 4.19b. In this case, there is no distortion in the outputs.

Figure 4.20 shows comparison results of neutral-point voltage oscillations when M_a is 0.77 and the power factor (*PF*) is 0.5. In this case, the magnitude of the neutral-point voltage oscillation is larger than when the *PF* is 1. This is because the magnitude of the neutral current under medium-voltage vectors when the *PF* is 0.5 is bigger than when the *PF* is 1. The voltage oscillations have their maximum magnitude when the *PF* is near zero. The neutral-point voltage oscillation is reduced after the time-offset injection method is applied. However, the oscillation

Table 4.11 Simulation parameters

DC-link	650 V	Output frequency	50 Hz
C_1, C_2	550 μF	Output current	43 A_{peak}
Control period	100 μs	Switching frequency	10 kHz

Table 4.12 Experiment parameters

DC-link	180 V(M_a = 0.83), 400 V(M_a = 0.365)	Output frequency	60 Hz
C_1, C_2	550 μF	Load (R,L)	10 Ω, 9 mH
Control period	100 μs	Switching frequency	10 kHz

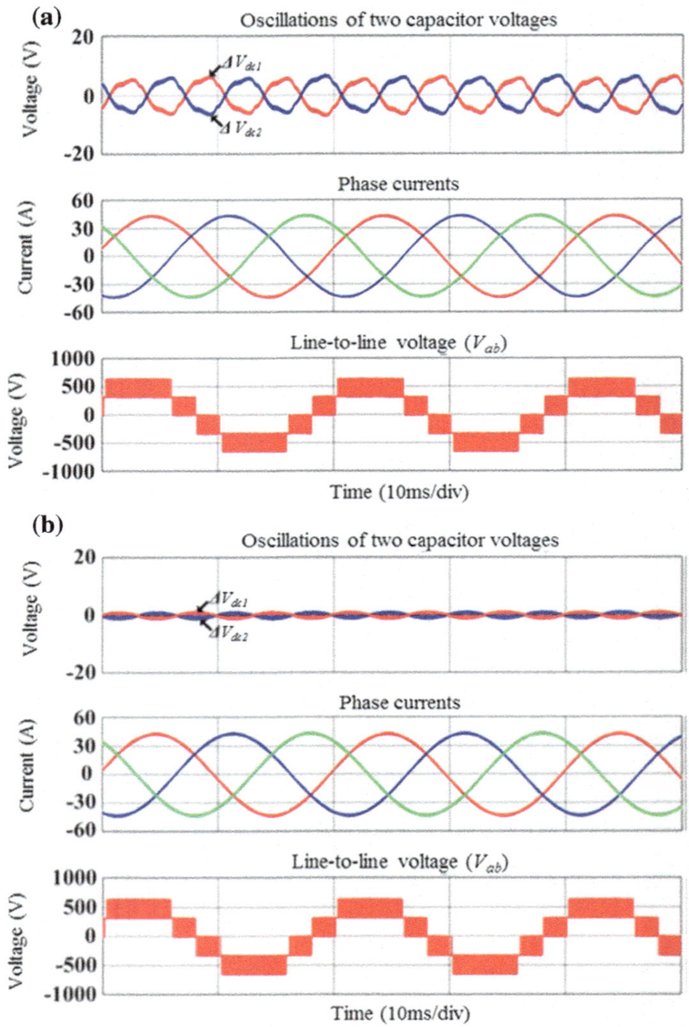

Fig. 4.18 Simulation results when M_a = 0.83 and PF = 1: **a** conventional SVM, **b** Neutral-point voltage ripple reduction method (reprinted from [9], Figs. 7 and 8)

still remains because there is a limitation to the injection of the time offset as mentioned in section III. To effectively decrease the neutral-point voltage oscillation, the switching sequences are changed in region 3-b. After the switching sequences are changed, the low voltage oscillations are significantly decreased as shown in Fig. 4.20.

Figure 4.21a shows experimental results before the neutral-point voltage ripple reduction method is applied when the modulation index is 0.83. After the method is

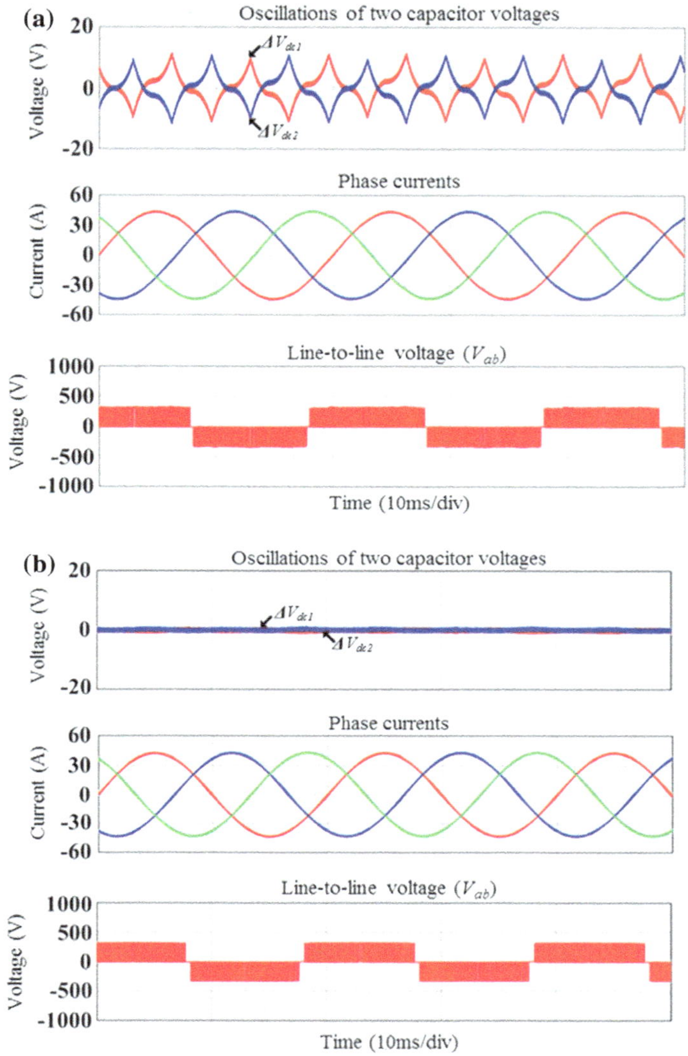

Fig. 4.19 Simulation results when M_a = 0.415 and PF = 1: **a** conventional SVM, **b** neutral-point voltage ripple reduction method (reprinted from [9], Figs. 9 and 10)

applied, the magnitudes of the neutral-point voltage oscillations are significantly reduced as shown in Fig. 4.21b. There are no distortions in the output line-to-line voltage or the phase current.

Figure 4.22a shows the two capacitor voltages, the line-to-line voltage, and the phase current when the modulation index is 0.365. The magnitudes of the neutral-point voltage oscillations are also decreased as shown in Fig. 4.22b. The

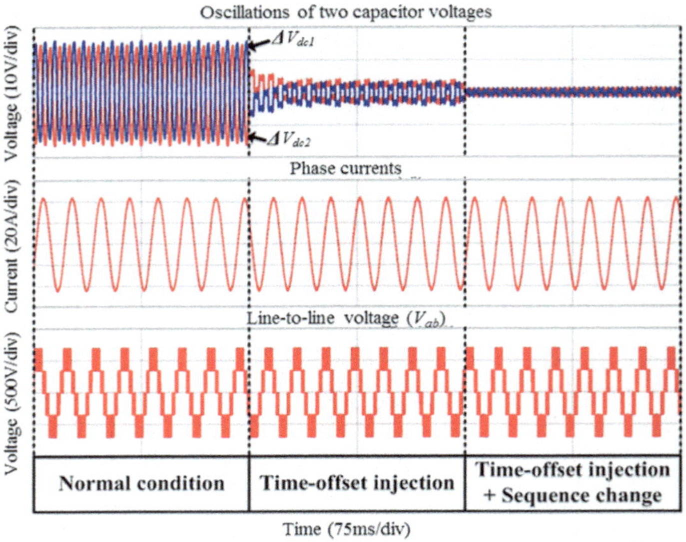

Fig. 4.20 Comparison of the neutral-point voltage oscillation and output phase currents when $M_a = 0.77$ and $PF = 0.5$ (reprinted from [9], Fig. 12)

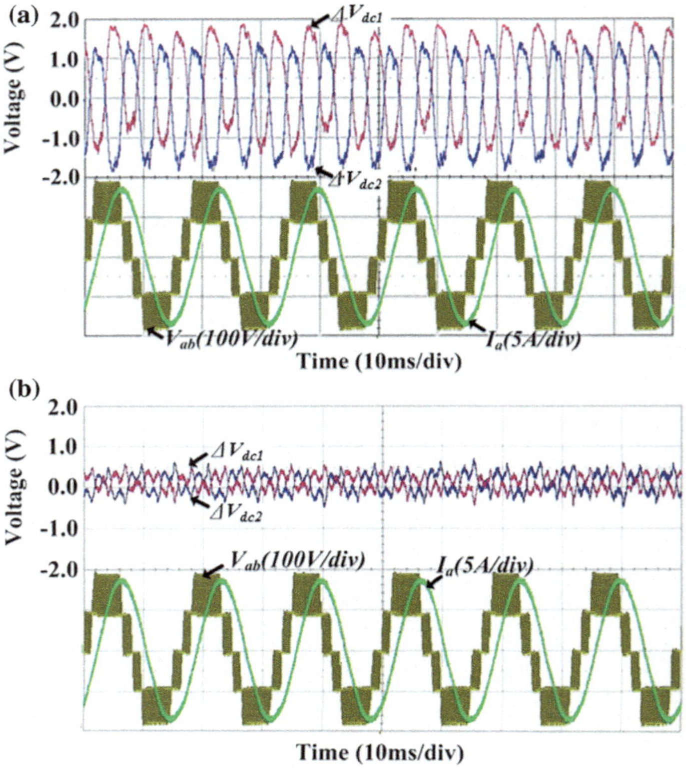

Fig. 4.21 Experimental results when $M_a = 0.83$: **a** conventional SVM, **b** neutral-point voltage ripple reduction method (reprinted from [9], Figs. 13 and 14)

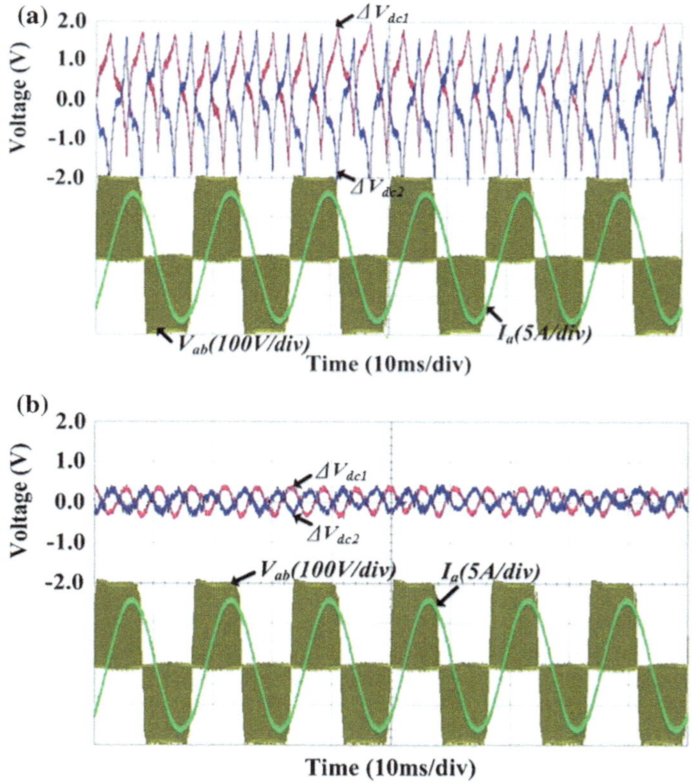

Fig. 4.22 Experimental results when $M_a = 0.365$: **a** conventional SVM, **b** neutral-point voltage ripple reduction method (reprinted from [9], Figs. 15 and 16)

oscillations of the neutral-point voltage in the experiments are smaller than those of the simulation results. This is due to the smaller output phase currents and larger output frequency when compared to those of simulations. The magnitude of the low voltage oscillations in the neutral-point is proportional to the magnitude of the output phase currents, and it is inversely proportional to the size of the capacitors and fundamental output frequency. Nevertheless, these experimental results show that the algorithm can reduce low-frequency voltage oscillations well without distortions in the output phase currents or the line-to-line voltages.

Figure 4.23 shows results of the neutral-point voltage balancing method. To make V_{dc1} and V_{dc2} unbalanced, a resistor is connected in parallel with the lower capacitor. Due to the neutral-point voltage deviation, the phase current is distorted as shown in Fig. 4.23a. After the balancing method is applied, the two capacitor voltages are converge at the same value, and the distortion of the phase current is eliminated as shown in Fig. 4.23b.

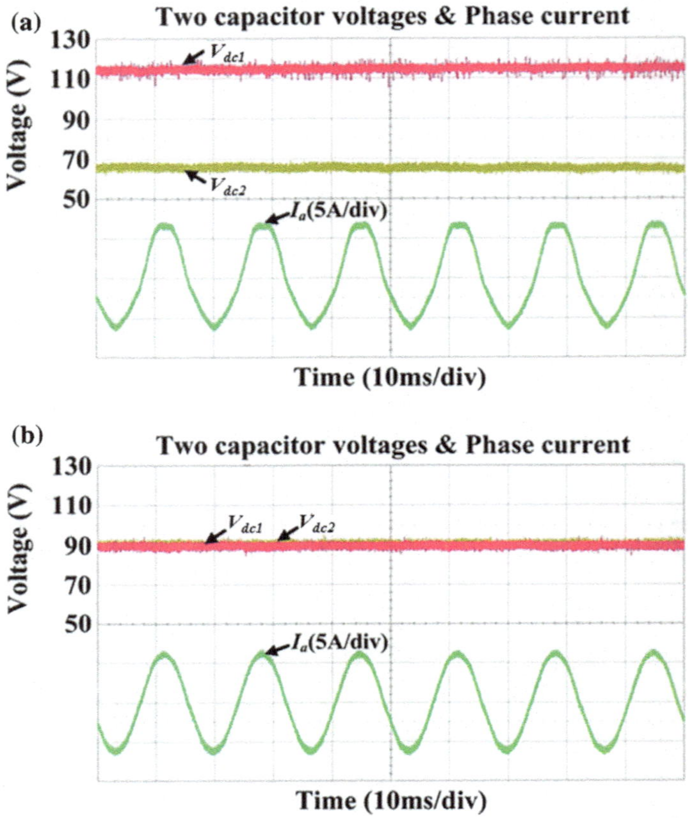

Fig. 4.23 Comparison of phase currents: **a** before the neutral-point voltage is balanced, **b** after the neutral-point voltage is balanced (reprinted from [9], Fig. 18)

References

1. Jiang, W.D., S.W. Du, L.C. Chang, Y. Zhang, and Q. Zhao. 2010. Hybrid PWM strategy of SVPWM and VSVPWM for NPC three-level voltage-source inverter. *IEEE Transactions on Power Electronics* 25 (10): 2607–2619.
2. Monge, S.B., S. Somavilla, J. Bordonau, and D. Boroyevich. 2007. Capacitor voltage balance for the neutral-point-clamped converter using the virtual space vector concept with optimized spectral performance. *IEEE Transaction on Power Electronics* 22 (4): 1128–1135.
3. Shen, J., S. Schröder, B. Duro, and R. Roesner. 2013. A neutral-point balancing controller for a three-level inverter with full power-factor range and low distortion. *IEEE Transaction on Industry Applications* 49 (1): 138–148.
4. Zabalza, J.C., P.I. Moreno, D. Madariaga, G. Calvo, and M.A. Rodríguez. 2013, October. Voltage balancing control in 3-level neutral-point clamped inverters using triangular carrier PWM modulation for FACTS applications. *IEEE Transaction on Power Electronics* 28 (10): 4473–4484.

5. Pou, J., J. Zaragoza, S. Ceballos, M. Saeedifard, and D. Boroyevich. 2012, A carrier-based PWM strategy with zero-sequence voltage injection for a three-level neutral-point-clamped converter. *IEEE Transactions on Power Electronics* 27 (2): 642–651.
6. Lee J.S., and K. B. Lee. 2014, A carrier-based PWM method for neutral-point ripple reduction of a 3-level inverter. In *Proceedings of IEEE Energy Conversion Congress and Exposition Conference* (pp. 2096–2100).
7. Choi, U.M., H.H. Lee, and K.B. Lee. 2013. Simple neutral-point voltage control for three-level inverters using a discontinuous pulse width modulation. *IEEE Transaction on Energy Conversion* 28 (2): 434–443.
8. Bendre, A., G. Venkataramanan, and D. Rosene. 2006. Modeling and design of a neutral-point voltage regulator for a three-level diode-clamped inverter using multiple-carrier modulation. *IEEE Transaction on Industrial Electronics* 53 (3): 718–726.
9. Choi, U.M., F. Blaabjerg and K.B. Lee. 2015. Method to Minimize the Low-Frequency Neutral-Point Voltage Oscillations with Time-Offset Injection for Neutral-Point-Clamped Inverters. *IEEE Transactions on Industry Applications* 51 (2): 1678–1691.
10. Lee, J.S., and K.B. Lee. 2016. Time-offset injection method for neutral-point AC ripple voltage reduction in a three-level inverter. *IEEE Transactions on Power Electronics* 31 (3): 1931–1941.
11. Maheshwari, R., S. Munk-Nielsen, and S. Busquets-Monge. 2013, May. Design of neutral-point voltage controller of a three-level NPC inverter with small DC-link capacitors. *IEEE Transaction on Industry Electronics* 60 (5): 1861–1871.
12. Wang, C., and Y. Li. 2010, Analysis and calculation of zero-sequence voltage considering neutral-point potential balancing in three-level NPC converters. *IEEE Transaction on Industrial Electronics* 57 (7): 2262–2271.

Chapter 5
Leakage Current Reduction

Leakage current can be defined as the current flowing through the ground in a lot of systems. The magnitude of the leakage current is defined based on the application and its topology. In particular, applications with transformerless topologies can be attacked by leakage current. Leakage current leads to safety problems and quality aggravation of the output current. Therefore, the leakage current has been restricted in a lot of systems such as wind turbine systems, photovoltaic systems, and energy storage systems by the regulation of the grid connection. A simple and normal solution for leakage current reduction is to add a transformer. However, adding a transformer decreases the total efficiency of the system and increases in terms of volume and cost [1–3]. It is a big disadvantage in the recent market situation due to a serious price war because transformers make up a large portion of the total cost of a system. Other solutions, switching devices and passive components are being used to reduce leakage current. Recently, modified PWM methods have been proposed to mitigate leakage current.

Most of the existing methods focus on the two-level converter topology. Recently, several leakage current reduction methods for small scale three-level converters have been proposed. This section analyzes the cause of leakage current in a photovoltaic system, which is a representative application suffering from leakage current. In addition, the PWM methods for leakage current reduction are introduced.

5.1 Leakage Current Analysis

5.1.1 Main Reason for Leakage Current

Leakage current and Common Mode Voltage (CMV) should be considered in PV systems. Leakage current is generated by the leakage capacitor C_{PV} which is

K.-B. Lee and J.-S. Lee, *Reliability Improvement Technology for Power Converters*, Power Systems, DOI 10.1007/978-981-10-4992-7_5

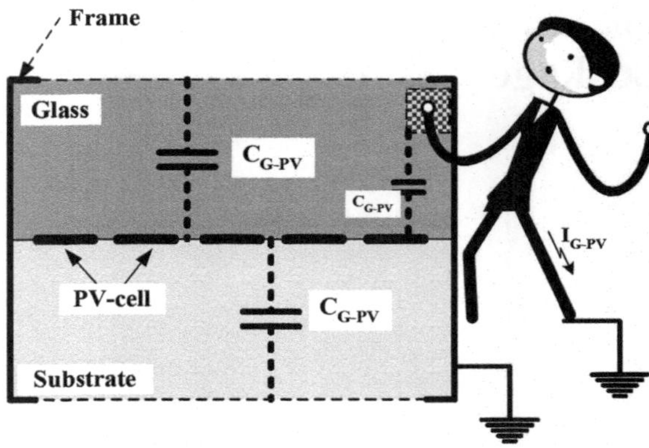

Fig. 5.1 Leakage capacitor existing in PV modules

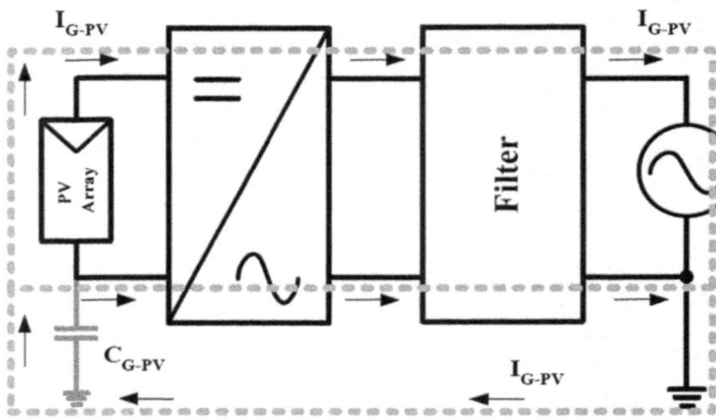

Fig. 5.2 Grid connected PV system

between the PV cells and the PV frame connected to the ground as shown in Figs. 5.1 and 5.2. If there is no voltage difference between the PV cells and the PV frame, leakage current does not flow. However, the P voltage and N voltage of the PV cells with respect to the grid ground have different voltage values depending on the inverter switching method. The generated P and N voltages are referred to as CMV and these voltages cause leakage current. Many countries restrict leakage current to low values, and rules about leakage current should be satisfied for the accreditation of manufactured goods.

5.1.2 Leakage Current Reduction Methods in H-Bridge and Two-Level Converters

A. H-bridge converters [4, 5]

It is not easy to change the PWM method in the H-bridge converters used in single-phase systems. Therefore, additional switching devices and passive

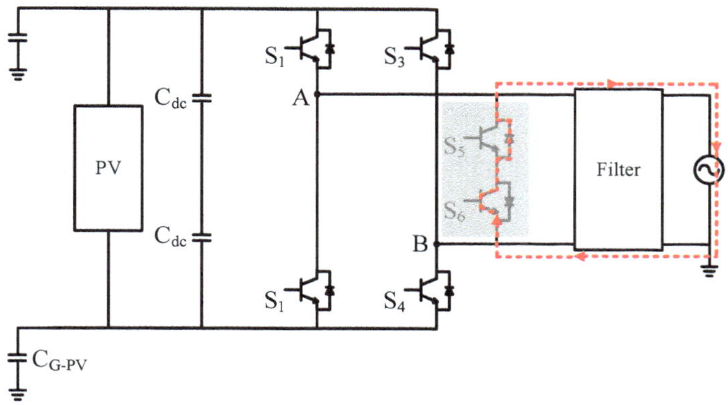

Fig. 5.3 Highly efficient and reliable inverter concept topology (SunWay)

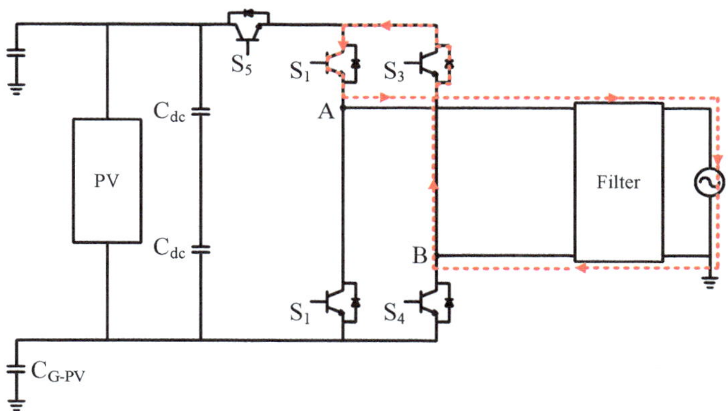

Fig. 5.4 H5 Topology (SMA)

components are used to reduce leakage current. Figures 5.3 and 5.4 show that additional switching devices isolate the DC-link with the PV from the grid. These topologies avoid building circuits that generate leakage current.

B. Two-level converters [6–12]

Not only new topologies based on two-level converter also a lot of PWM methods have been proposed for the leakage current reduction of two-level converters. The discontinuous PWM, active zero state PWM, and near state PWM methods are representative methods for the leakage current reduction of two-level converters. Figure 5.5 shows the PWM pattern and common-mode voltage leading to the leakage current of the representative methods compared to the conventional PWM method. The three PWM methods have a magnitude of common-mode voltage that is lower than that of the conventional PWM method.

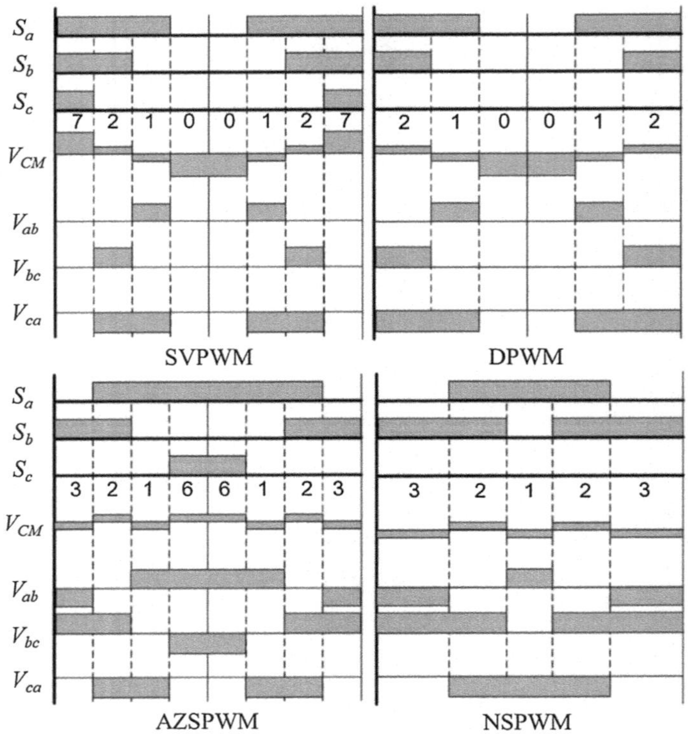

Fig. 5.5 Comparison of PWM methods for leakage current reduction

5.1.3 Three-Level Converters

Three-level converters have more voltage vectors than two-level converters. Therefore, a lot of PWM methods can be considered in three-level converters. This section explains the reason for leakage current in three-level converters using circuit analysis and introduces a PWM method for the leakage current reduction of three-level converters.

A. **Common-mode voltage of three-level converters**

A grid-connected three-level inverter with PV is shown in Fig. 5.6. This figure shows the leakage capacitor C_{PV}, three level inverter, and the L-filter. It also shows the leakage inductance L_{PV} and leakage resistor R_{PV} between the leakage capacitor C_{PV} and the grid ground (n). S_{a1} and S_{a3} of a leg are operated in reverse at the same time. S_{a2} and S_{a4} also perform this operation. The leakage capacitor C_{PV} is influenced by weather conditions. Therefore, the leakage current value can be changed due to the environmental factors.

Figure 5.7 represents a simple model of Fig. 5.6 by considering the leakage capacitor C_{PV} that is connected at N. The three-phase voltages to N (V_{AN}, V_{BN}, and

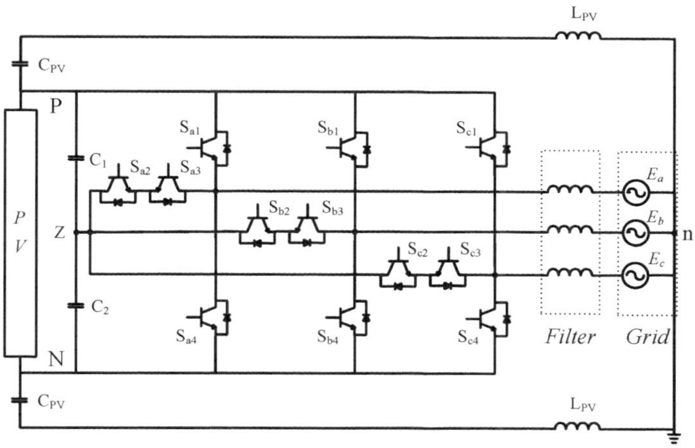

Fig. 5.6 Three-phase grid-connected transformerless PV three-level inverter (reprinted from [13], Fig. 1)

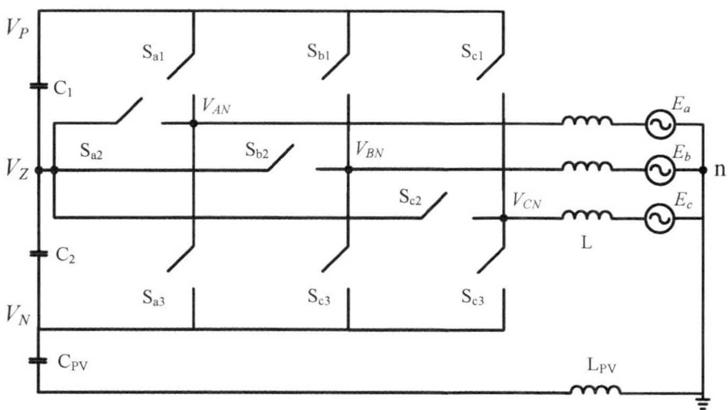

Fig. 5.7 Simple model of a transformerless PV inverter (reprinted from [13], Fig. 2)

V_{CN}) are considered in this model and V_{dc} is sum of V_{C1} and V_{C2}. In a three-level inverter, the output voltages of the inverter's legs are determined by the connection with one point among V_P, V_Z, and V_N. If the grid ground (n) is the standard, the output voltages can be divided into three cases depending upon the connection. In the first case, when the inverter's output is connected to V_P, the output voltage is $V_{dc} + V_N$. In the second case, the output voltage is $V_{dc}/2 + V_N$ when the output is connected at V_Z. In the last case, V_N is the output voltage when the output is attached to V_N.

V_N is defined by using the fact that the sum of the grid currents is zero. In addition, the output voltages are denoted as V_{AN}, V_{BN}, and V_{CN}. The grid currents flow due to the voltage difference between the output voltages of the inverter and the grid voltages in each phase. Therefore, the sum equation of the grid currents is expressed as:

$$\frac{V_{AN}+V_N-E_a}{Z_L}+\frac{V_{BN}+V_N-E_b}{Z_L}+\frac{V_{CN}+V_N-E_c}{Z_L}=0 \tag{5.1}$$

$$E_a+E_b+E_c=0 \tag{5.2}$$

where Z_L is the impedance of the L-filter. Equation (5.1) can be represented by using (5.2), and the represented equation is expressed as:

$$\frac{V_{AN}+V_{BN}+V_{CN}+3V_N}{Z_L}=0 \tag{5.3}$$

V_N can be calculated from (5.3) and this voltage is expressed as:

$$V_N=-\frac{V_{AN}+V_{BN}+V_{CN}}{3} \tag{5.4}$$

Likewise, V_P is defined using the same method.

V_N is determined according to the output voltages of the inverter as (5.4). This means that V_N can be influenced due to the inverter switching state. Figure 5.8 shows the circuit for the output voltage levels (V_{dc} + V_N, $V_{dc}/2$ + V_N, and V_N) and the values of V_{AN}, V_{BN}, and V_{CN} which are the ones among V_{dc}, $V_{dc}/2$, and 0.

$$\mathrm{S_{a,b,c}}(V_{AN,BN,CN})=\left\{\begin{array}{c}\mathrm{S_a}=1,\ V_{AN}=V_{dc}\\ \mathrm{S_a}=\frac{1}{2},\ V_{AN}=\frac{V_{dc}}{2}\\ \mathrm{S_a}=0,\ V_{AN}=0\\ \\ \mathrm{S_b}=1,\ V_{BN}=V_{dc}\\ \mathrm{S_b}=\frac{1}{2},\ V_{BN}=\frac{V_{dc}}{2}\\ \mathrm{S_b}=0,\ V_{BN}=0\\ \mathrm{S_c}=1,\ V_{CN}=V_{dc}\\ \mathrm{S_c}=\frac{1}{2},\ V_{CN}=\frac{V_{dc}}{2}\\ \mathrm{S_c}=0,\ V_{CN}=0\end{array}\right\} \tag{5.5}$$

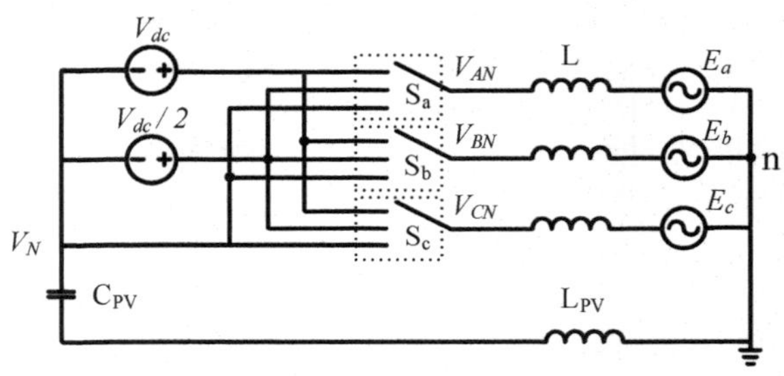

Fig. 5.8 Circuit for the output voltages of an inverter depending on the switching states (reprinted from [13], Fig. 3)

Equation (5.1) is represented with switching states as:

$$\frac{V_{dc}(S_a) + V_N - E_a}{Z_L} + \frac{V_{dc}(S_b) + V_N - E_b}{Z_L} + \frac{V_{dc}(S_c) + V_N - E_c}{Z_L} = 0 \qquad (5.6)$$

Then, (5.6) is simply represented as:

$$V_N = -\frac{V_{dc}(S_a) + V_{dc}(S_b) + V_{dc}(S_c)}{3} \qquad (5.7)$$

Therefore, the value of V_N can be selected among $-V_{dc}$, $-5V_{dc}/6$, $-4V_{dc}/6$, $-3V_{dc}/6$, $-2V_{dc}/6$, $-1V_{dc}/6$, and 0 depending on the switching states, and V_N includes an AC component. The ac-component of V_N generates leakage current through the line including C_{PV}.

The current equation is calculated by considering C_{PV}. In Fig. 5.9, there are four paths and the sum of the four currents through these four paths is zero. Therefore, the current equation should be expressed as:

$$\frac{V_{AN} + V_{CM} - E_a}{Z_L} + \frac{V_{BN} + V_{CM} - E_b}{Z_L} + \frac{V_{CN} + V_{CM} - E_c}{Z_L} + \frac{V_{CM}}{Z_{PV}} = 0 \qquad (5.8)$$

where V_{CM} is the voltage of V_N with consideration of Z_{PV}, and Z_{PV} is the combined impedance of C_{PV}, R_{PV}, and L_{PV}.

From (5.8), using (5.4), V_{CM} is expressed as:

$$V_{CM} = \frac{Z_{PV}}{3Z_{PV} + Z_L} 3V_N \qquad (5.9)$$

V_{CM} is proportional to V_N. The leakage current can be calculated using V_{CM} because the voltages of the circuit in Fig. 5.9 are defined from the grid ground (n). The magnitude of the leakage current is calculated as:

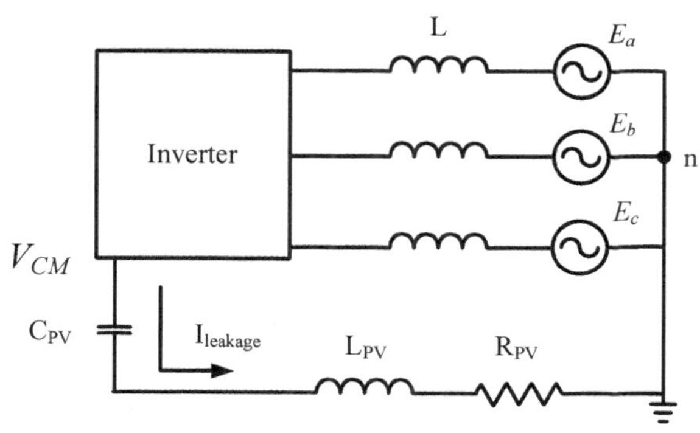

Fig. 5.9 Model circuit for the output voltage of an inverter (reprinted from [13], Fig. 4)

$$|I_{CM}| = \frac{|V_{CM}|}{|Z_{PV}|} = \frac{3|V_N|}{\left|3(\mathrm{R_{PV}} + j2\pi f \mathrm{L_{PV}} + \frac{1}{j2\pi f \mathrm{C_{PV}}}) + j2\pi f \mathrm{L}\right|} \tag{5.10}$$

The leakage current is the same as the current of the LC resonant circuit because the resistor is normally small. The current reaches its peak at the resonant frequency according to L and C.

Figure 5.10 shows the leakage current magnitude depending on the switching frequency. These waveforms are calculated using (5.11). There are three waveforms depending on the leakage resistor $\mathrm{R_{PV}}$. The leakage resistor $\mathrm{R_{PV}}$ acts as a damping. Therefore, the leakage current decreases due to a large $\mathrm{R_{PV}}$.

The total leakage current is the sum of each of the currents due to the odd multiples of the switching frequency because CMV is a square wave. Therefore, the total leakage current is expressed as:

$$\left|I_{CM,total}\right| = \sum_{n=1,3,5,\ldots}^{\infty} |I_{CM}(nf)| \tag{5.11}$$

The total leakage current can increase if the switching frequency increases to the value of the resonant frequency. However, the total leakage current is not proportional to the switching frequency due to the damping resistor, which is the leakage resistor $\mathrm{R_{PV}}$.

The leakage current is affected by the magnitude and frequency of CMV, the leakage inductance $\mathrm{L_{PV}}$, leakage resistor $\mathrm{R_{PV}}$, and the leakage capacitor $\mathrm{C_{PV}}$ from (5.10). The leakage inductance $\mathrm{L_{PV}}$ and the leakage resistor $\mathrm{R_{PV}}$ are excluded from the leakage current calculation, and the leakage capacitor $\mathrm{C_{PV}}$ is decided based on environmental factors. To decrease the leakage current, the magnitude and frequency of V_N or CMV should be low. In addition, (5.10) is valid when the grid is replaced with a load.

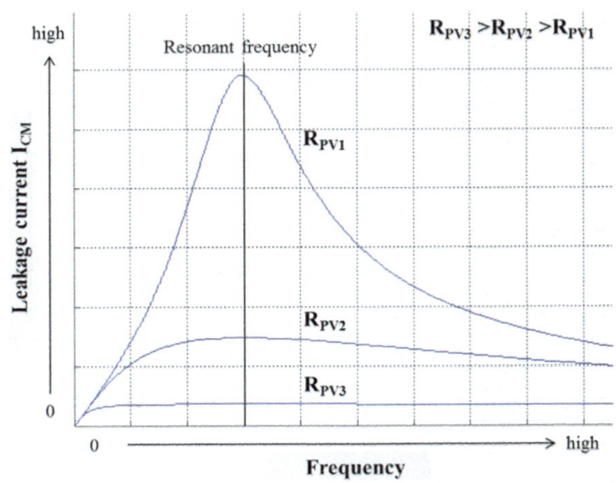

Fig. 5.10 Leakage current magnitude depending on the frequency and R_{PV} value (reprinted from [13], Fig. 5)

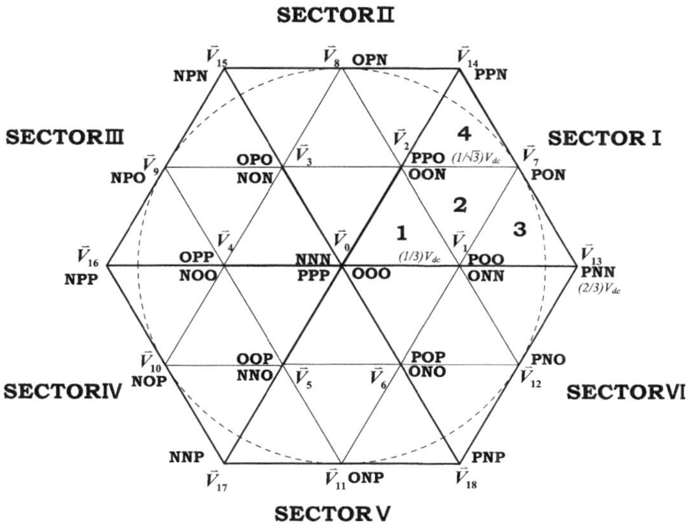

Fig. 5.11 Space vector diagram of a three-level inverter

B. V_N of the conventional SVPWM

The conventional PWM is commonly used in three-level inverters. A space vector diagram of a three-level inverter including four types of vectors is shown in Fig. 5.11. There are zero vectors, small vectors, medium vectors, and large vectors. There are three combinations for a zero vector, two combinations for a small vector, and one combination for a medium and a large vector as shown in Fig. 5.11.

Depending on the vector, the magnitude of V_N for all of the vectors is defined from (5.7) in Table 5.1. The conventional PWM for a three-level inverter uses all of the vectors to control its output voltage. Therefore, V_N can be from 0 V to V_{dc}.

Solutions to reduce the leakage current are a small magnitude and a low switching frequency of V_N. These two solutions can be identified through (5.10). The magnitude of V_N should be small, and its frequency should be low to reduce the leakage current.

In each switching period T_T, the modulation index of an inverter is usually greater than 0.44. For this condition, the zero vectors NNN and PPP are not used to generate the output voltage of the inverter. Therefore, the magnitude of V_N is from $V_{dc}/6$ to $4V_{dc}/6$ or from $2V_{dc}/6$ to $5V_{dc}/6$ according to the sector in Fig. 5.11. The conventional PWM requires a low frequency of V_N because this modulation switches only once for each switching period T_T in V_N as shown in Fig. 5.12. As a result, the conventional PWM shows the best sequence that does not generate a high frequency.

Table 5.1 V_N value according to the switching states

V_{Small}	S_a	S_b	S_c	$\|V_N\|$	V_{Medium}	S_a	S_b	S_c	$\|V_N\|$
V_0(NNN)	0	0	0	0	V_7(PON)	1	1/2	0	$3V_{dc}/6$
V_0(OOO)	1/2	1/2	1/2	$3V_{dc}/6$	V_8(OPN)	1/2	1	0	$3V_{dc}/6$
V_0(PPP)	1	1	1	V_{dc}	V_9(PNO)	1	0	1/2	$3V_{dc}/6$
V_1(POO)	1	1/2	1/2	$4V_{dc}/6$	V_{10}(NOP)	0	1/2	1	$3V_{dc}/6$
V_1(ONN)	1/2	0	0	$1V_{dc}/6$	V_{11}(ONP)	1/2	0	1	$3V_{dc}/6$
V_2(PPO)	1	1	1/2	$5V_{dc}/6$	V_{12}(PNO)	1	0	1/2	$3V_{dc}/6$
V_2(OON)	1/2	1/2	0	$2V_{dc}/6$	$\mathbf{V_{Large}}$	$\mathbf{S_a}$	$\mathbf{S_b}$	$\mathbf{S_c}$	$\mathbf{\|V_N\|}$
V_3(OPO)	1/2	1	1/2	$4V_{dc}/6$	V_{13}(PNN)	1	0	0	$2V_{dc}/6$
V_3(NON)	0	1/2	0	$1V_{dc}/6$	V_{14}(PPN)	1	1	0	$4V_{dc}/6$
V_4(OPP)	0	1	1	$5V_{dc}/6$	V_{15}(NPN)	0	1	0	$2V_{dc}/6$
V_4(NOO)	1	1/2	1/2	$2V_{dc}/6$	V_{16}(NPP)	0	1	1	$4V_{dc}/6$
V_5(POO)	1	1/2	1/2	$4V_{dc}/6$	V_{17}(NNP)	0	0	1	$2V_{dc}/6$
V_5(NNO)	0	0	1/2	$1V_{dc}/6$	V_{18}(PNP)	1	0	1	$4V_{dc}/6$
V_6(POP)	1	1/2	1	$5V_{dc}/6$					
V_6(ONO)	1/2	0	1/2	$2V_{dc}/6$					

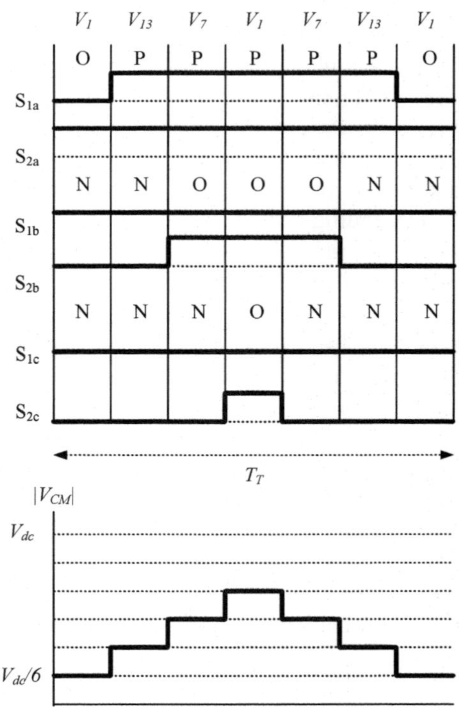

Fig. 5.12 Switching order and V_N for the conventional PWM at the region 3 of sector 1 (reprinted from [13], Fig. 7)

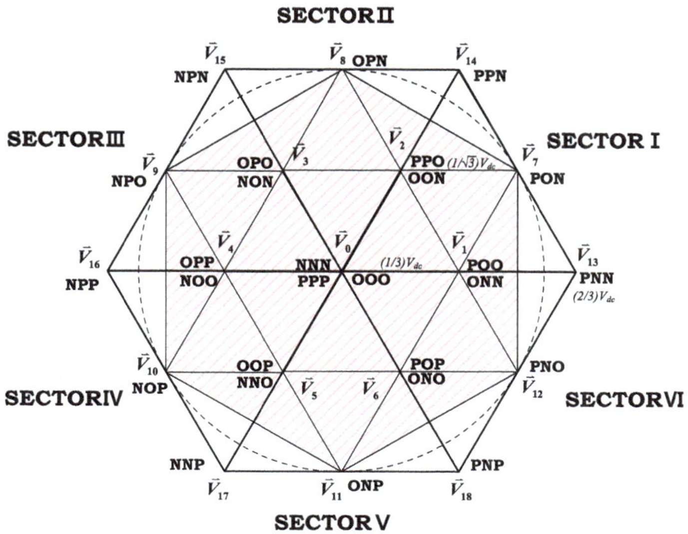

Fig. 5.13 Space vector diagram of a modulation technique using medium and zero vectors

C. **Modulation technique using medium and zero vectors** [12, 14]

Many combinations can produce a switching sequence because many vectors exist in the space vector diagram with three levels. Therefore, the magnitude and frequency of V_N are changed depending on the selected switching sequence. Modulation using only medium and zero vectors has been proposed as shown in Fig. 5.13. This modulation can generate an output voltage by $V_{dc}/2$. However, it requires a higher DC-link voltage compared to that of the conventional PWM. Furthermore, it switches twice in a switching period T_T. This makes the leakage current high and increases the THD of the grid currents.

D. **LMZVM (Large, Medium and Zero Vector Modulation) method** [13]

The first discussed modulation method has the best sequence in terms of frequency, which is similar to that of the conventional PWM. In addition, it has a lower magnitude of V_N when compared to the conventional PWM. Using small vectors critically increases or decreases V_N. Therefore, this modulation does not use small vectors to generate the output voltages of the inverter. Instead, it uses only large vectors, medium vectors, and zero vectors. In addition, only OOO is used as the zero vector because the fluctuation of V_N is minimized. Therefore, the space vector diagram of a three-level inverter is divided into 12 sectors as shown in Fig. 5.14.

When the output vector is in sector 1, V_{13}, V_7, and V_0 are used to make the output vector. In Fig. 5.15, the switching time for the output vector is calculated. The medium vector's switching time T_M and the large vector's switching time T_L are

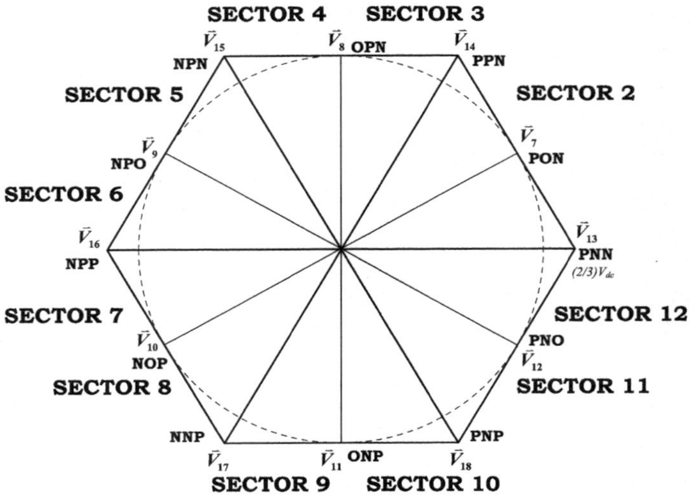

Fig. 5.14 Space vector diagram of a three-level inverter for LMZVM (reprinted from [13], Fig. 8)

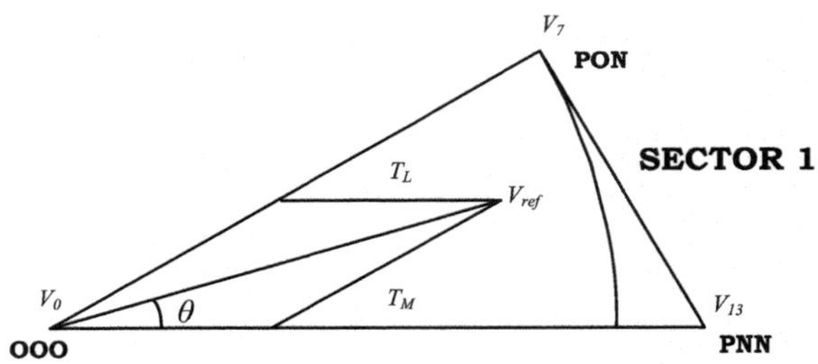

Fig. 5.15 Output vector times in sector 1 (reprinted from [13], Fig. 9)

calculated in advanced according to θ. Then the remaining time T_Z is allocated for the zero vector. As in sector 1, the adjacent vectors are used in N sector.

Another consideration is the switching sequence. This sequence is important for reducing the leakage current. Thus, only one switching should be executed in a switching period T_T. If frequencies higher than the switching frequency are generated, the leakage current increases. Therefore, the selected sequence is shown in Fig. 5.16.

The output vector of sector 1 consists of a large vector, a medium vector, and a zero vector in this order. This sequence switches only once per switching period T_T, and the magnitudes of V_N are from $2V_{dc}/6$ to $3V_{dc}/6$ and from $3V_{dc}/6$ to $4V_{dc}/6$ depending on the sector. In summary, the frequency of V_N is the same as that of the conventional PWM, and the magnitude of V_N is half that of the conventional PWM.

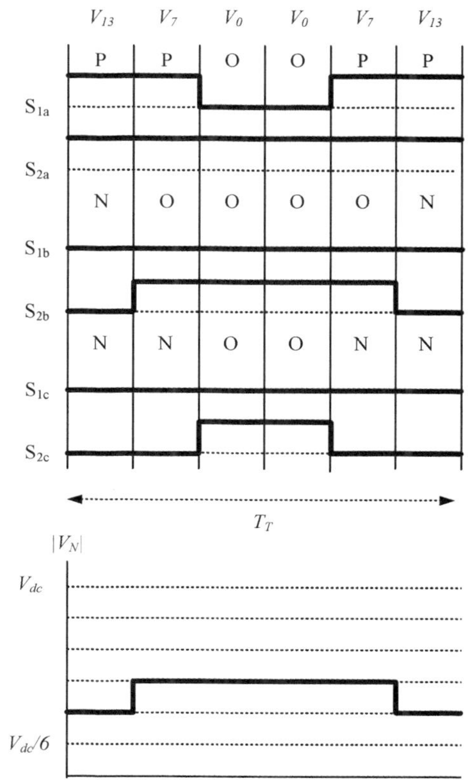

Fig. 5.16 Switching order and V_N for LMZVM in sector 1 (reprinted from [13], Fig. 10)

Table 5.2 Simulation and experimental parameters

DC-link voltage	600 V (Simulation)/300 V (Experiment)	Filter	2 mH
DC-link capacitor	1100 μF	C_{PV}	0.01 μF, 0.015 μF
Switching frequency	10 kHz	R_{PV}	15 Ω

- Simulation and experimental results of the LMZVM method.

Table 5.2 shows the parameters used in the simulation and experiment.

Figure 5.17 shows V_N and the leakage current waveforms when the conventional PWM is used. V_N has a magnitude from 200 to 500 V and from 100 to 400 V depending on the sector. At that time, the leakage current ranges in value from –500 to 500 mA.

Figure 5.18a shows the dropped V_N and the leakage current waveforms when the LMZVM is used. The small vectors that generate V_N of $-5V_{dc}/6$ and $-V_{dc}/6$ are not used. In comparison with Fig. 5.18a, V_N is half that in Fig. 5.17. For this reason, the leakage current decreases approximately by half. Figure 5.18b shows the ON “1”/

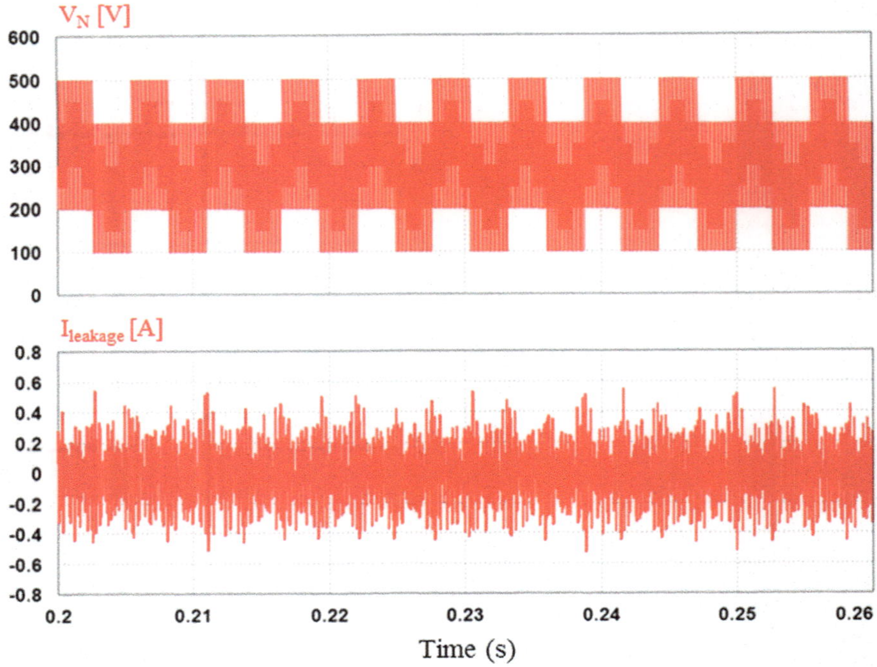

Fig. 5.17 Simulation waveforms of VN and leakage current with the conventional PWM (reprinted from [13], Fig. 14)

OFF "0" signals of the switches and V_N when θ is in the sector 1 of Fig. 5.14. The switching order is V_{13}(PNN)-V_7(PON)-V_0(POO)-V_7(PON)-V_{13}(PNN), which is the same as that of Fig. 5.16. Also, the magnitude of V_N is from 200 to 300 V and switching occurs once in T_T.

Figure 5.19a, b show V_N and the leakage current when the conventional PWM is used in the experiment. The magnitude ranges of V_N are from 100 to 250 V and from 50 to 200 V. The peak values of the leakage current are 200 and 280 mA as shown in Fig. 5.19a, b. V_N and the leakage current are shown in Fig. 5.20a, b when the LMZVM is applied. The leakage currents decrease and the peak values are 100 and 180 mA because V_N decreases with a magnitude from 150 to 200 V and from 100 to 150 V. When the magnitude of V_N is reduced by one-third, the leakage current also is reduced. These results indicate the same characteristics as the simulation results.

E. **LMSVM (Large, Medium and Small Vector Modulation) method** [13]

During the LMZVM operation, it is not possible to control the neutral-point voltage because this technique uses large, medium, and zero vectors. For neutral-point voltage balance, the small vectors have to be used. This is available to the N-type small vectors for top side DC-link charging and to the P-type small vectors for bottom side dc-link charging. Therefore, the second technique, based on

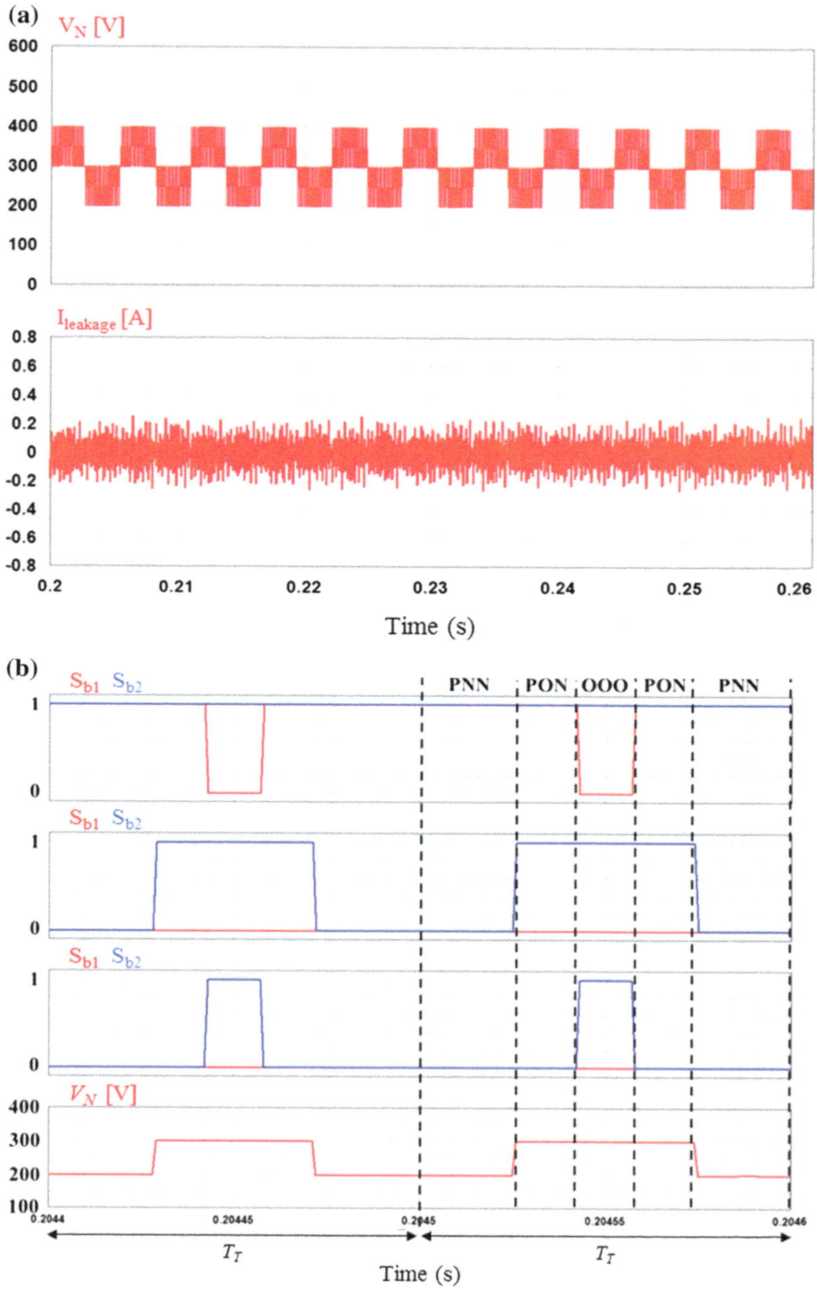

Fig. 5.18 Simulation waveforms of LMZVM: **a** VN and leakage current, **b** with the ON "1"/OFF "0" signals and V_N (reprinted from [13], Figs. 15 and 16)

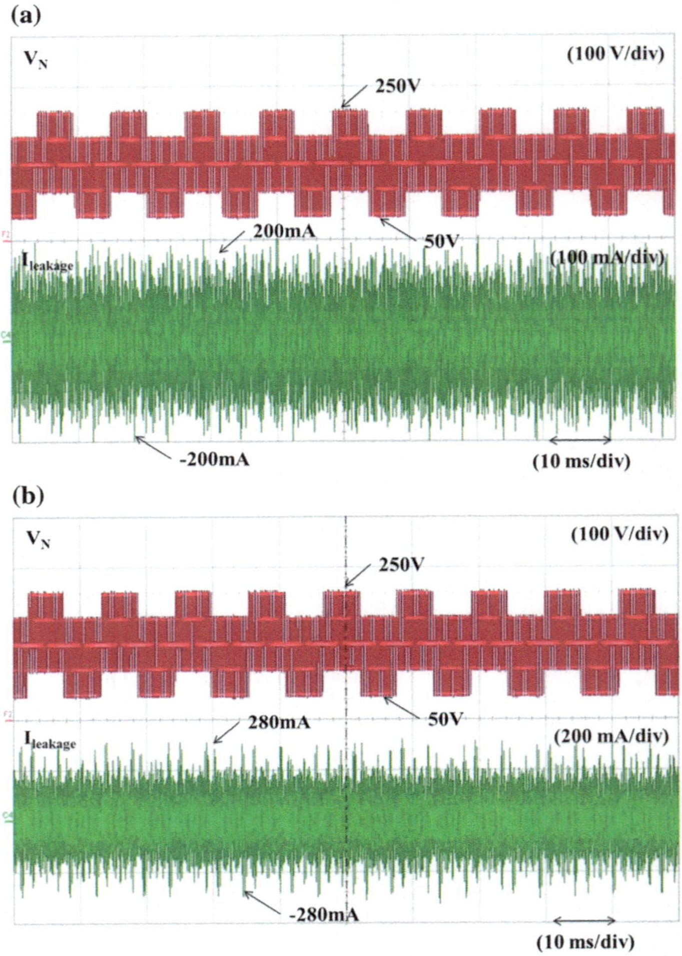

Fig. 5.19 Experimental result of V_N and the leakage current using the conventional PWM: **a** $C_{PV} = 0.01$ μF, **b** $C_{PV} = 0.015$ μF (reprinted from [13], Fig. 22)

LMZVM, substitutes the zero vector with the small vector (LMSVM), which is the closest to the present large and medium vectors to prevent the THD of output current from increasing.

In sector 1, V_2[OON], one of the N-type vectors, is selected for neutral-point voltage control. Then, the switching order is V_2[OON]-V_7[PON]-V_{13}[PNN]-V_7[PON]-V_2[OON] which is shown in Fig. 5.21. The top side capacitor voltage increases due to the N-type small vector V_2[OON]. The magnitude of V_N is maintained as the magnitude of the LMZVM. In sector 2, V_1[POO], one of the P-type vectors, is used and the switching order is V_1[POO]- V_7[PON]-V_{14}[PPN]-

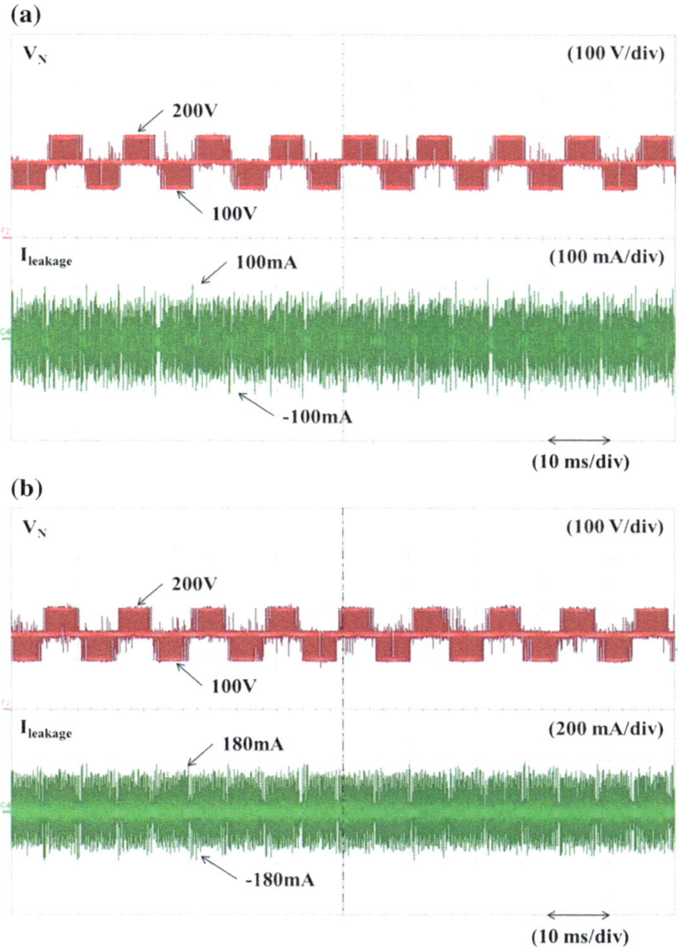

Fig. 5.20 Experimental result of V_N and the leakage current using LMZVM: **a** C_{PV} = 0.01 μF, **b** C_{PV} = 0.015 μF (reprinted from [13], Fig. 23)

V_7[PON]-V_1[POO]. The bottom side capacitor voltage increases due to the P-type small vector V_1[POO], and the magnitude of V_N is the same as that of the LMZVM. In sectors 1, 4, 5, 8, 9, and 12, the N-type small vectors are chosen and these vectors increase the top side capacitor voltage. The chosen P-type small vectors increase the bottom side capacitor voltage in sectors 2, 3, 6, 7, 10, and 11.

The switching times in the odd sectors and even sectors are calculated as shown in the following equations:

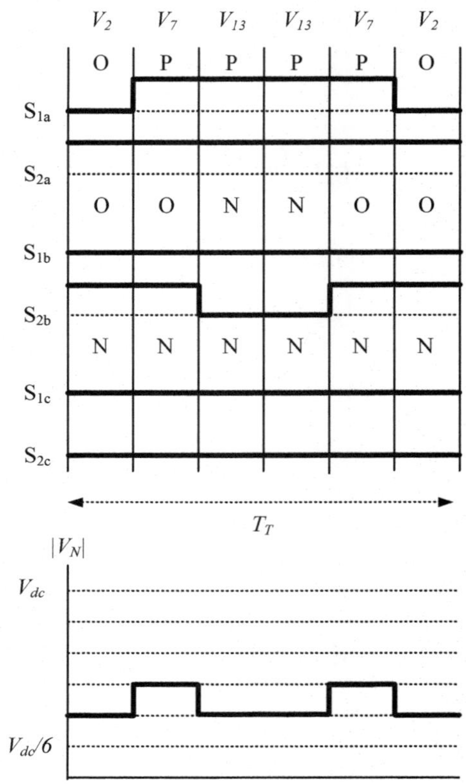

Fig. 5.21 Switching order and V_N for LMSVM in sector 1 (reprinted from [13], Fig. 11)

Even sectors:

$$\begin{aligned} T_L &= \left[1 - 2\sqrt{3}m_i \sin\theta\right] T_T \\ T_M &= \left[3m_i(\sqrt{3}\sin\theta + \cos\theta) - 2\right] T_T \\ T_S &= \left[2 - m_i(\sqrt{3}\sin\theta + 3\cos\theta)\right] T_T \end{aligned} \tag{5.12}$$

Odd sectors:

$$\begin{aligned} T_L &= [1 + m_i(\sqrt{3}\sin\theta - 3\cos\theta)]T_T \\ T_M &= [2m_i(3cos\theta) - 2]T_T \\ T_S &= [2 + m_i(\sqrt{3}\sin\theta - 3\cos\theta)]T_T \end{aligned} \tag{5.13}$$

where T_L is the large vector's switching time, T_M is the medium vector's switching time, T_S is the small vector's switching time, θ is the grid's angle, M_a is the modulation index (V_{ref}/V_{dc}), and T_T is the switching period.

The switching times, which are calculated by (5.12) and (5.13), have to be positive values. Therefore, the θ range to obtain positive values for T_L and T_M is considered because T_T can be calculated by the following equation:

$$T_S = T_T - (T_M + T_L) \tag{5.14}$$

- Feasible range of the LMSVM method

The θ range to obtain positive values for T_L and T_M is related to M_a, and is defined depending on the sectors. In sectors 1 and 2, the relationship between θ and M_a is expressed as

$$\begin{aligned}&\text{Sector 1 } (0^\circ < \theta < 30^\circ):\\ &\sin\theta < \frac{1}{2\sqrt{3}M_a}, \quad \text{for } \mathrm{T}_L > 0\\ &\sin(\theta - 30^\circ) > \frac{1}{3M_a}, \quad \text{for } \mathrm{T}_M > 0\end{aligned} \tag{5.15}$$

$$\begin{aligned}&\text{Sector 2 } (30^\circ < \theta < 60^\circ):\\ &\sin(\theta - 30^\circ) > -\frac{1}{2M_a}, \quad \text{for } \mathrm{T}_L > 0\\ &\cos\theta > \frac{1}{3M_a}, \quad \text{for } \mathrm{T}_M > 0\end{aligned} \tag{5.16}$$

The M_a value is calculated as 0.518 by setting V_{dc} as 600 V and V_{ref} as 311.14 V, which are the parameters used at the grid connection. To satisfy the two equations in (5.15) and (5.16), each θ range is calculated as

$$\begin{aligned}&\text{sector 1}: 10.05^\circ < \theta < 30^\circ\\ &\text{sector 2}: 30^\circ < \theta < 49.95^\circ\end{aligned} \tag{5.17}$$

The LMSVM is applied in the range of (5.17). The θ range from (5.17) can be expanded to all of the sectors and these ranges are expressed as:

$$\begin{aligned}&30(S-1) + 10.05 < \theta < 30S, \quad \text{for odd sectors}\\ &30(S-1) < \theta < 60S - 10.05, \quad \text{for even sectors}\end{aligned} \tag{5.18}$$

where S is a sector.

All of the ranges from (5.18) are shown as Fig. 5.22. The gray parts are the ranges for the LMSVM. In the gray parts, the parts with patterns indicate the range where the P-type small vectors are chosen to increase the bottom side capacitor voltage, and the N-type small vectors are chosen for the top side capacitor voltage in the parts without a pattern. In the rest of the parts, the LMZVM is always applied.

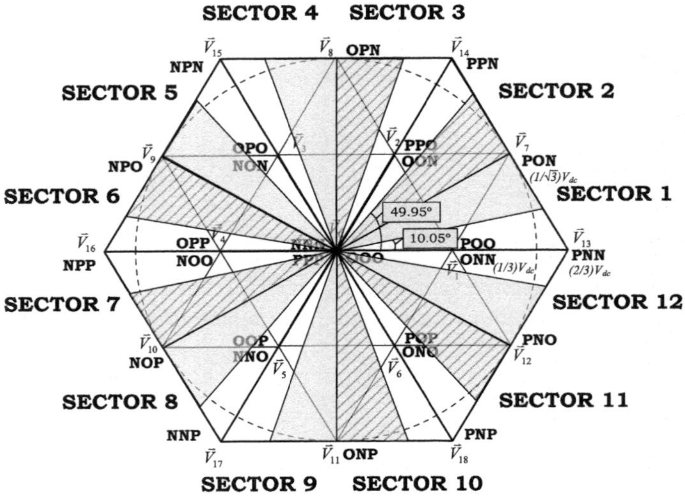

Fig. 5.22 Space vector diagram of the θ range for LMSVM when M_a is 0.518 (reprinted from [13], Fig. 12)

If the bottom side capacitor voltage is larger than the top side capacitor voltage, the LMSVM is applied in the gray parts without a pattern, and the LMZVM is used both in the gray parts with a pattern and in the rest of the parts. On the other hand, if the top side capacitor voltage is larger than the bottom side capacitor voltage, the LMSVM is used in the gray parts with a pattern, and the LMZVM is used in the rest of the gray parts without a pattern.

F. **Hybrid operation of the LMZVM and LMSVM methods** [13]

The proposed modulation technique is based on the SVM technique. In the SVM technique, the switching times are calculated by using θ and M_a. The calculation amount of the first proposed modulation technique is similar to that of the conventional PWM based on the SVM technique. However, to apply the second proposed modulation technique, the sector calculation of the neutral-point voltage balancing and the (5.15) and (5.16) calculations are needed.

Figure 5.23 shows a control block diagram to balance the dc-link capacitor voltages by using the LMSVM. From Fig. 5.23, both leakage current reduction and neutral-point voltage balancing control are possible. When the dc-link voltages are balanced, only the LMZVM is applied. If unbalancing of the DC-link voltages occurs, the LMSVM is substituted for the LMZVM depending on the sectors.

- Simulation and experimental results of the hybrid method.

Table 5.3 shows the parameters used in the simulation and experiment.

Neutral-point voltage balancing control is conducted by the control block of Fig. 5.23, and the results are shown in Fig. 5.24a. The switching mode 0 means the

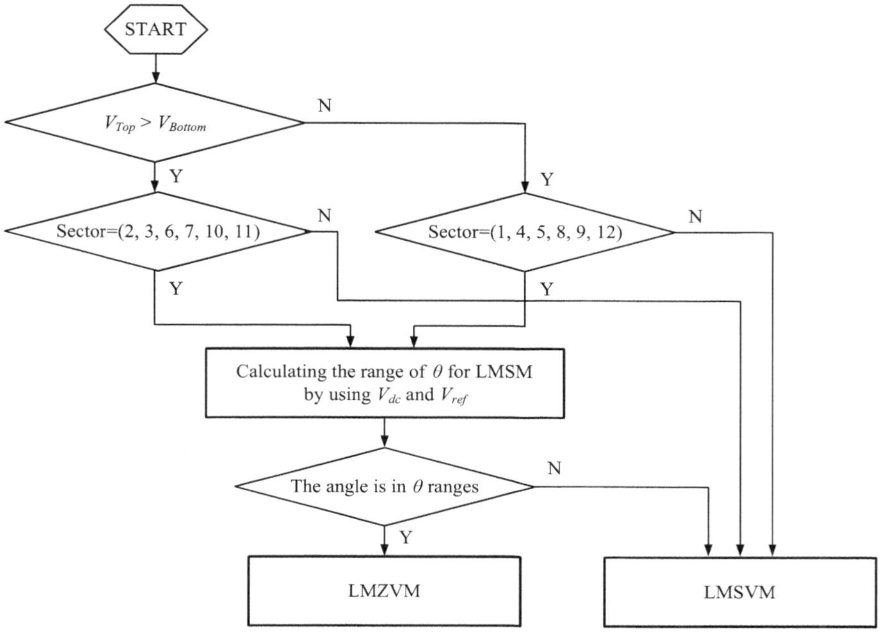

Fig. 5.23 Block diagram of the leakage current reduction and the neutral-point voltage balancing control (reprinted from [13], Fig. 13)

Table 5.3 Simulation and experimental parameters

DC-link voltage	600 V (Simulation)/300 V (Experiment)	Filter	2 mH
DC-link capacitor	1100 uF	C_{PV}	0.01 μF, 0.015 μF
Switching frequency	10 kHz	R_{PV}	15 Ω

LMZVM and switching mode 1 means the LMSVM. The bottom side capacitor voltage is higher than the top side before 0.8 s. From 0.8 s, the LMSVM is applied to decrease the bottom side capacitor voltage in sectors 1, 4, 5, 8, 9, and 12. Sectors 1, 4, 5, 8, 9, and 12 contain the N-type vectors. The LMSVM is applied in the valid range of these sectors that is calculated from (5.14).

When the LMZVM and LMSVM are applied together, the magnitude of V_N is maintained from 200 to 300 V and from 300 to 400 V, which is the same as when only the LMZVM is applied, and the results are shown in Fig. 5.24b. However, compared to when only the LMZVM is applied, the leakage current becomes bigger in the LMSVM operation parts because the switching number in V_N is doubled.

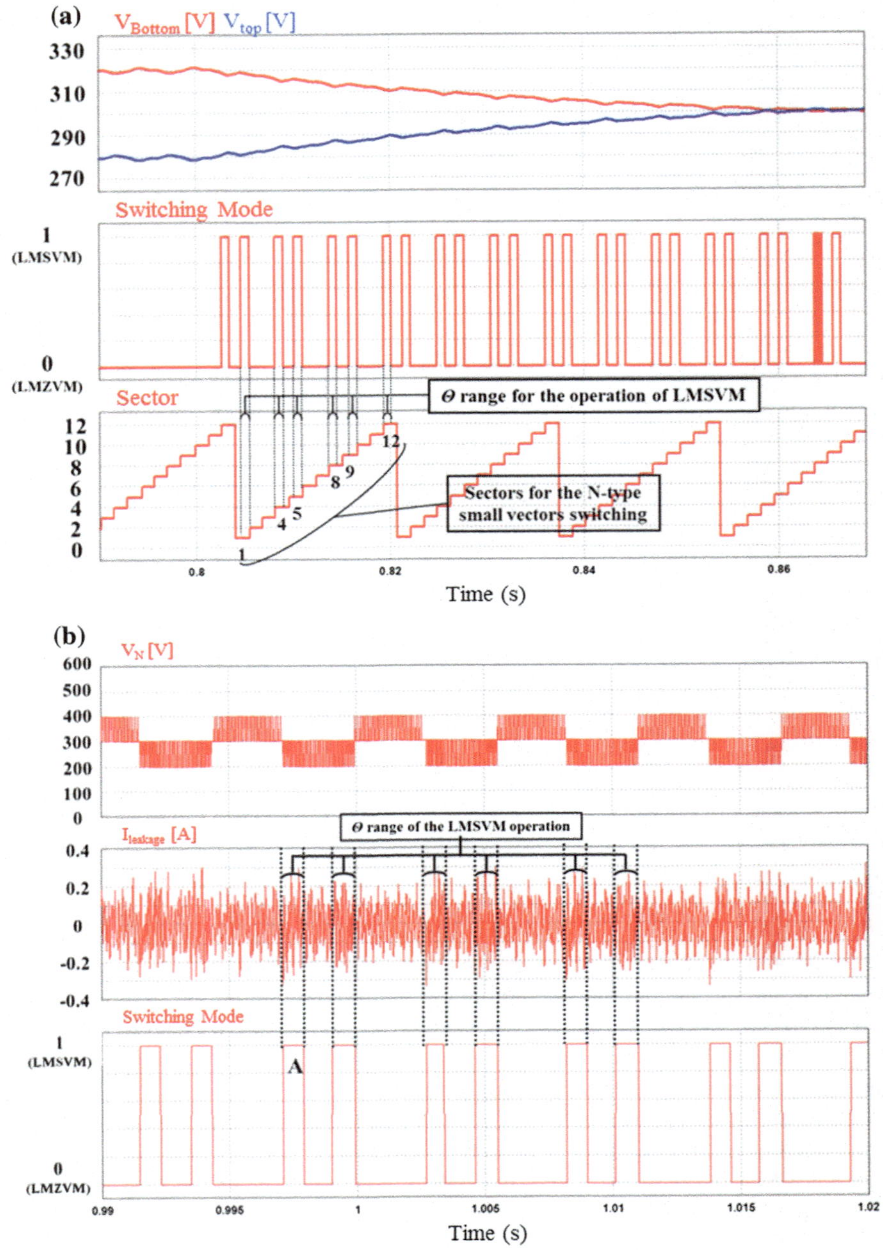

Fig. 5.24 Simulation waveforms of V_N and leakage current with a hybrid method of LMZVM and LMSVM (reprinted from [13], Fig. 17 and 18)

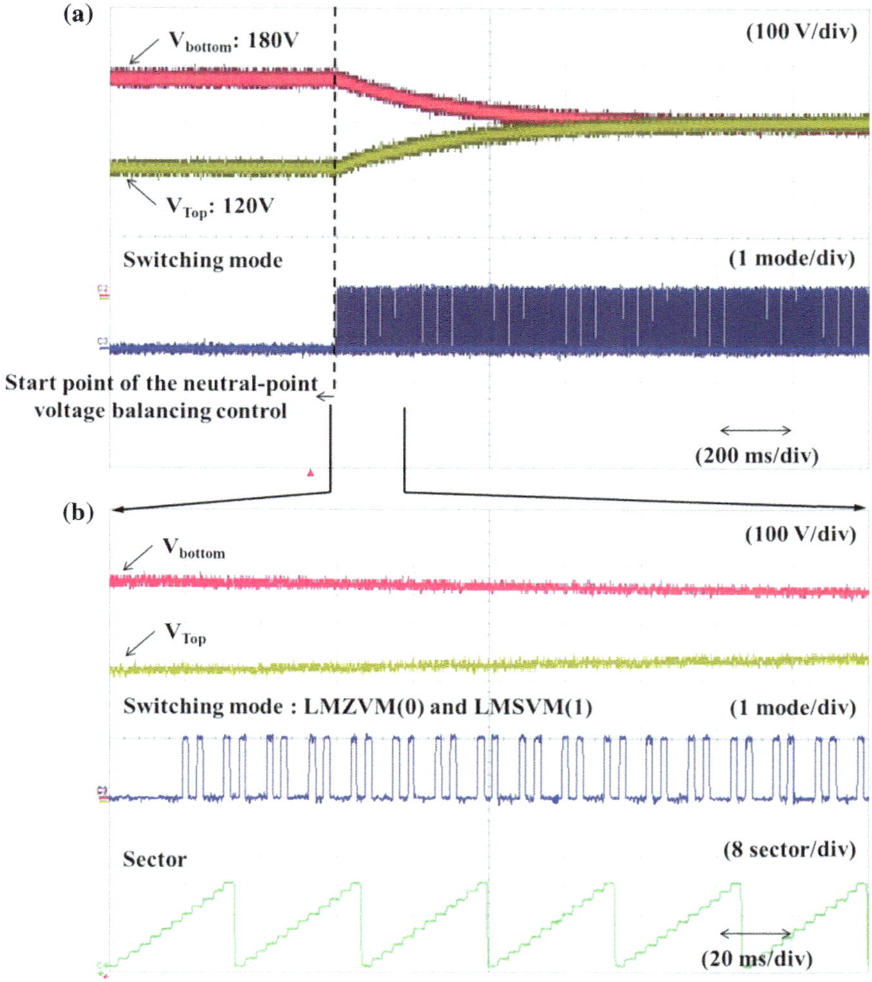

Fig. 5.25 Experimental result of the neutral-point voltage balancing control with a hybrid method of LMZVM and LMSVM (reprinted from [13], Fig. 24)

Figure 5.25 shows the transition of the neutral-point voltage balancing control in the experiment. Switching modes 0 and 1 represent the LMZVM and LMSVM. The values of V_{Top} and V_{Bottom} are different before the neutral-point voltage balancing control is applied. After the start point in Fig. 5.25a, the switching mode is continuously changed. The switching mode becomes 1 in the ranges that contain the N-type small vectors. From Fig. 5.25b, it can be seen that the switching mode is changed to 1 in the valid range of sectors 1, 4, 5, 8, 9, and 12. This valid value is from (5.14) and such simulations. In the rest of these ranges which are represented as the switching mode 0, the LMZVM is used. The difference between V_{Top} and V_{Bottom} progressively decreases and becomes zero with the LMSVM.

In the LMSVM operation range, V_N is maintained at the same value as the LMZVM as shown in Fig. 5.26a, b. Like the analysis in section III-B, the LMSVM causes a double frequency in these operation ranges, and this double frequency is identified in V_N of Fig. 5.26a, b. Due to the increased frequency, the leakage current is raised during the LMSVM operation. In Fig. 5.26a, the leakage current is from −100 to 100 mA during the LMZVM operation and from −130 to 130 mA during the LMSVM operation.

Both the LMSVM and the LMSVM with LMZVM raise the THD of the output currents. Figure 5.27a shows the line-to-line voltage of an inverter using the

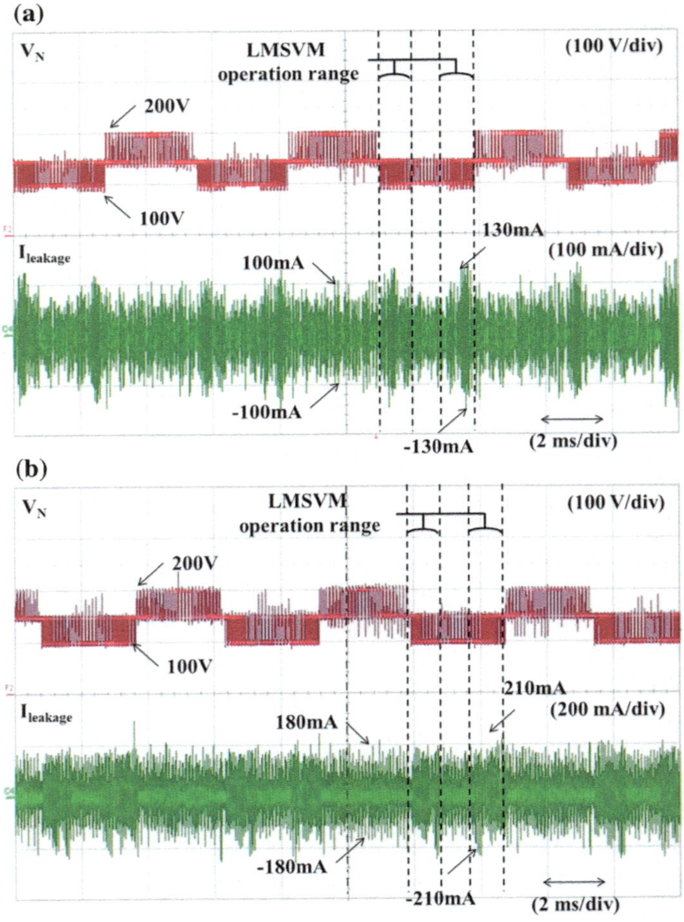

Fig. 5.26 Experimental result of VN and the leakage current using a hybrid method of LMZVM and LMSVM: **a** $C_{PV} = 0.01$ μF, **b** $C_{PV} = 0.015$ μF (reprinted from [13], Fig. 25)

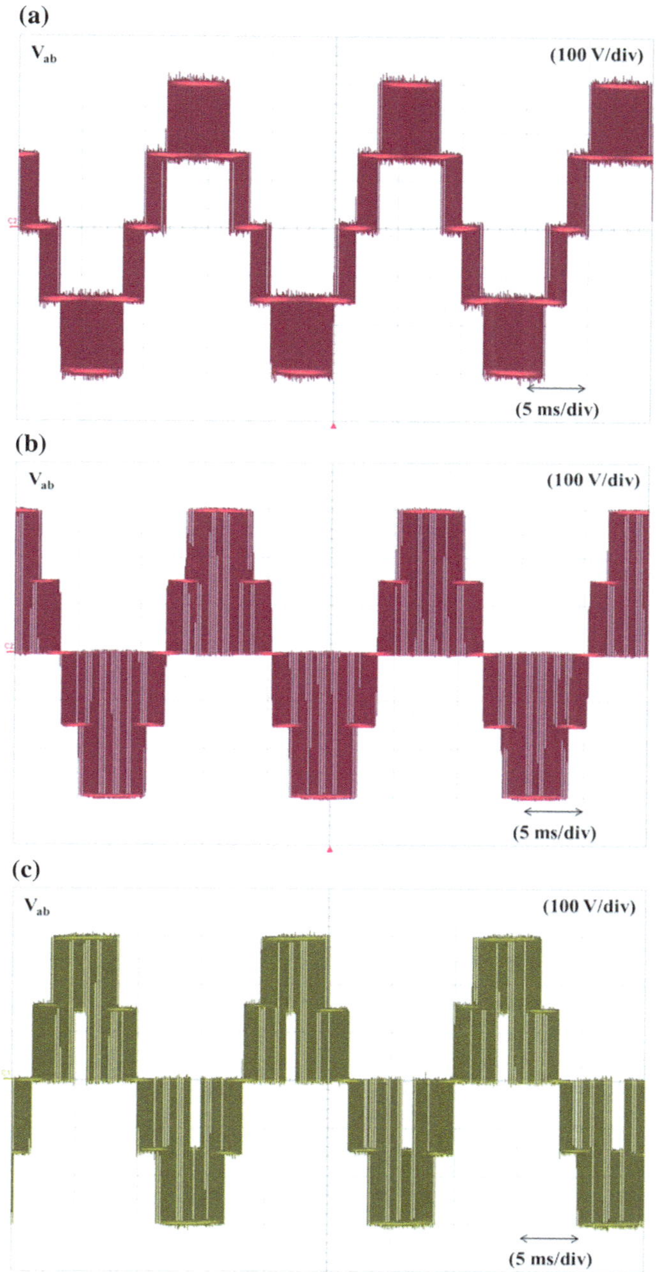

Fig. 5.27 Experimental result of the line-to-line voltage: **a** SVPWM, **b** LMZVM, **c** hybrid method (LMZVM and LMSVM) (reprinted from [13], Figs. 26, 27 and 28)

conventional PWM. The level of the line-to-line voltage is five levels. This is a characteristic of three-level inverters, and the cause of the THD is decreased than that of two-level inverters.

In the LMZVM and LMSVM, the THD increase is identified from Fig. 5.27b, c. The waveform of Fig. 5.27b is the line-to-line voltage using the LMZVM. The peak part of the line-to-line voltage has values of V_{dc} and 0. During the LMSVM operation, the line-to-line voltage has the same value as that of the convectional PWM in the LMSVM operation part as shown in Fig. 5.27c. Therefore, the THD decreases more in the LMZVM operation with LMSVM than with the LMZVM operation alone.

References

1. González, R., J. López, P. Sanchis, and L. Marroyo. 2007. Transformerless inverter for single-phase photovoltaic systems. *IEEE Transactions on Power Electronics* 22 (2): 693–697.
2. Koutroulis, E., and F. Blaabjerg. 2013. Design optimization of transformerless grid-connected PV inverters including reliability. *IEEE Transactions on Power Electronics* 28 (1): 325–335.
3. Zhang, L., K. Sun, L. Feng, H. Wu, and Y. Xing. 2013. A family of neutral point clamped full-bridge topologies for transformerless photovoltaic grid-tied inverters. *IEEE Transactions on Power Electronics* 28 (2): 730–738.
4. Saridakis, S., E. Koutroulis, and F. Blaabjerg. 2013. Optimal Design of Modern Transformerless PV Inverter Topologies. *IEEE Transactions on Energy Conversion* 28 (2): 394–404.
5. Zhao, Z., J. Lai, and Y. Cho. 2013. Dual-Mode Double-Carrier-Based Sinusoidal Pulse Width Modulation Inverter With Adaptive Smooth Transition Control Between Modes. *IEEE Transactions on Industrial Electronics* 60 (5): 2094–2103.
6. Kerekes, T., R. Teodorescu, M. Liserre, C. Klumpner, and Mark Sumner. 2009. Evaluation of three-phase transformerless photovoltaic inverter topologies. *IEEE Transactions on Power Electronics* 24 (9): 2202–2211.
7. Heldwein, M.L., H. Ertl, J. Biela, and J.W. Kolar. 2010. Implementation of a transformerless common-mode active filter for offline converter systems. *IEEE Transactions on Industrial Electronics* 57 (5): 1772–1786.
8. Dong, D., F. Luo, D. Boroyevich, and P. Mattavelli. 2012. Leakage current reduction in a single-phase bidirectional AC–DC full-bridge inverter. *IEEE Transactions on Power Electronics* 27 (10): 4281–4291.
9. Yang, B., W. Li, Y. Gu, W. Cui, and X. He. 2013. Improved transformerless inverter with common-mode leakage current elimination for a photovoltaic grid-connected power system. *IEEE Transactions on Power Electronics* 27 (2): 752–762.
10. Hava, A.M., and E. Ün. 2011. A high-performance PWM algorithm for common-mode voltage reduction in three-phase voltage source inverters. *IEEE Transactions on Power Electronics* 26 (7): 1998–2008.
11. Kimball, J.W., and M. Zawodniok. 2011. Reducing common-mode voltage in three-phase sine-triangle PWM with interleaved carriers. *IEEE Transactions on Power Electronics* 26 (8): 2229–2236.
12. Cavalcanti, M.C., K.C. de Oliveira, A.M. de Farias, F.A.S. Neves, G.M.S. Azevedo, and F.C. Camboim. 2010. Modulation techniques to eliminate leakage currents in transformerless three-phase photovoltaic systems. *IEEE Transactions on Industrial Electronics* 57 (4): 1360–1368.

13. Lee, J.S., and K.B. Lee. 2014. New modulation techniques for a leakage current reduction and a neutral-point voltage balance in transformerless photovoltaic systems using a three-level inverter. *IEEE Transactions on Power Electronics* 29 (4): 1720–1732.
14. Cavalcanti, M.C., A.M. Farias, K.C. Oliveira, F.A.S. Neves, and J.L. Afonso. 2012. Eliminating leakage currents in neutral point clamped inverters for photovoltaic system. *IEEE Transactions on Industrial Electronics* 59 (1): 435–443.

Chapter 6
Switch Fault Diagnosis and Tolerant Control in Cascaded H-Bridge Multilevel Inverters

As previous mentioned, multilevel inverters have become an increasingly attractive topology for industrial high-voltage applications. They have enabled important development in the ability to reach high power with increasing voltage levels. Multilevel inverters have particular advantages and are used in many applications such as pipeline pumps, rolling machines, and railway electrical traction drive systems [1–4]. Multilevel inverters have many topologies, such including diode-clamped multilevel inverters like the NPC topology, flying capacitor multilevel inverters, and CHMIs. Among these topologies, the CHMI is one of the most popular inverter topologies used in high-power and medium-voltage (MV) drives [5, 6]. CHMIs are simple and modular in structure, and do not require extra hardware such as clamp diodes or capacitors. As a result, the manufacturing process is fast and economical. The modularized structure of this topology as shown in Fig. 6.1 enables the number of output voltage levels to be easily increased by adding more H-bridge inverter cells.

A common characteristic of all multilevel topologies is that a large number of semiconductors and passive components are required. With this increase in the number of switches, the fault probability also increases. As a result, reliability decreases. The reliability of power electronics equipment is very important because multilevel inverter systems are utilized in high-power applications. For example, industrial applications such as industrial manufacturing are dependent on induction motors and their inverter systems for process control. A fault can cause an unbalanced output voltage and current while a traction motor is operating. The results can be serious damage to the traction motor if it runs in this state for a long period of time. In addition, if a system has to be shut down, the downtime can lead to a great financial loss per hour. Therefore fault tolerance is vital to companies. Many engineers and researchers have focused their efforts on fault detection and preventive maintenance to mitigate this issue. Fortunately, the CHMI provides a large number of redundant switching states, which allows an inverter to maintain normal operation under internal fault conditions. Therefore, the several studies on switch fault detection and tolerant control for multilevel inverters are still in

K.-B. Lee and J.-S. Lee, *Reliability Improvement Technology for Power Converters*, Power Systems, DOI 10.1007/978-981-10-4992-7_6

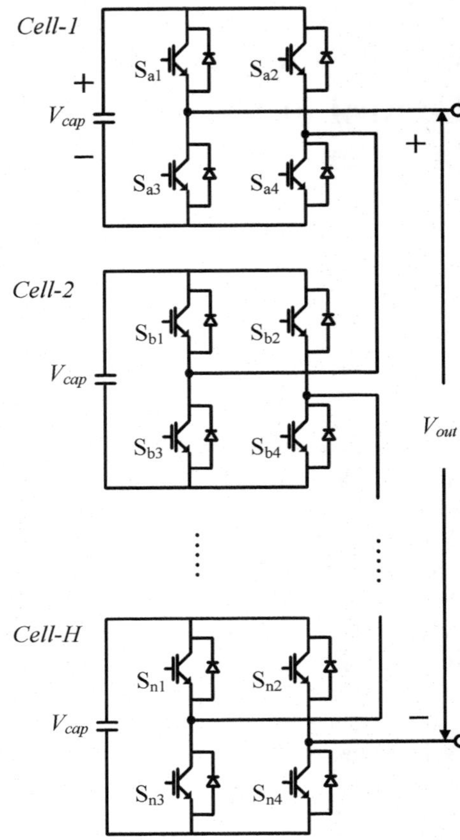

Fig. 6.1 Structure of a CHMI (reprinted from [7], Fig. 1)

progress. This chapter introduces switching device fault detection methods and tolerant controls for the CHMI.

6.1 Cascaded H-Bridge Multilevel Inverters

6.1.1 Configuration

The general a CHMI is composed of several H-bridge cells that are connected in series and it is expanded to three-phase systems as shown in Fig. 6.2. In the CHMI, the minimum variation between the voltage levels is the DC-link voltage (V_{cap}) connected to each H-bridge cell, and the number (m) of voltage levels of a phase is determined using the number (H) of H-bridge cells connected in series. The relationship between them is expressed as:

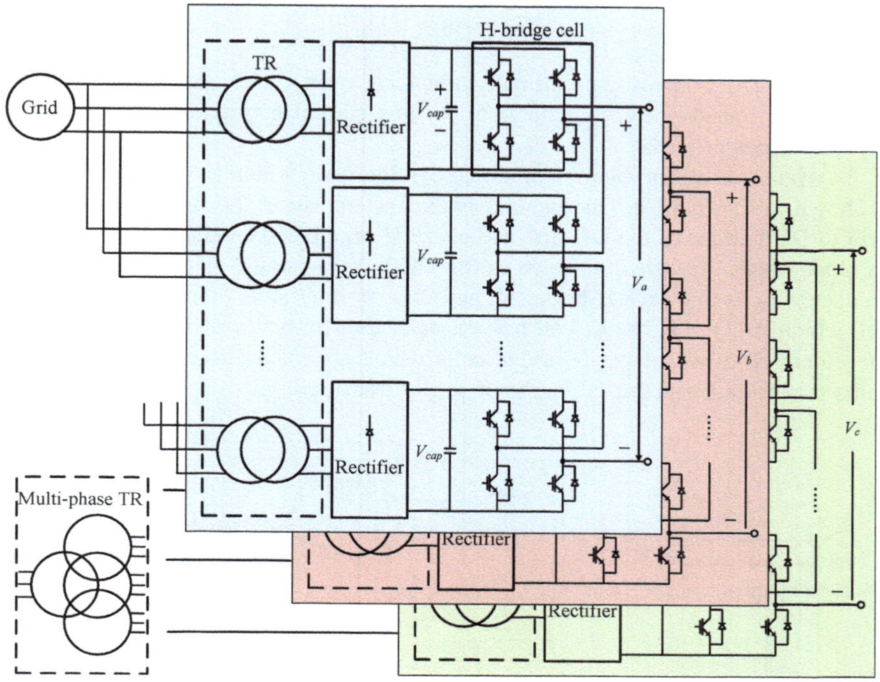

Fig. 6.2 Three-phase CHMI system

$$m = 2H + 1. \tag{6.1}$$

At the inverter side, the switching state of *Cell-H* (H = 1, 2, 3) is determined by the logical value of two signals (S_{k1}, S_{k3}), which can be 1 and 0 representing the ON and OFF state of each switch device, respectively. S_{k2} and S_{k4} are switched complementarily to S_{k1} and S_{k3}. This leads to four different binary combinations that generate three different output voltages $+V_{cap}$, 0, and $-V_{cap}$. Because the cells are connected in series, the total inverter output voltage of the u-phase, for example, is given by:

$$v_{uM} = \sum_{k=1}^{N} v_{uk} = \sum_{k=1}^{N} V_{cap}(S_{k1} - S_{k3}) \tag{6.2}$$

The voltage level, m, is always an odd number for the CHMI and can be extended to any number of voltage levels. The total number (N_{sw}) of active switches (IGBTs) used in the CHMI can be calculated from

$$N_{sw} = 6(L - 1). \tag{6.3}$$

Isolated DC sources are normally obtained from diode rectifiers, including multi-phase transformers, to achieve a low line current harmonic distortion and a high input power factor.

To overcome this problem, a modified configuration which is a network inverter is shown in Fig. 6.3 [6]. One rectifier is used to transform the AC voltages of the grid to DC voltage of the DC-link and every H-bridge cell should have a DC-link. In other words they are connected to the DC-link in parallel. The most important characteristic is that each H-bride cell has a transformer at its output. One H-bridge cell which is a sub H-bridge cell has a special transformer which is different from the others. The rest of the H-bridge cells which are main H-bridge cells use the same transformer and are in charge of power transmission.

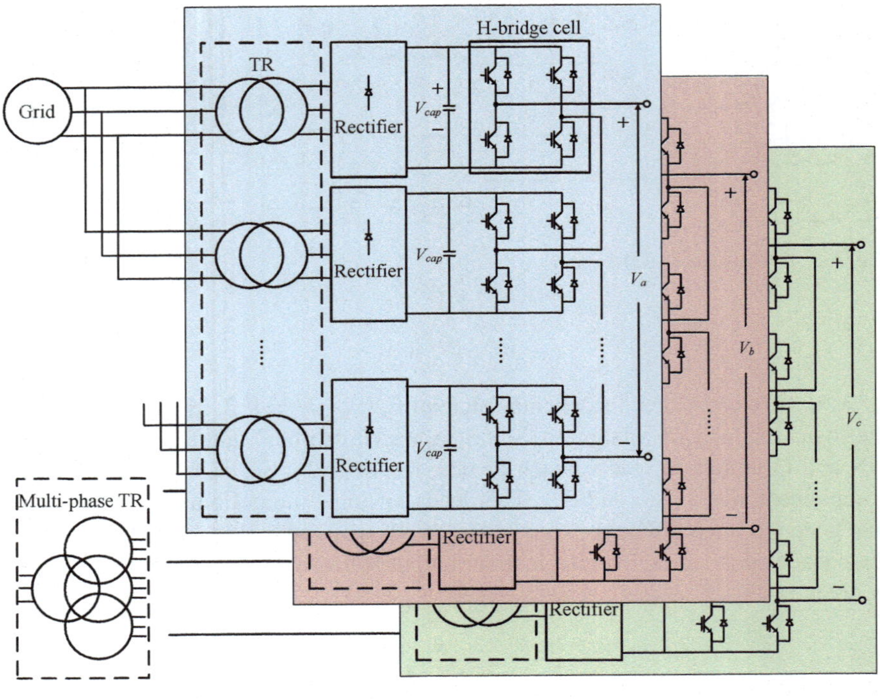

Fig. 6.3 Three-phase modified CHMI system

6.1.2 PWM Methods

Various switching methods exist for multilevel inverters, including space vector pulse-width modulation (PWM), multicarrier-based PWM, selective harmonic elimination, and nearest level control. Among these methods, the multicarrier-based PWM is the most popular due to its simple realization and applicability to various topologies. Multicarrier PWM methods can be classified into Phase-Shifted PWM (PS-PWM) and Level-Shifted PWM (LS-PWM) methods.

A. **Level-shifted PWM**

In LS-PWM, the m-level CHMI needs (m − 1) carriers which have the same frequency and magnitude, and are vertically disposed. The operating principle of the LS-PWM method, which is a seven-level CHMI, is shown in Fig. 6.4. The seven-level multilevel inverter requires six carriers with the same frequency and amplitude. Each carrier is vertically disposed. To determine the switching states of each H-bridge cell, the six carriers are compared with their corresponding reference signals. The phase voltage of the CHMI is generated through a summation of the output voltage for each of the H-bridge cells. The amplitude of the voltage level is the separated DC source, V_{cap}. By generating the output voltage of each H-bridge cell, a seven voltage-level scan is generated by the LS-PWM switching process.

In the LS-PWM, it is identified that only one of the H-bridge cells conducts the switching operation and the other H-bridge cells keep their switching state every one switching period. This means that the main switching frequency of the phase voltage is the same as that of the H-bridge cell.

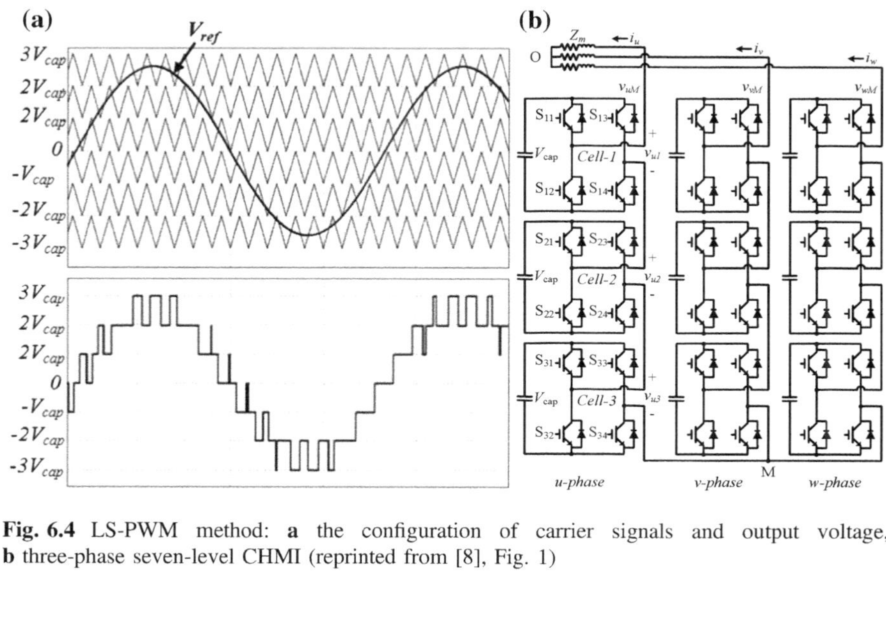

Fig. 6.4 LS-PWM method: **a** the configuration of carrier signals and output voltage, **b** three-phase seven-level CHMI (reprinted from [8], Fig. 1)

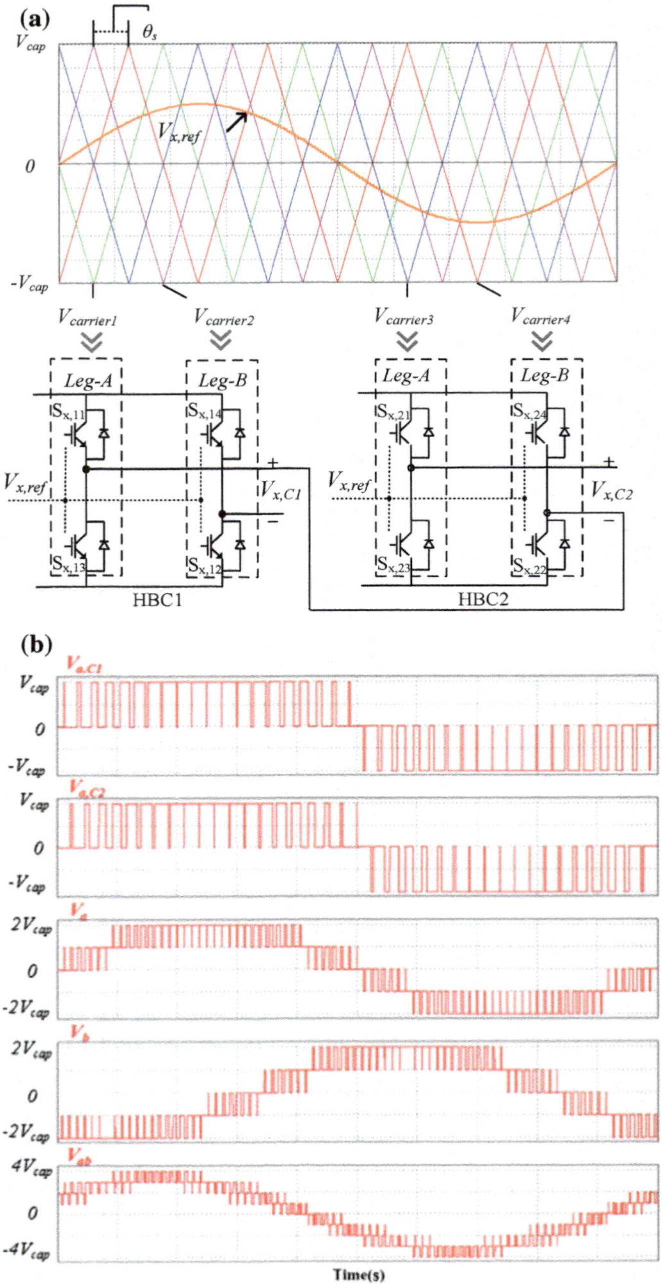

Fig. 6.5 Three-phase CHB 5-level inverter: **a** the configuration of carrier signals in a phase, **b** voltage waveforms

B. **Phase-shifted PWM**

Like the level-shifted PWM, the m-level CHMI needs (m − 1) phase-shifted carriers. The carriers have the same frequency and magnitude. The phase shift (θ_c) between two adjacent carriers depend on the voltage level and is represented as:

$$\theta_c = \frac{360^\circ}{m-1}. \tag{6.4}$$

Each carrier determines the switching state of each leg. The phase voltage of the CHMI is the summation of all of the H-bridge cells connected in series. In the PS-PWM, all of the H-bridge cells conduct the switching operation. Therefore, the main frequency (f_{inv}) of the phase voltage is increased depending on the number (H) of H-bridge cells. This relationship is expressed as:

$$f_{inv} = (m-1)f_{HBC} = 2Hf_{HBC}, \tag{6.5}$$

where f_{HBC} is the switching frequency applied to the H-bridge cells.

The operating principle of the PS-PWM method, which is a five-level CHMI, is shown in Fig. 6.5. Four carriers are used and the phase shift between any two adjacent carriers is 90°. The two carrier signals which are symmetric with respect to the time-axis are used together in one HBC and the switches at each HBC operate by the unipolar PWM method. As a result, a different output voltage is generated in each HBC, and the synthesized phase voltage of the CBHI has a frequency that is 4-times larger than the switching frequency.

6.2 Fault Diagnosis

The fault diagnosis methods for multilevel inverters deal with both the open-circuit fault and short-circuit faults in a switch. Like the fault diagnosis methods of three-level converters, the methods using additional devices will be introduced first. Then the state of the art fault diagnosis method which use only currents will be shown.

6.2.1 Fault Diagnosis Using Additional Devices

Voltage sensors are installed at the output of H-bridge cell as shown in Fig. 6.6 [9]. This method focuses on short-circuit fault and uses the fault characteristic where the output voltage of the H-bridge cell containing the short-circuit fault switch becomes

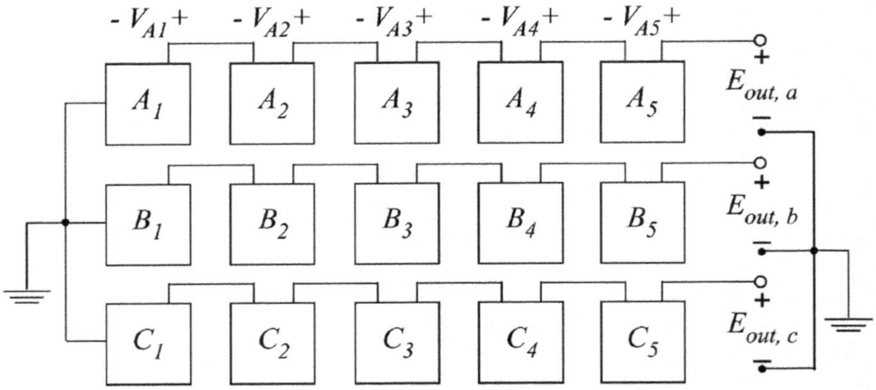

Fig. 6.6 Fault diagnosis using voltage sensors installed at the output of each H-bridge cell (reprinted from [9], Fig. 4)

zero. The measured output voltages are compared with the corresponding desired output voltages. If they are not equal, the fault signal is generated at every H-bridge cell.

Voltage sensors installed at the output of a phase are used to diagnosis fault switches as shown in Figs. 6.7 and 6.8 [10, 11]. The method shown in Fig. 6.7 can diagnose both open-circuit and short-circuit faults. The position of the fault switch is detected through a frequency analysis of the measured output voltage of the phase. The switching components of the H-bridge cells are extracted from the output voltage of the phase. The magnitude and phase for each of the switching components are used to distinguish the position of the fault switch. Therefore, the performance of this method is guaranteed in a CHMI using the PS-PWM method. The method shown in Fig. 6.8 uses an artificial intelligence algorithm to detect a switch fault depending on the output voltage pattern caused by the switch fault, and the position of the fault switch is determined.

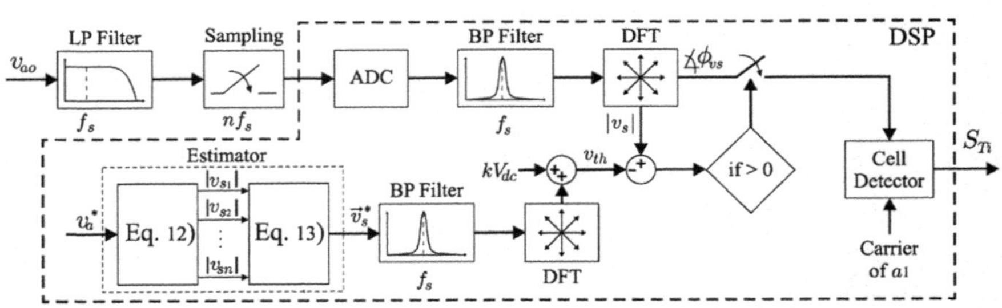

Fig. 6.7 Frequency based fault diagnosis using voltage sensors installed at the output of a phase (reprinted from [10], Fig. 7)

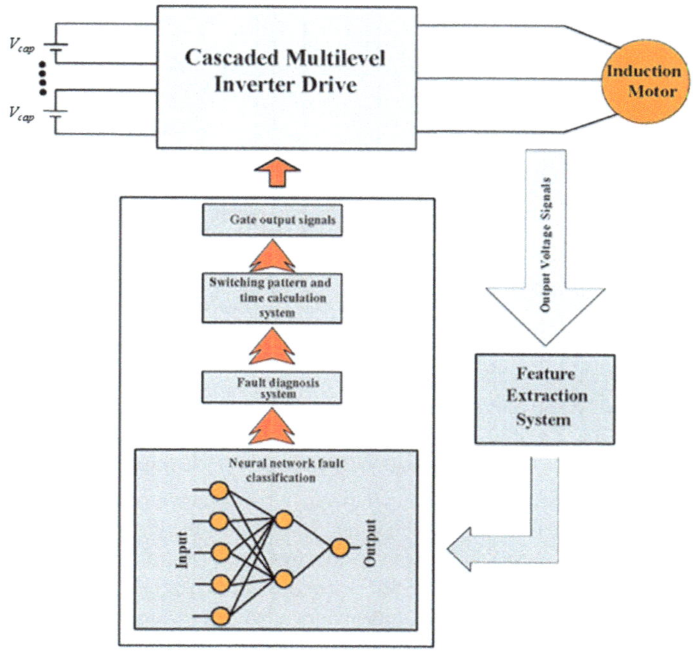

Fig. 6.8 AI-based diagnosis using voltage sensors installed at the output of phase (reprinted from [11], Fig. 8)

6.2.2 Fault Diagnosis Using Currents [7]

The current shape is influenced by the position of the fault switch. Therefore, it is possible to detect the position of a fault switch from an analysis of the relationship between the current shape and the fault switch. To implement it, the overall operation conditions such as the switching method applied in the multilevel inverter, the current path depending on the switching state, the power factor, and so on, must be known. This chapter introduces an open-circuit fault switch diagnosis method for the CHMI using the LS-PWM method. This method only uses output current which is already used to control the CHMI. Therefore, it does not need additional devices.

A. **Zero-voltage switching states**

When an open-switch fault occurs in a CHMI using the LS-PWM method, the location of the current distortion is different depending on the faulty cell. If the open-switch fault occurs in *Cell-2*, for instance, the current distortion is generated in the second voltage level ($+V_{cap} \sim +2V_{cap}$ or $-V_{cap} \sim -2V_{cap}$). Therefore, if the output voltage level of the location of the current distortion is detected, the faulty cell can be found. However, to apply this method, zero-voltage switching states

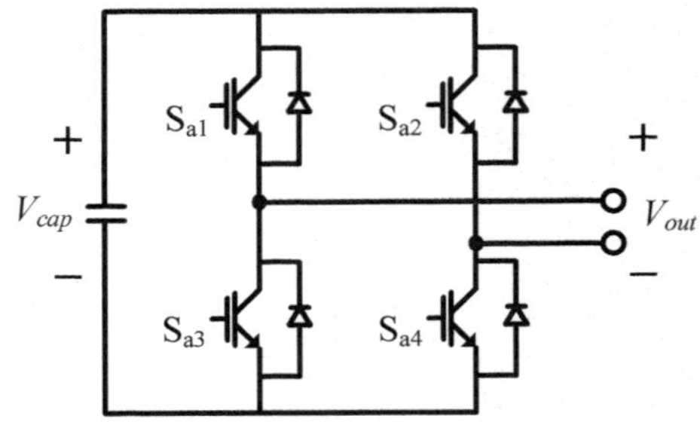

Fig. 6.9 An example of one H-bridge cell

must be considered, because the location of the current distortion differs depending on zero-voltage switching states.

There are two types of zero-voltage switching states in an H-bridge cell. In Fig. 6.9, one type is when S_{x1} and S_{x2} are on while S_{x3} and S_{x4} are off. The other is the exact opposite. This is summarized in Table 6.1. This basic theory is important for the open-switch fault detection method. In this chapter, the current paths depending on the zero-voltage switching states are presented. In addition, an analysis of the current paths is done considering open-switch fault conditions.

The current path, according to the two types of zero-voltage switching states in a seven-level CHMI with a low modulation index M_a, is shown in Fig. 6.10. When M_a is less than 0.33, only a single cell generates output voltage. In this case, *Cell-1* generates the output voltage, while the other cells make the current path by zero-voltage switching. The output currents flow through the on-state switches and diodes. Therefore, the current path is changed according to the zero-voltage switching states. In addition, the zero-voltage switching states affect the current path when an open-switch fault occurs.

B. **Current path with open-switch faults**

In each H-bridge cell, both S_{x1} and S_{x4} are used for the positive output voltage and current. However, when an open-switch fault occurs, the characteristic of the current distortion differs depending on the zero-voltage switching states (this also applies to S_{x2} and S_{x3}).

The current paths of *Cell-3* under normal and open-switch fault conditions, when S_{x3} and S_{x4} are on for zero-voltage switching and the output current $I > 0$, are shown in Fig. 6.11. The solid line indicates the current path under normal conditions, while the dashed line indicates faulty conditions. If an open-switch fault occurs on switch S_{c1}, the current path is the same as the normal condition, which does not cause a distortion in the current waveform. However, if the modulation index is greater than 0.66, the current distortion is generated by an S_{c1} fault, as

Table 6.1 Zero-voltages switching states

S_{x1}	S_{x2}	S_{x3}	S_{x4}	AC terminal voltage
On	On	Off	Off	0 V
Off	Off	On	On	0 V

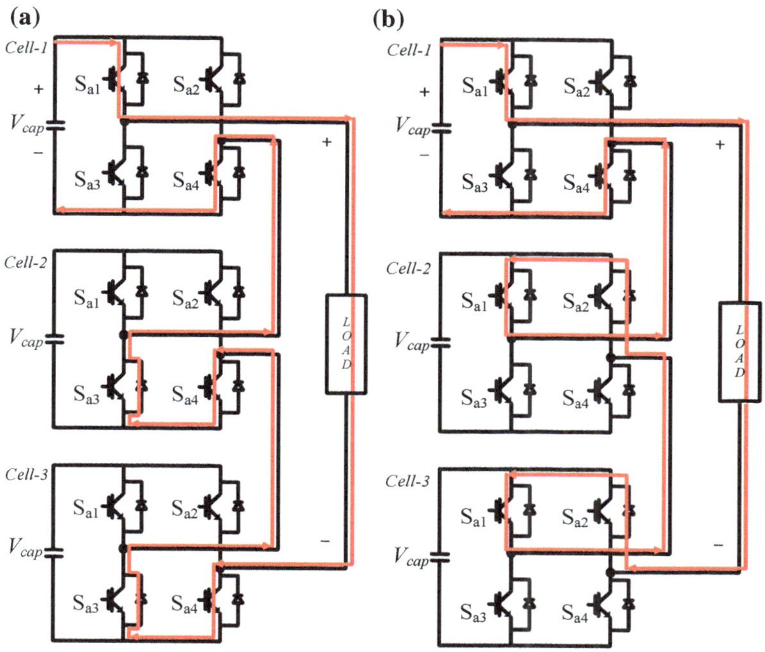

Fig. 6.10 Current path according to two types of zero-voltage switching states for ($M_a < 0.33$): **a** S_{x1}, S_{x2}-On, **b** S_{x3}, S_{x4}-On (reprinted from [7], Fig. 4)

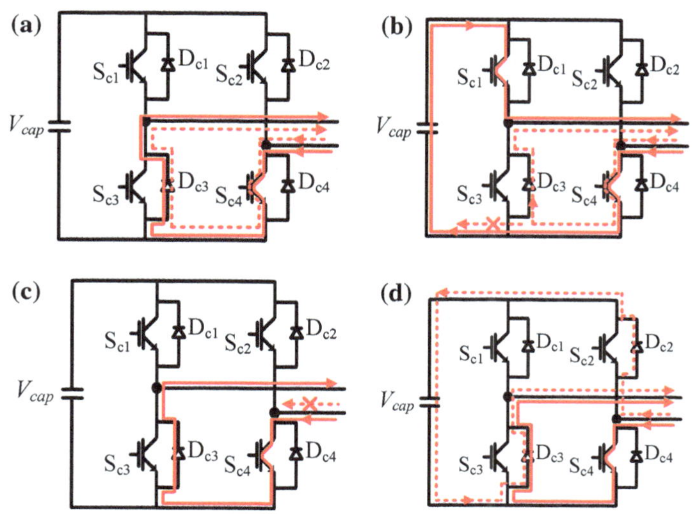

Fig. 6.11 Current path of *Cell-3* under normal and open-switch fault conditions (S_{x3} and S_{x4}-On, $I_o > 0$): **a** S_{c1} fault ($M_a < 0.66$), **b** S_{c1} fault ($M_a > 0.66$), **c** S_{c4} fault ($M_a < 0.33$), **d** S_{c4} fault ($M_a > 0.33$). (reprinted from [7], Fig. 5)

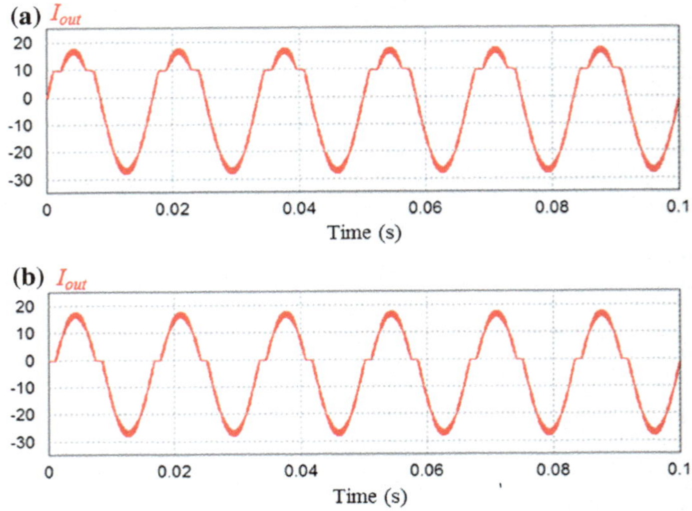

Fig. 6.12 Output current waveforms under open-switch fault conditions (S_{x3} and S_{x4}-On): **a** S_{b1} fault, **b** S_{b4} fault (reprinted from [7], Fig. 6)

shown in Fig. 6.11b. On the other hand, if S_{c4} has an open-switch fault, the current path is disconnected by the faulty switch when the modulation index $M_a < 0.33$. When *Cell-2* generates the output voltage by increasing modulation index M_a, the output node voltage of *Cell-3* becomes larger than the supply voltage. Therefore, current flows through the diode D_{c2}, as shown in Fig. 6.11d. In this case, the current distortion is generated at the lowest voltage level. Therefore, the location of the current distortion differs depending on S_{c1} and S_{c4} faults, even though they generate the same output voltage in the on states at the same time.

Waveforms of the current distortion, under different fault conditions and zero-voltage switching states, are shown in Figs. 6.12 and 6.13. *Cell-2* starts to operate at the second voltage level. If an open fault occurs on Sb1 during zero-voltage switching through the use of S_{x3} and S_{x4}, current distortion is generated at the second voltage level ($+V_{cap} \sim +2V_{cap}$). Likewise, if an open fault occurs on S_{b3} during zero-voltage switching through S_{x1} and S_{x2}, current distortion is generated at the second voltage level ($-V_{cap} \sim -2V_{cap}$). On the other hand, if an open fault occurs on the switch of the zero-voltage switching state, current distortion is only generated at the lowest voltage level, as shown in Figs. 6.12b and 6.13b.

C. Asymmetric zero-voltage switching for fault diagnosis

In the "Current path with open-switch faults" section, the current waveforms caused by each switch fault according to the zero-voltage switching states were discussed. If an open fault occurs on a switch that performs zero-voltage switching, regardless of the cell, it is impossible to detect faults, because distortions of all the current waveforms are generated at the equivalent voltage level between $-V_{cap}$ and

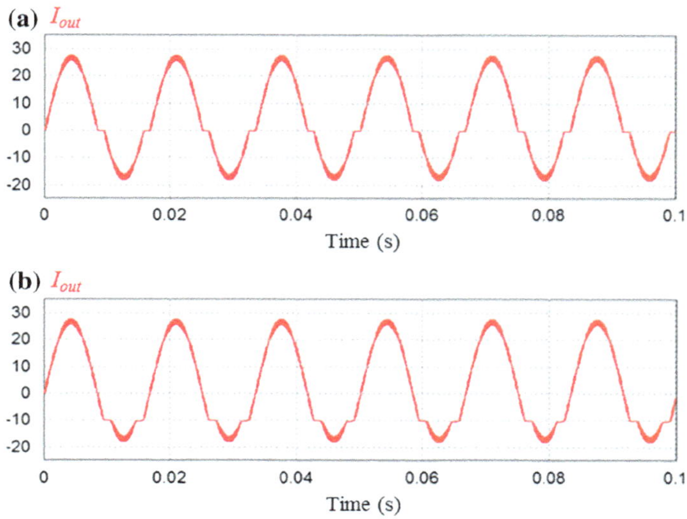

Fig. 6.13 Output current waveforms under open-switch fault conditions (S_{x1} and S_{x2}-On): **a** S_{b2} fault, **b** S_{b3} fault (reprinted from [7], Fig. 7)

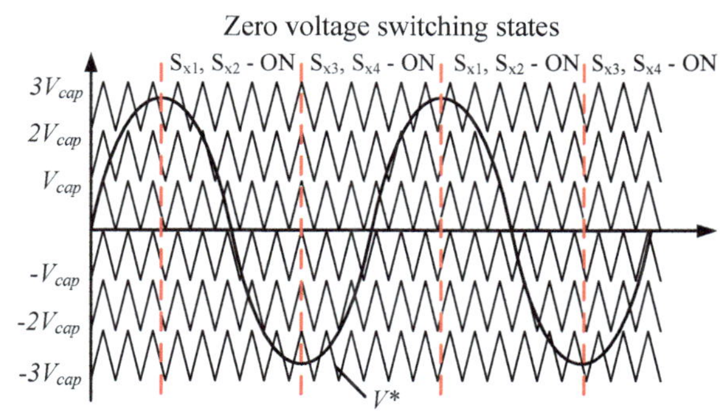

Fig. 6.14 Asymmetric zero-voltage switching (reprinted from [7], Fig. 8)

$+V_{cap}$. Therefore, as shown in Fig. 6.14, the zero-voltage switching states of the increasing interval and the decreasing interval of the reference voltage are changed as follows:

$$\begin{cases} S_{x1},\ S_{x2} - ON,\ \frac{dV_{ref}}{dt} < 0 \\ S_{x3},\ S_{x4} - ON,\ \frac{dV_{ref}}{dt} < 0 \end{cases}, \tag{6.6}$$

In this case, when a fault occurs on a single switch, the location of the current distortion for every half period varies as shown in Fig. 6.15. If the open fault occurs on S_{b1} of *Cell-2*, the distortion is generated in the second voltage level because S_{b1} is not related to the zero-voltage switching. In the decreasing interval, distortion is

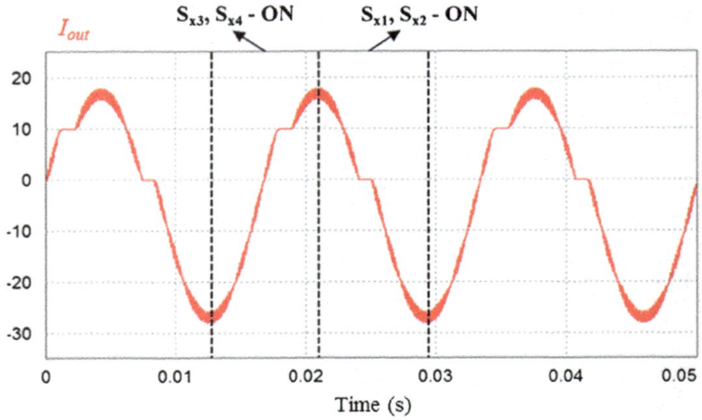

Fig. 6.15 Current distortion waveform with asymmetric zero-voltage switching under the S_{b1} open-fault condition (reprinted from [7], Fig. 9)

generated at the lowest voltage level because of the disconnection of the current path, which is caused by the S_{b1} fault. Therefore, current distortion is generated at the lowest voltage level during a half cycle of the current. In the other half cycle, the current distortion is generated at a voltage level depending on the faulty cell.

During one cycle of the current, one of the half cycles is used to detect faulty cells and to determine which set of switches, (S_{x1}, S_{x4}) or (S_{x2}, S_{x3}), produces a positive output voltage and which set produces a negative output voltage. The other half cycle is used to identify which set of switches, (S_{x1}, S_{x2}) or (S_{x3}, S_{x4}), is used for zero-voltage switching. Using a combination of the two half cycles, it is possible to accurately detect the faulty cell and switch.

D. **Evaluating the slope of the current for fault diagnosis**

To diagnose an open-switch fault using the distortion of the current waveform, the slope of the current is used. If an open fault occurs, the slope of the current is decreased compared to the normal condition. Therefore, the fault can be detected when the slope decreases below K_{ratio} compared with the previous state, as shown in (6.7). The value of K_{ratio} is important in detecting an open-switch fault. For a unity power factor, the range of K_{ratio} is not critical because the current slope is almost zero when an open-switch fault occurs. However, for a non-unity power factor, as the power factor decreases, the value of K_{ratio} increases.

$$\frac{I_{slope,N}|_k}{I_{slope,N}|_{k-1}} < K_{ratio}, \ (0 < K_{ratio} < 1), \tag{6.7}$$

$$I_{slope} = \frac{di}{dt}, I_{slope,N} = \frac{I_{slope}}{|I_o| \times \omega_s}, \tag{6.8}$$

where I_{slope} is the slope of the output current, $I_{slope,N}$ is the normalized slope, I_o is the magnitude of the output current, K_{ratio} is the current slope ratio compared with the previous state, ω_s is the angular frequency, and k is the period index.

To reduce noise effects, the normalized slope is used, as expressed in (6.8). By comparing this value with that of the normalized slope of the previous state, the location of the faulty cell and switch can be determined.

Figure 6.16 shows current waveforms depending on the PWM method under the switch S_{a1} fault condition. The LS-PWM method has an unbalanced conduction time for each cell. To distribute the conduction time evenly, a switching pattern should rotate among the cells.

When the conventional PWM method is used, it is impossible to accurately detect faulty switches. This is due to the fact that the location of the current distortion is the same regardless of the zero-voltage switching states, as shown in Fig. 6.16a. In this case, it is possible to distinguish either (S_{x1}, S_{x4}) or (S_{x2}, S_{x3}) fault. However, it is impossible to know exactly which switch is faulty. On the other hand, in the rotating PWM method, the location of the current distortion is shifted in every fundamental cycle. Since a seven-level inverter is composed of three cells, the fault-detection cycle comprises three cycles of current, as shown in Fig. 6.16b. The slope of the current is compared with those in previous states during the

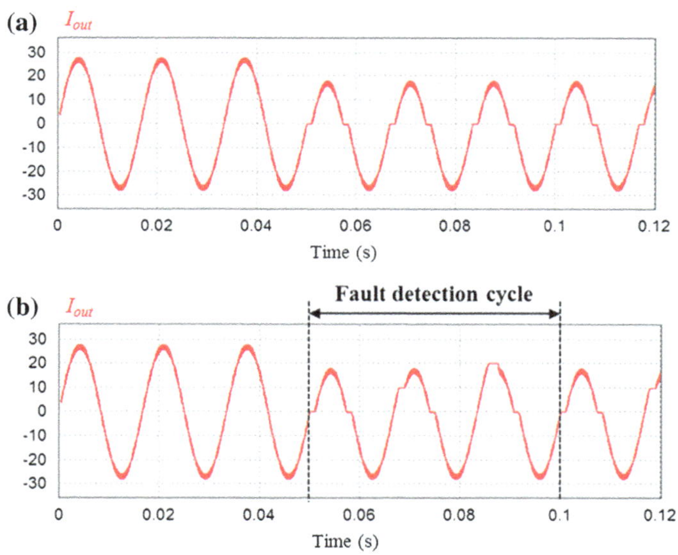

Fig. 6.16 Current waveform depending on the PWM method under a switch S_{a1} fault in *Cell-1*: **a** the conventional LS-PWM, **b** the rotating LS-PWM (reprinted from [7], Fig. 10)

fault-detection cycle. Through the location of the current distortion, depending on the rotating cells during the fault-detection cycle, the faulty cells and switches can be detected.

In the rotating PWM method, the detection cycle increases with an increasing number of levels. Therefore, this method is more effective with a low number of levels. However, it can be applied to a large number of levels because open-switch faults do not have a significant impact on a system in a short period of time.

E. **Considering a non-unity power factor**

Faulty cells can be detected by comparing the position of current distortions with their corresponding reference voltage level. Therefore, if a system has a non-unity power factor, it is impossible to accurately determine an open-switch fault. A non-unity power factor system is shown in Fig. 6.17, where the difference in phase between the current and the voltage is θ_{PF}. To compensate the lagging output current, the virtual reference of the voltage $V_{ref(PF)}$, as expressed in (6.9). Therefore, it is possible to detect an open-switch fault by comparing the distorted current point and the virtual reference of the voltage $V_{ref(PF)}$.

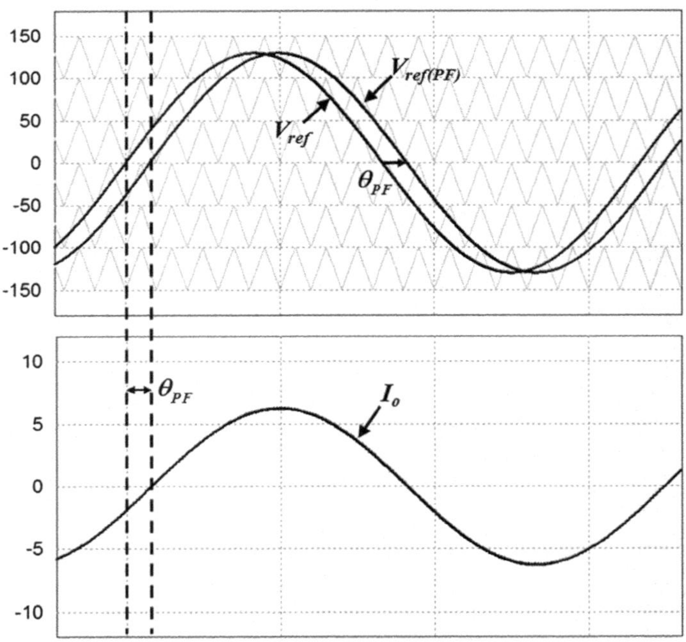

Fig. 6.17 Method for considering a non-unity power factor (reprinted from [7], Fig. 11)

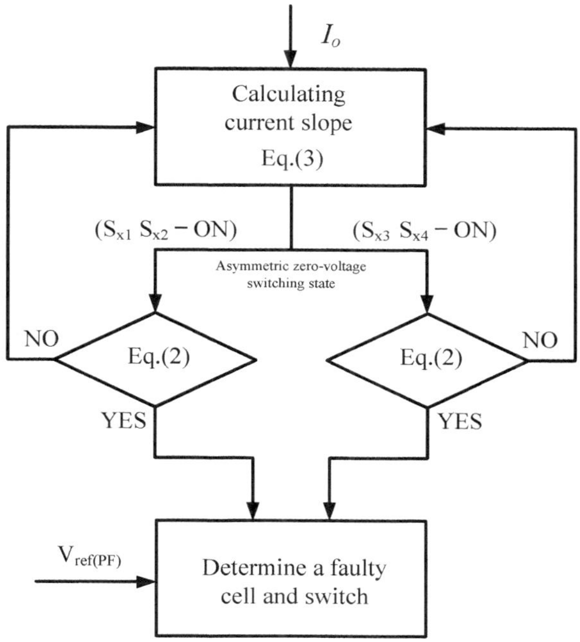

Fig. 6.18 Flowchart of the open-switch fault-detection method (reprinted from [7], Fig. 12)

$$V_{ref} = V_a \cos(\theta), \tag{6.9}$$

$$V_{ref(PF)} = V_a \cos(\theta - \theta_{PF}), \tag{6.10}$$

A flowchart of the open-switch fault detection method is shown in Fig. 6.18.

F. **Simulation results**

To effectively verify the fault-detection method, Simulations are implemented and the simulation parameters are presented in Table 6.2.

Simulation results concerning the rotated switching pattern for each cell are shown in Fig. 6.19. The switching pattern of each cell is shifted every period while maintaining a seven level output voltage. In comparison with the conventional LS-PWM method, the output voltages of each cell have the same switching frequencies and conduction times.

Table 6.2 Simulation parameters

Parameter	Value
Individual DC-link voltage	50 (V)
R_{load}	50 (Ω)
L_{load}	1.5 (mH)
Switching frequency	10 (kHz)
Power factor	≈1
Fundamental frequency	60 (Hz)

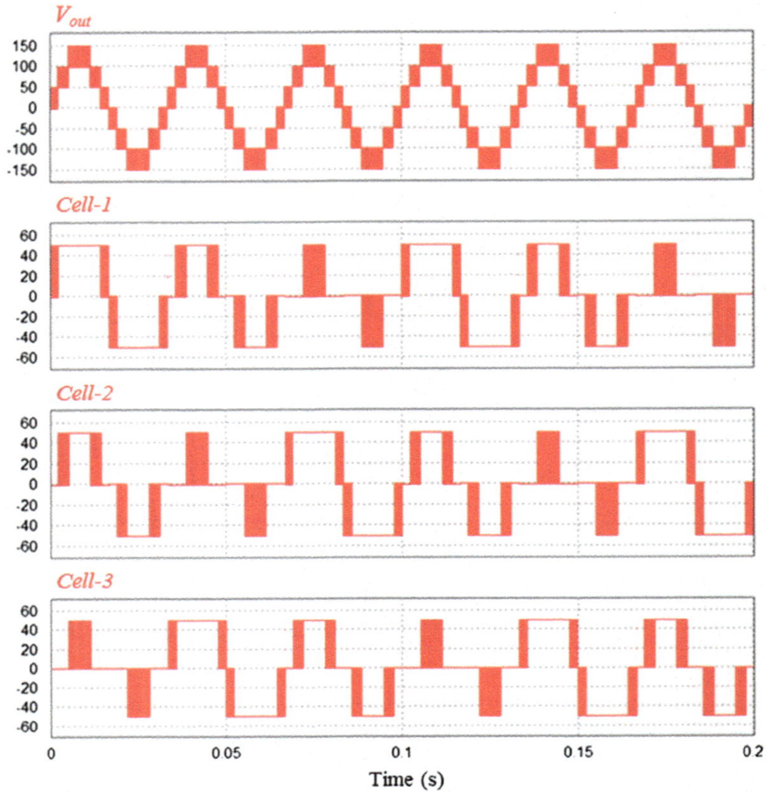

Fig. 6.19 Simulation results of the seven-level output voltage and each H-bridge cell output voltage (reprinted from [7], Fig. 13)

Figure 6.20 shows simulation results of the fault-detection method under open-switch faults in *Cell-2*. It presents two cases of open-switch faults on S_{b1} and S_{b4}. It can be seen that the current waveforms have different shapes after an open-switch fault occurs at 0.1 s. The slope of the current sharply decreases when the distortion is generated by the open-switch fault. By analyzing the slope of the current distortion during three cycles of the current, the fault can be detected. In the non-faulty state, each fault-detection signal is set as zero. When an S_{b1} fault occurs in *Cell-2*, the fault-detection signal of *Cell-2* becomes one, which means that the open fault occurred on S_{b1} in *Cell-2*. In addition, the fault-detection signal of *Cell-2* becomes four, as shown in Fig. 6.20b, when an S_{b4} open-switch fault occurs. For the two cases of open-switch faults in *Cell-2*, the faulty cells and switches are precisely distinguished.

Figure 6.21 shows simulation results of the fault detection method under open-switch faults in *Cell-3*. Likewise, it presents two cases of open-switch faults

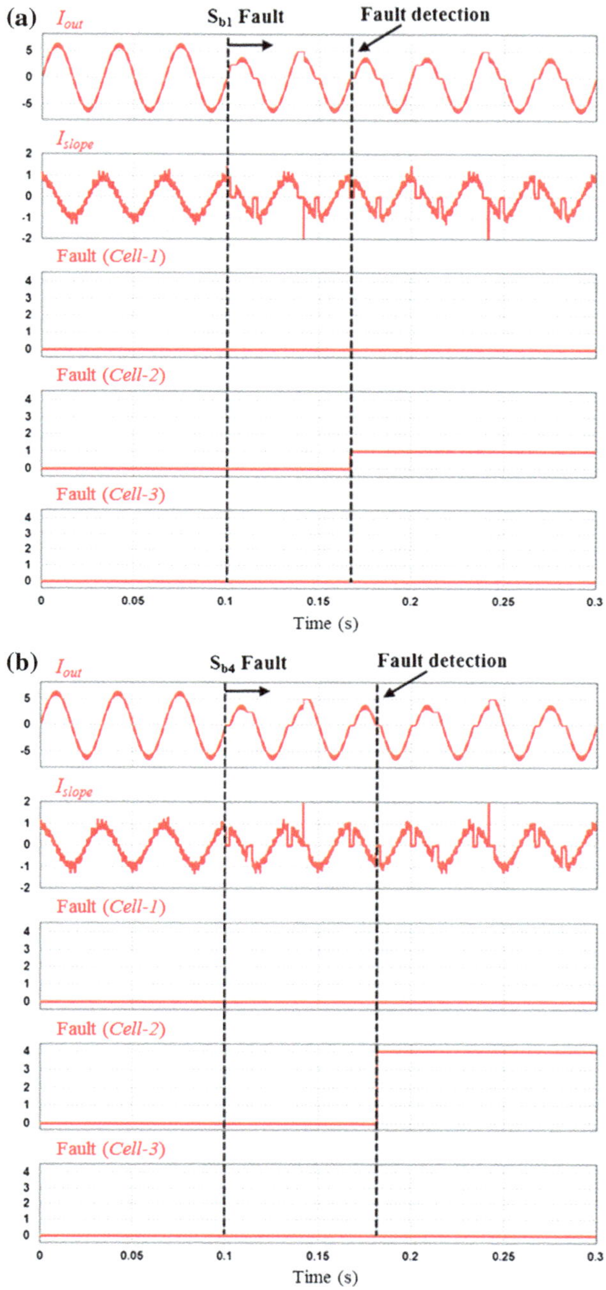

Fig. 6.20 Simulation results of the fault detection under an open-switch faults in *Cell-2*: **a** S_{b1} fault in *Cell-2*, **b** S_{b4} fault in *Cell-2* (reprinted from [7], Fig. 14)

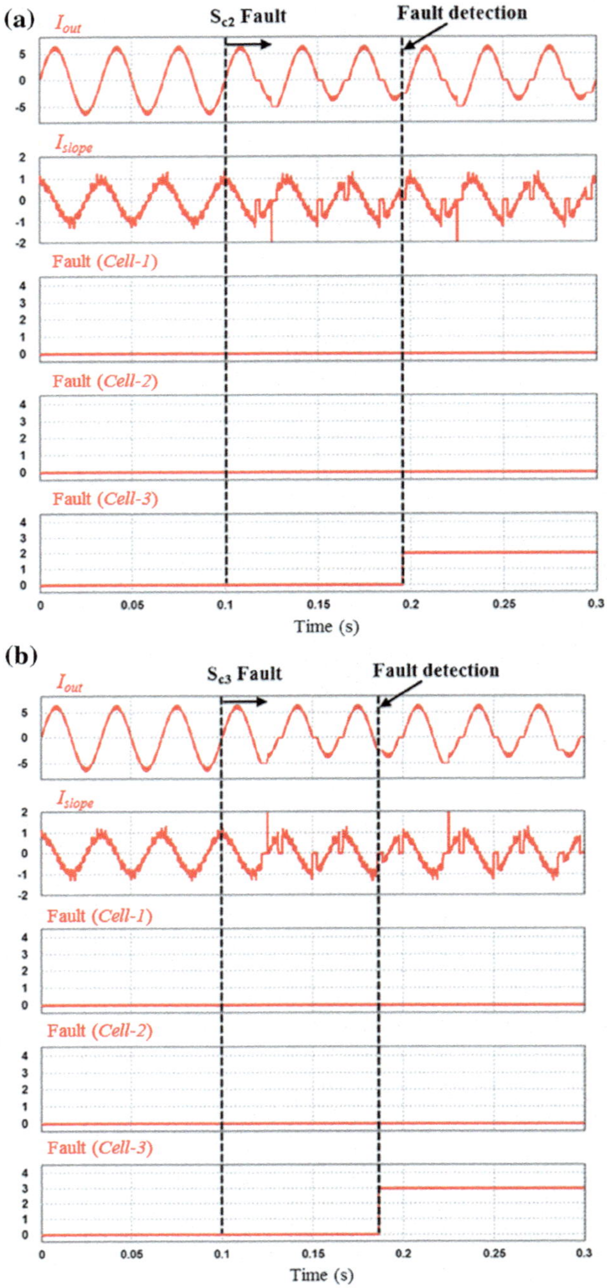

Fig. 6.21 Simulation results of the fault detection under an open-switch faults in *Cell-3*: **a** S_{c2} fault in *Cell-3*, **b** S_{c3} fault in *Cell-3* (reprinted from [7], Fig. 15)

on S_{c2} and S_{c3}. In this case, the fault-detection signal of *Cell-3* is given a value when a fault is detected. Each value of the fault signal indicates the number of faulty switches. Therefore, the faulty cells and switches are identified. Consequently, the simulation results demonstrate that the fault-detection method is useful when an open-switch fault occurs in the CHMI.

G. **Experimental results**

A 6-kW three-phase, four-layer CHMI prototype was manufactured, as shown in Fig. 6.22. The system is composed of a main control board, sensor boards, a SMPS, and several H-bridge modules. To verify the open-switch fault detection method, a one-phase seven-level inverter is used. Instantaneous values of the voltage reference are transmitted from the main control board to each H-bridge control board by a controller area network communication protocol. The experiments were carried out under the same conditions as the simulation, as shown in Table 6.3.

Figure 6.23 shows the seven-level output voltage of the CHMI using the rotating LS-PWM. This figure also includes the output voltage of one H-bridge cell that is rotated every period. Therefore, it leads to balancing power for each of the H-bridge cells.

Figure 6.24 shows the output voltage and current under the R–L load and non-fault state. In this case, the load inductance L_{load} is 1.5 mH, which is a unity power factor. On the other hand, the peak value of the current is 6 A. Thus, the current slope is normalized with 6 A. In the fault-detection method, faulty cells can be detected by comparing the position of the current distortions with their corresponding reference voltage level.

Figure 6.25 shows output voltage and current distortion waveforms when an open-switch fault occurs in *Cell-2*. Due to the characteristics of the LS-PWM method, current distortion is generated at the corresponding voltage level, and the slope of the distorted current is almost zero. By comparing the slope of the distorted current with the previous value, the open-switch fault can be detected. On the other hand, S_{x1} and S_{x4} open-switch faults prevent the generation of positive voltage, which can lead to a reduction of the positive voltage level. Therefore, current distortion is generated at a positive current interval in both cases. However, the positions of the current distortion are different due to the different zero-voltage switching states. In the case of an S_{x1} open-switch fault, current distortion occurs on the rising edge of the current, whereas an S_{x4} open-switch fault generates current distortion on the falling edge of the current. Similarly, the same principle is applied for open-switch faults in S_{x2} and S_{x3}. Therefore, it is confirmed through experimental results that the current distortion waveforms are different from each other due to the asymmetric zero-voltage switching state. In addition, by rotating the switching pattern among the H-bridge cells, the position of the current distortion is shifted every period.

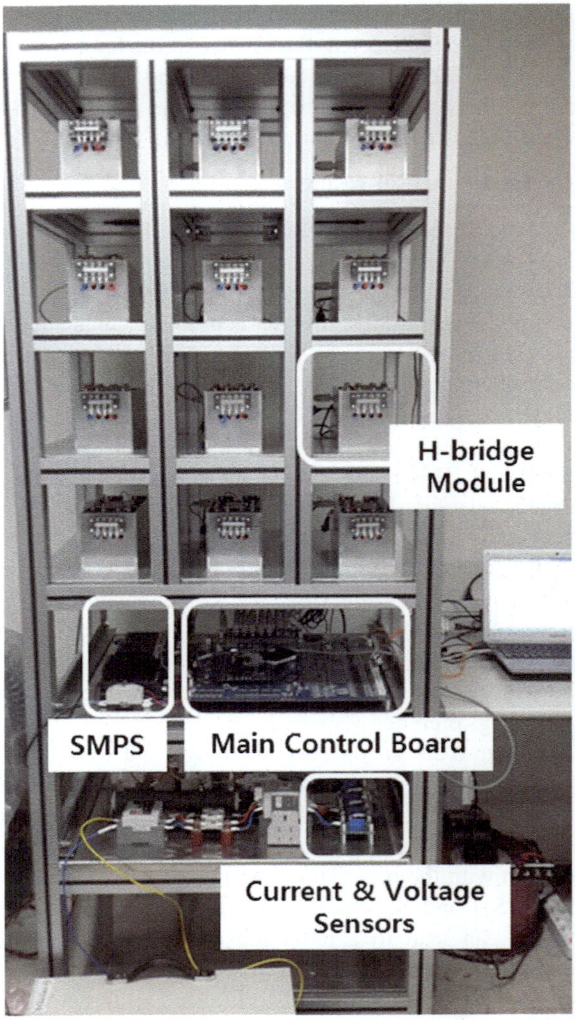

Fig. 6.22 Experimental setup of the CHMI (reprinted from [7], Fig. 16)

Table 6.3 Experiment parameters

Parameter	Value
Individual DC-link voltage	50 (V)
R_{load}	20 (Ω)
L_{load}	1.5/15 (mH)
Switching frequency	10 (kHz)
Fundamental frequency	60 (Hz)

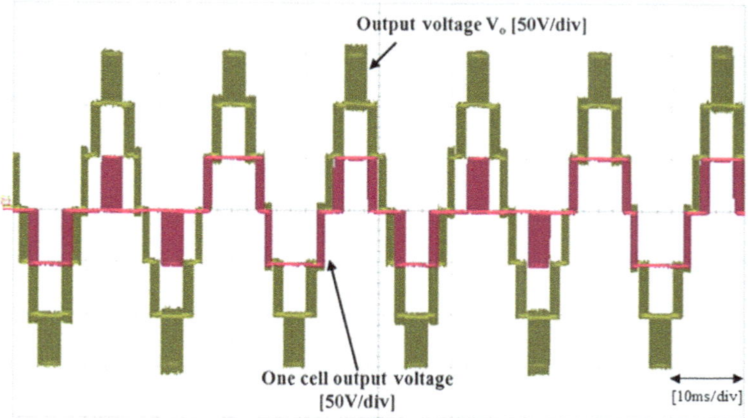

Fig. 6.23 Output phase voltage of the seven-level CHMI and the output voltage of one H-bridge cell (reprinted from [7], Fig. 17)

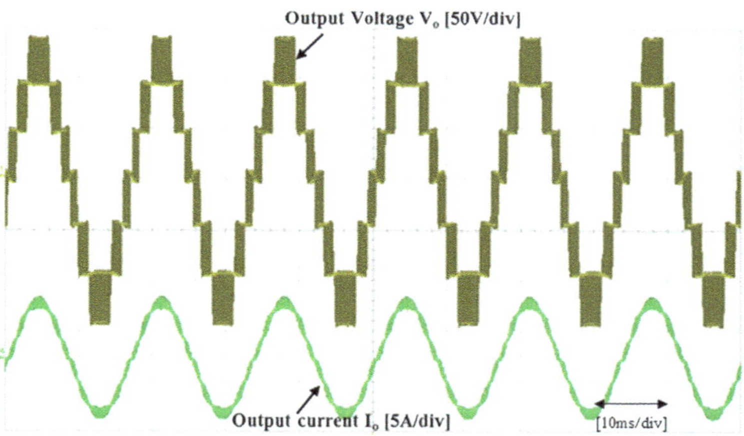

Fig. 6.24 Output voltage and current waveform under the non-fault state. (reprinted from [7], Fig. 18)

Figure 6.26a, b show output current waveforms and fault-detection signal when S_{b1} and S_{b4} open-switch faults occur, respectively. The open-switch fault-detection method identifies both the faulty cells and the switches. In Fig. 6.26a, when an open-switch fault occurs in S_{b1}, the fault-detection signal becomes one after the detection cycle, which is three periods of the reference voltage. In addition, when an

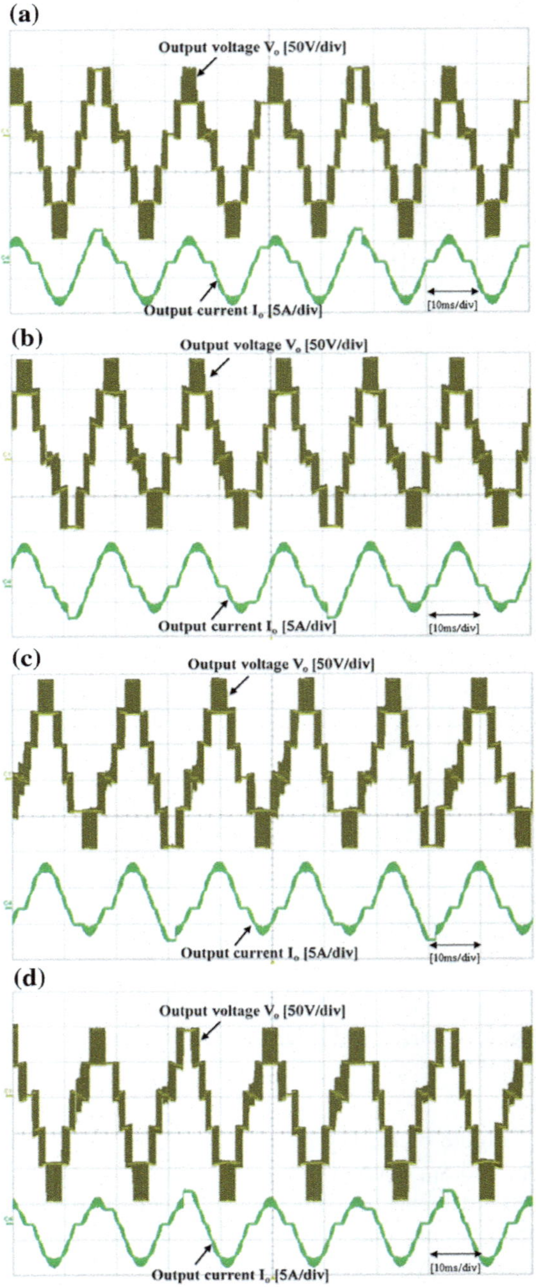

Fig. 6.25 Output voltage and current waveforms under the open-switch fault condition in *Cell-2*: **a** S_{b1} open-switch fault, **b** S_{b2} open-switch fault, **c** S_{b3} open-switch fault, **d** S_{b4} open-switch fault (reprinted from [7], Fig. 19)

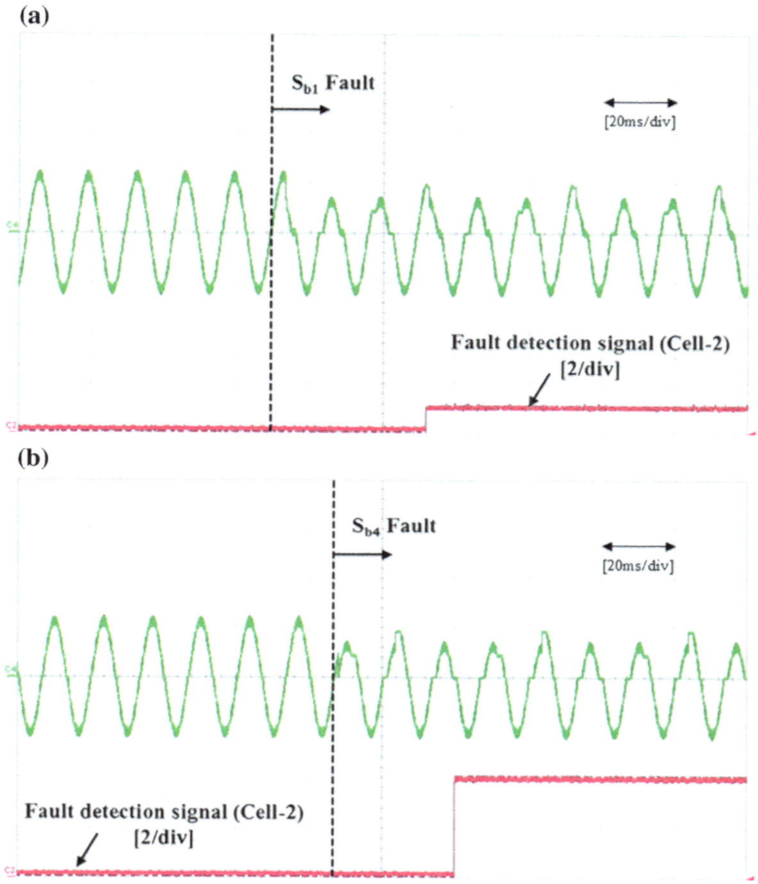

Fig. 6.26 Experimental results under the open-switch fault conditions in *Cell-2* ($PF \approx 1$): **a** S_{b1} open-switch fault, **b** S_{b4} open-switch fault (reprinted from [7], Fig. 20)

open-switch fault occurs in S_{b4}, the fault-detection signal becomes four. This fault-detection signal indicates an open-fault switch in *Cell-2*. Furthermore, an additional experiment was performed for a non-unity power factor. Figure 6.27 shows experimental results at a non-unity power factor when S_{c1} and S_{c2} open-switch faults occur. In this case, S_{c1} and S_{c2} in *Cell-3* have open-switch faults. By applying the detection algorithm of the power factor, it can be seen that the open-switch faults are accurately detected.

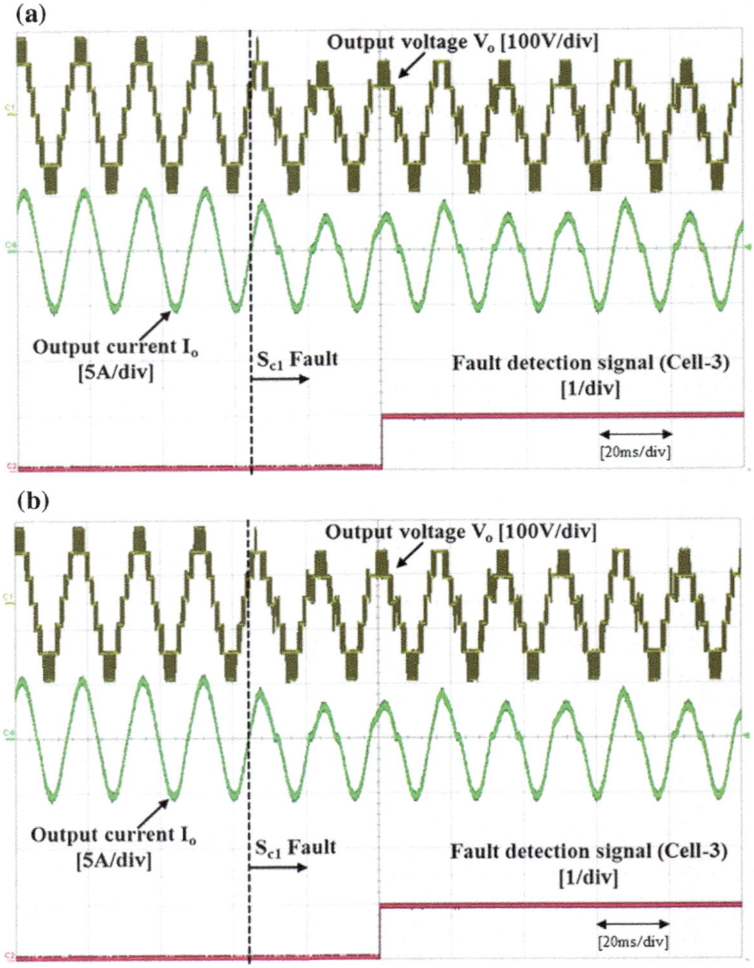

Fig. 6.27 Experimental results for a non-unity power factor ($PF \approx 0.9$): **a** S_{c1} open-switch fault, **b** S_{c2} open-switch fault (reprinted from [7], Fig. 21)

6.3 Fault Tolerant Control

Many engineers and researchers have focused their attention on fault detection and preventive maintenance. Fortunately, the CHMI provides a large number of redundant switching states, which allows an inverter to maintain normal operation under internal fault conditions.

Several fault-tolerant methods use additional devices such as thyristor switches and a bypass contactor as shown in Fig. 6.28 [12]. Fault cell isolation methods cope with the case of an open or short fault in switches larger than two. However, it is

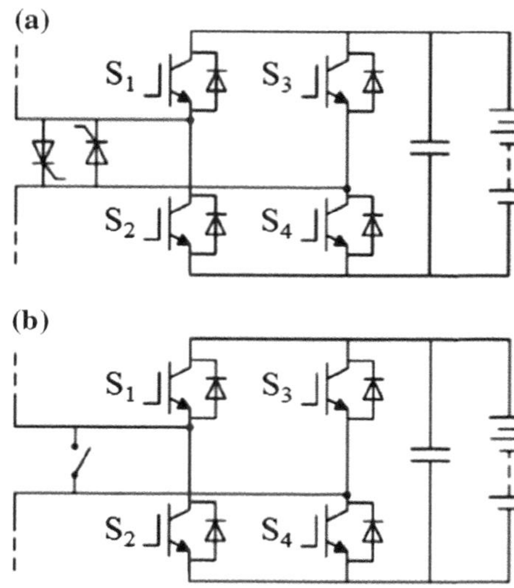

Fig. 6.28 Methods for bypassing a faulty converter cell using: **a** thyristor switches and **b** a bypass contactor (reprinted from [12], Fig. 3)

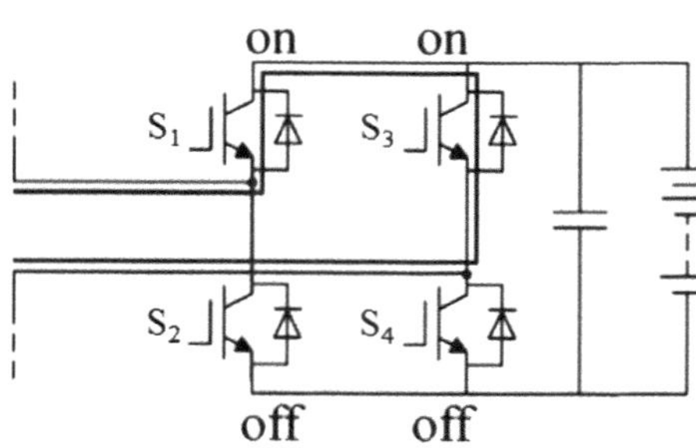

Fig. 6.29 Method for bypassing a faulty cell (reprinted from [12], Fig. 4)

possible to make the current path by bypassing the fault cell as shown in Fig. 6.29. This requires no additional bypass switches. For example, if S_1 fails in the short mode, S_3 is ON, while the two bottom switches S_2 and S_4 are OFF, which effectively bypasses the faulty converter cell. This chapter focuses on tolerant control methods using no additional devices.

A. **Fault condition of PS-PWM and LS-PWM** [1]

Tolerant control must be influenced by the PMW method. PS-PWM is mainly employed for multi-cell topologies, because all of the power cells can operate under the same switching conditions. This makes the power distribution among inverter power cells equal. However, it is known when compared to other multicarrier methods, LS-PWM has a better output voltage quality with a lower distortion. Because the carrier signals of PS-PWM need to have a phase displacement between any two adjacent carriers, the output line-to-line and load voltages have some additional dv/dT_s. These distortion factors are not produced with LS-PWM where all the carriers are in phase. LS-PWM methods are based on the amplitude shifts between carriers. As shown in Fig. 6.30, there are three schemes for LS-PWM: Phase Disposition (PD), Alternative Phase Opposite Disposition (APOD) and Phase Opposite Disposition (POD).

Figure 6.31 shows the modulation results of the CHMI based on PS-PWM and LS-PWM under a faulty single-inverter cell of the u-phase. Each voltage level has an amplitude of V_{cap}. Both of the output voltage waveforms contain only five voltage levels instead of seven. However, the waveforms are different in the central portion of the positive and negative half-cycles of v_{uM}.

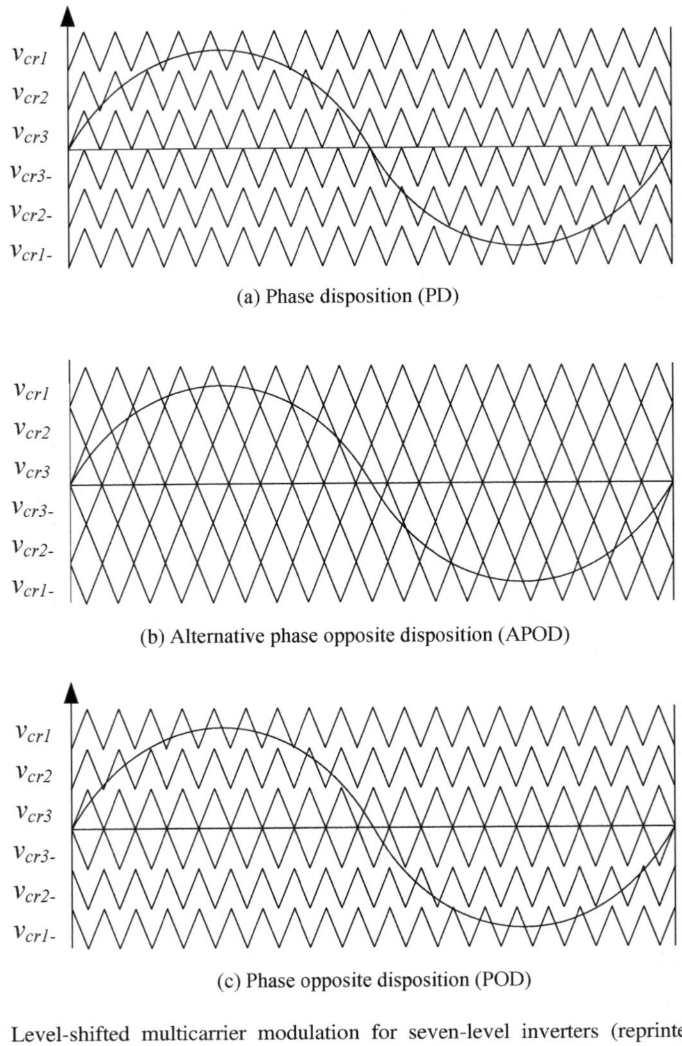

Fig. 6.30 Level-shifted multicarrier modulation for seven-level inverters (reprinted from [8], Fig. 2)

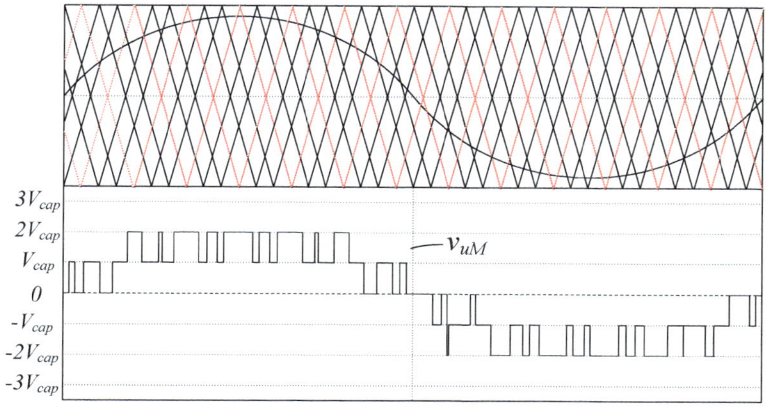

(a)Phase-shifted modulation

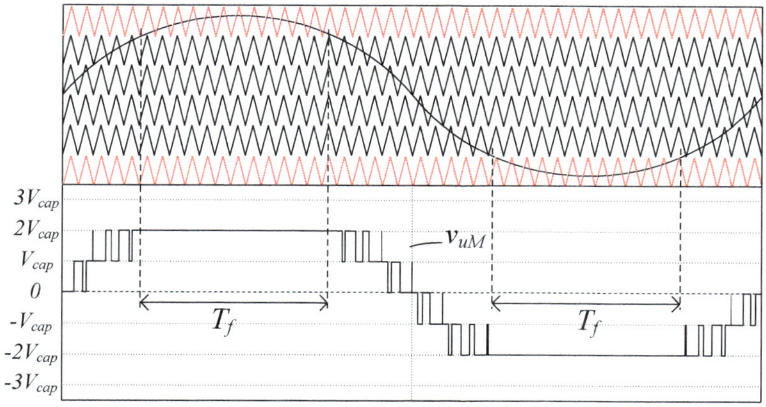

(b) Level-shifted modulation

Fig. 6.31 Level-shifted multicarrier modulation for seven-level inverters (reprinted from [8], Fig. 3)

In PS-PWM, the amplitude modulation index (M_a) is always in the range of 0–1 even if the CHMI has faulty inverter cells. This means that the CHMI based on PS-PWM can afford to modulate a modified reference signal with carrier bands for fault-tolerant control.

However, M_a of the CHMI based on LS-PWM can be represented as follows:

$$0 < M_a < \frac{H - n_f}{H}, \ (n_f = 0, 1, 2, \ldots, H) \tag{6.11}$$

where H is the number of operating inverter cells, and nf is the number of faulty inverter cells.

For the CHMI based on LS-PWM, M_a is also in the range of 0 to 1 under normal operation conditions. However, if one of the inverter cells in the u-phase of the seven-level CHMI is out of order, the available M_a for the faulty u-phase is reduced to the range of 0 to 2/3 by (6.4). As shown in Fig. 6.31, the output voltage v_{uM} in the central portion (T_f), where the desired voltage ranges from $2V_{cap}$ to $3V_{cap}$, maintains an amplitude of $2V_{cap}$ because there are no available carrier bands to be modulated with the reference signal in this portion, T_f.

The general basic strategy for achieving fault tolerance of CHMIs is increasing M_a for the healthy inverter cells in the faulty phase, and many studies are based on this strategy. However, CHMIs based on LS-PWM cannot apply these methodologies because the range of the available M_a is reduced when an inverter cell malfunctions, as described previously. The fundamental wave of the output voltage cannot be increased in CHMIs based on LS-PWM even if M_a of the reference signal is increased for fault tolerance. For the faulty region (T_f), where the reference signal should be modulated with the outermost carrier bands in the normal operation mode, v_{uM} is always $2V_{cap}$. Thus the fundamental wave of v_{uM} cannot be increased under the faulty S_{11} condition.

B. **Neutral-shift method** [12]

The neutral shift control method is used to maximize the available voltage after a bypass to increase the number of module failures that can be tolerated. For the neutral shift of this study, a fundamental-frequency zero-sequence voltage is injected. This section briefly introduces the neutral shift method. It should be noted that this fault-tolerant method is also based on PS-PWM as its switching scheme.

In the following, it is assumed that the u-phase inverter *Cell-1* has malfunctioned. Therefore, faulty *Cell-1* is bypassed.

$$v_{u1} = 0 \tag{6.12}$$

As shown in Fig. 6.32b, the modulation index M_a of the remaining healthy inverter cells in the u-phase need to be increased by a factor of $H/(H - 1)$ so as to maintain three-phase balanced line-to-line voltages.

$$\begin{aligned} v_{uM} &= H/(H-1)M_a V_{dc} \sin \omega t \\ v_{vM} &= M_a V_{dc} \sin(\omega t - 2\pi/3) \\ v_{wM} &= M_a V_{dc} \sin(\omega t + 2\pi/3) \end{aligned} \tag{6.13}$$

where N is the cascaded number, V_{dc} is the total DC-link voltage which equals $H{\cdot}V_{cap}$.

Figure 6.32b shows a phasor diagram of (6.13) for $H = 3$. If *Cell-1* of the u-phase is out of order, only M_a of *Cell-2* and *Cell-3* should be increased to maintain the balanced line-to-line voltage while the healthy inverter cells in the v-phase and w-phases are working normally, which is similar to the condition in Fig. 6.32a.

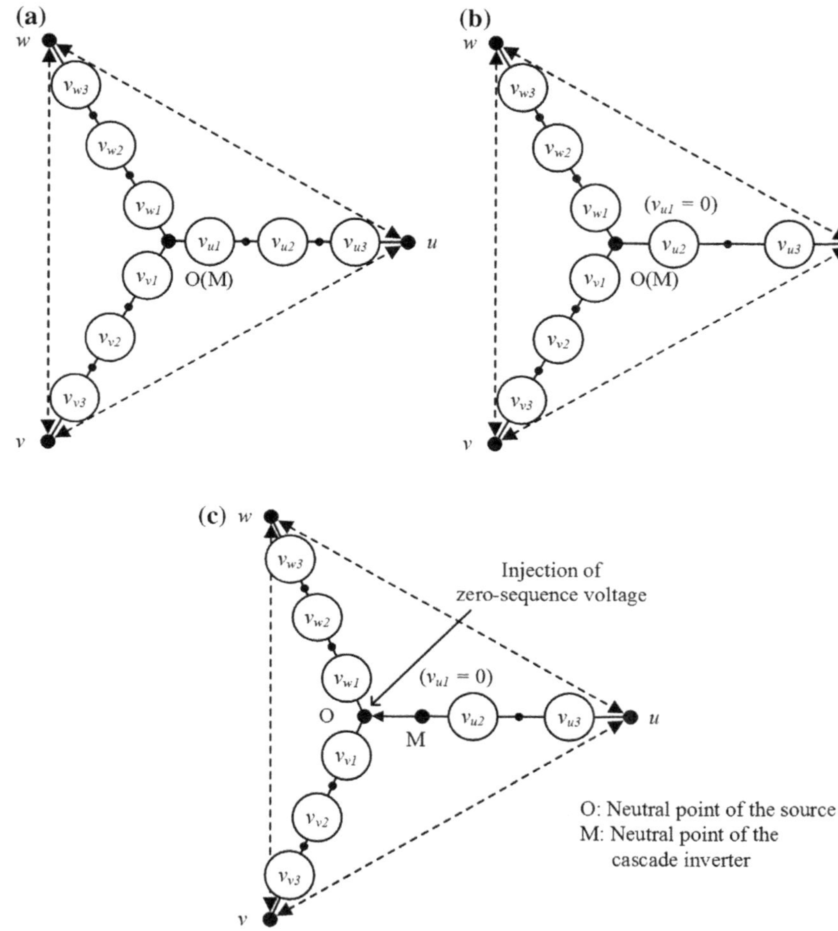

Fig. 6.32 Phasor diagrams of inverter-cell voltages with a cascade number H = 3 during a fault in u-phase inverter cell number 1 with an injection of zero-sequence voltage (reprinted from [8], Fig. 4)

After the neutral shift by a zero-sequence voltage injection, the reference voltage of the healthy inverter cells can be expressed as:

$$
\begin{aligned}
v_{uM} &= H/(H-1)\left(M_a V_{dc}\sin\omega t - v_{0f}\sin\omega t\right) \\
v_{vM} &= M_a V_{dc}\sin(\omega t - 2\pi/3) - v_{0f}\sin\omega t \\
v_{wM} &= M_a V_{dc}\sin(\omega t + 2\pi/3) - v_{0f}\sin\omega t.
\end{aligned}
\tag{6.14}
$$

The second term on the right-hand side of each equation represents the injected zero-sequence voltage. The amplitude of the zero-sequence voltage v_{0f} is derived as:

$$v_{0f} = \frac{2}{3H-1} M_a V_{dc}. \tag{6.15}$$

For a cascaded number $H = 3$, as shown in Fig. 6.32c, the zero-sequence voltage is calculated as:

$$v_{0f} = \frac{1}{4} M_a V_{dc}. \tag{6.16}$$

Substituting (6.16) into (6.14) yields:

$$\begin{aligned} v_{uM} &= 1.125 M_a V_{dc} \sin \omega t \\ v_{vM} &= 1.145 M_a V_{dc} \sin(\omega t - 130.9) \\ v_{wM} &= 1.145 M_a V_{dc} \sin(\omega t + 130.9) \end{aligned} \tag{6.17}$$

When the neutral shift is applied, the modulation index M_a is increased by a factor of 1.125 in the u-phase and by a factor of 1.145 in the v-phase and w-phases during a single-inverter-cell fault in the u-phase. Without the neutral shift, an increase by a factor of 1.5 in the u-phase would be necessary to maintain the three-phase balanced line-to-line voltages and currents. In other words, the neutral shift can share the additional burden of the increasing M_a with all of the $3H - 1$ healthy inverter cells in the three phases.

If the u-phase and v-phase fail simultaneously, this condition can be tolerated in the same way. Because the u-phase and v-phase have failed, the two zero-sequence voltages should be calculated, and these two offset voltages are added to each phase. Thus, the three-phase reference voltage signals of the healthy inverter cells are expressed as:

$$\begin{aligned} v_{uM} &= H/(H-1)\left(M_a V_{dc} \sin \omega t - v_{0f} \sin \omega t - v_{0f} \sin(\omega t - 2\pi/3)\right) \\ v_{vM} &= H/(H-1)(M_a V_{dc} \sin(\omega t - 2\pi/3) - v_{0f} \sin \omega t \\ &\quad - v_{0f} \sin(\omega t - 2\pi/3)) \\ v_{wM} &= M_a V_{dc} \sin(\omega t + 2\pi/3) - v_{0f} \sin \omega t - v_{0f} \sin(\omega t - 2\pi/3). \end{aligned} \tag{6.18}$$

For a cascaded number N = 3, substituting (6.16) into (6.18) yields:

$$\begin{aligned} v_{uM} &= 1.352 M_a V_{dc} \sin(\omega t + 13.89^\circ) \\ v_{vM} &= 1.352 M_a V_{dc} \sin(\omega t - 133.89^\circ) \\ v_{wM} &= 1.25 M_a V_{dc} \sin(\omega t + 120^\circ) \end{aligned} \tag{6.19}$$

In this faulty case, the increasing factor of M_a is smaller than 1.5 in the u-phase and v-phase. In a medium-voltage system, a high number of H-bridge cells, i.e., six or more, may be used. By increasing the cascade number H, the increasing factor of M_a is further decreased.

C. **Tolerant control based on LS-PWM** [8]

This section presents an LS-PWM scheme for fault-tolerant operation of a CHMI. First, the bypass method is introduced briefly. The tolerant control is then described with two types of schemes so as to cover both open-switching and short-switching device faults. Bypassing the faulty inverter cell is used. If an open-fault of S_1 is detected, a fault signal is generated and transmitted to the controller. In addition, S_3 is given the OFF signal, while the two bottom switches S_2 and S_4 are given the ON signal. Thus, the faulty inverter cell is effectively bypassed. On the other hand, if a short-fault of S_1 is detected, S_3 is given the ON signal, and S_2 and S_4 are given the OFF signal. This method can be expanded to all of the switch fault cases.

Type 1: For both open-switching and short-switching device faults

Figure 6.33 shows the reference signal of the u-phase and the triangular wave carrier bands. It is assumed that the switching device S_{11} of *Cell-1* in the u-phase has failed. In this first scheme, *Cell-1* and the faulty S_{11} are bypassed. Therefore, both open-fault and short-fault cases of S_{11} can be effectively dealt with.

From the left side of Fig. 6.33, it can be seen that faulty *Cell-1* is bypassed. Because *Cell-1* has lost its switching ability, the two carrier bands v_{cr1} and v_{cr1-} cannot be compared with the reference signal (V_{ref_u}) of the u-phase. Therefore, the gating signals for S_{11} and S_{13} are not generated. Even if M_a is increased by neutral

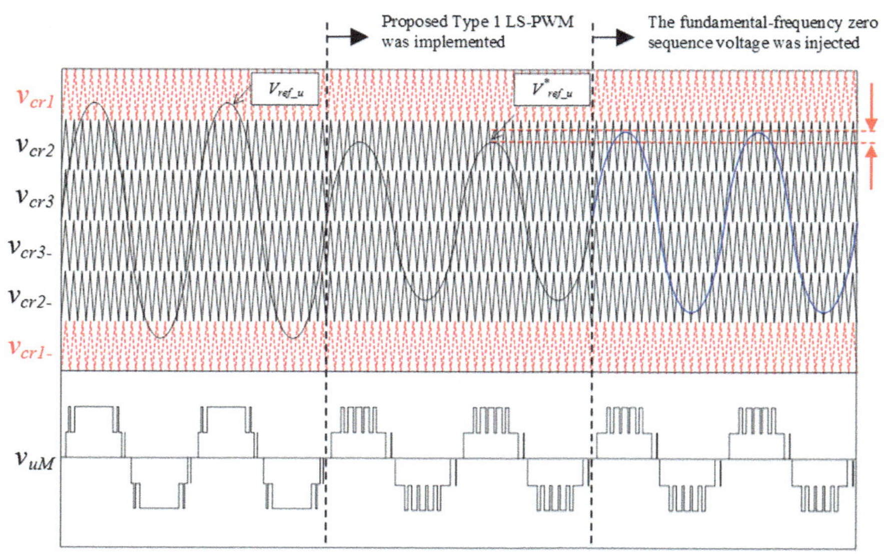

Fig. 6.33 Output voltage waveforms when a reference signal is adjusted by the Type 1 LS-PWM (reprinted from [8], Fig. 6)

shift for fault tolerance as discussed in the previous section, the upper-most and lower-most portions of V_{ref_u} still cannot be modulated with the carrier bands. To allow the reference signal to be modulated linearly with carrier bands, the reference signal is adjusted in this PWM method. In the middle of the time scale of Fig. 6.33, M_a of the reference signal is decreased, and the adjusted reference signal is marked by $V^*_{ref\_u}$. The factor of this decrement can be derived as follows:

$$V^*_{ref\_x} = \frac{N - n_f}{N} V_{ref\_x} \tag{6.20}$$

where x represents the phase including the faulty inverter cells.

For a cascade number $H = 3$ and a number of faulty inverter cells $n_f = 1$, the adjusted reference signal $V^*_{ref\_u}$ is given by:

$$V^*_{ref\_u} = \frac{2}{3} V_{ref\_u} \tag{6.21}$$

As shown on the right-side of Fig. 6.33, even though the reference signal $V^*_{ref\_u}$ has been increased by an injection of zero-sequence voltage for the neutral shift, $V^*_{ref\_u}$ can be modulated linearly from its peak-to-peak value with the carrier bands.

Type 2: For open-switching device faults only

In the explanation of Type 1, the faulty *Cell-1* of the u-phase is bypassed, and the u-phase can generate a five-level output voltage when the cascade number $H = 3$. Even though only one switching device has failed, the corresponding H-bridge cell is only used as a current path and cannot generate any voltage level.

If the faulty inverter cell is not bypassed, the remaining healthy switching devices in the faulty cell can be used, and it is possible to generate more output voltage levels than when the faulty inverter cell is bypassed.

Figure 6.34 shows the reference signal of the u-phase and the triangular wave carrier bands when the second type of LS-PWM scheme is applied. Since only switching device S_{11} has failed, the reference signal is modified during the original modulation period of S_{11}. For a cascade number N = 3 and a number of faulty inverter cells $n_f = 1$, the adjusted reference signal $V^*_{ref\_u}$ is expressed as follows:

$$\begin{aligned} &\text{where } -1 < V_{ref\_u} < 1, \\ &\text{(i) if } V_{ref\_u} > 2/3,\ V^*_{ref\_u} = \frac{2}{3} V_{ref\_u} \\ &\text{(ii) if } V_{ref\_u} < 2/3,\ V^*_{ref\_u} = V_{ref\_u} \end{aligned} \tag{6.22}$$

The lowermost portion of $V^*_{ref\_u}$ can still be compared with the carrier band v_{cr1}. Thus, shows that the healthy switching device S_{13} can generate a voltage level in the negative domain. Finally, the output voltage of the u-phase can be formed with six levels.

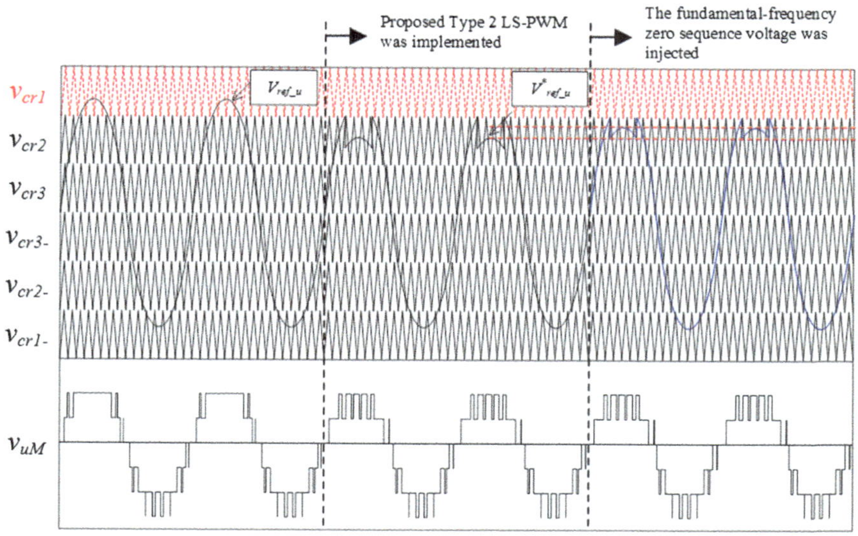

Fig. 6.34 Output voltage waveforms when a reference signal is adjusted by the Type 2 LS-PWM (reprinted from [8], Fig. 7)

The fundamental-frequency zero-sequence voltage is also only injected in the period where the reference signal is adjusted. Since all of the remaining healthy switching devices can be operated normally, it is possible to minimize the effect of the switching device failure on the output voltage waveform.

D. **Rotation scheme of the switching pattern** [8]

In LS-PWM, the power delivery of each H-bridge cell in a phase is not evenly distributed because the conduction times of the devices are unequal. In addition, the switching frequency is not the same for devices in different H-bridge cells. These unequal power delivery and switching frequencies cause unequal switch device losses. To evenly distribute power delivery, the switching pattern should rotate among the H-bridge cells.

In Fig. 6.35a, it is assumed that the switching device S_{11} of *Cell-1* in the u-phase has failed. The type 2 LS-PWM is applied with the conventional rotation scheme. The faulty *Cell-1* should be fixed for the modulation of the outermost voltage level, and *Cell-2* and *Cell-3* rotate their switching patterns mutually.

The faulty *Cell-1* conducts power for a very short time when compared to the other healthy inverter cells. This means that the healthy inverter cells, *Cell-2* and *Cell-3*, experience more conduction losses than *Cell-1*. If the faulty *Cell-1* can take on more power delivery, the switching device losses can be distributed among the H-bridge inverter cells. In this section, a modified rotation scheme for LS-PWM is introduced.

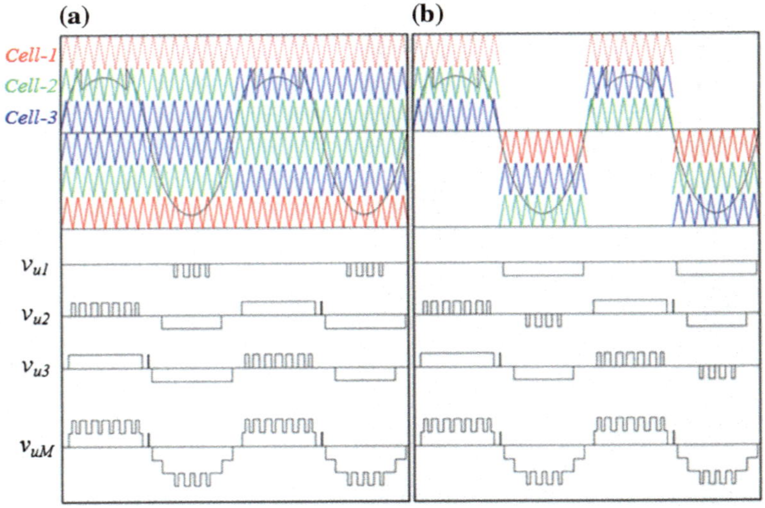

Fig. 6.35 Output voltage waveforms under the faulty *Cell-1* condition after the type 2 LS-PWM is applied with: **a** conventional rotation scheme, **b** proposed rotation scheme (reprinted from [8], Fig. 8)

Figure 6.35b shows the output waveforms after the type 2 LS-PWM is applied with the rotation scheme under the faulty switching device S_{11} of the *Cell-1* condition. The faulty *Cell-1* is responsible for the inner-most modulation level of the negative region. The healthy inverter cells, *Cell-2* and *Cell-3*, are responsible for the inner-most or middle level in the positive region and middle and outer-most level in the negative region. With this modified rotation scheme, the total amount of power delivered by *Cell-2* and *Cell-3* is still equal during the two cycles of the fundamental wave. However, the faulty *Cell-1* can take on more power-conduction time, and *Cell-2* and *Cell-3* can share the burden of their power delivery with *Cell-1*. The improved power distribution among the H-bridge cells can extend the life-cycle of the switching devices.

E. **Simulation results**

To ensure that all of the H-bridge cells implement the switching operation under normal operating conditions, the magnitude of the three-phase ac reference voltages is set to 750 V. Therefore, the amplitude modulation index M_a is 0.83. The simulation conditions are shown in Table 6.4.

Table 6.4 Simulation parameters

Parameter	Value
Individual DC source voltage	300 V
Total DC bus voltage	900 V
Amplitude modulation index	0.8333
Reference signal frequency	60 Hz
Switching frequency	2.5 kHz

Figure 6.36 shows simulation waveforms when the switching device S_{11} of *Cell-1* in the *u*-phase has failed. In this simulation, the faulty *Cell-1* is bypassed, and only five-level output voltage is generated in the *u*-phase. On the left side of Fig. 6.36, the *u*-phase reference signal cannot be modulated at the outermost region, and the output voltage retains a $2V_{cap}$ magnitude (600 V) in this region. The type 1 LS-PWM scheme is applied at the middle of the time scale. The output voltage waveform of the *u*-phase still has five levels. However, the reference signal can be modulated from the peak to peak linearly with the four carrier bands.

The neutral shift is implemented on the right side of the time scale by an injection of zero-sequence voltage. Due to the injection of offset voltage, the three-phase reference voltage signals are increased. Finally, the balanced three-phase line-to-line voltages and currents are generated. Note that LS-PWM and the neutral shift should be implemented at the same time. The waveforms in Fig. 6.36 help explain the fault-tolerant method step by step.

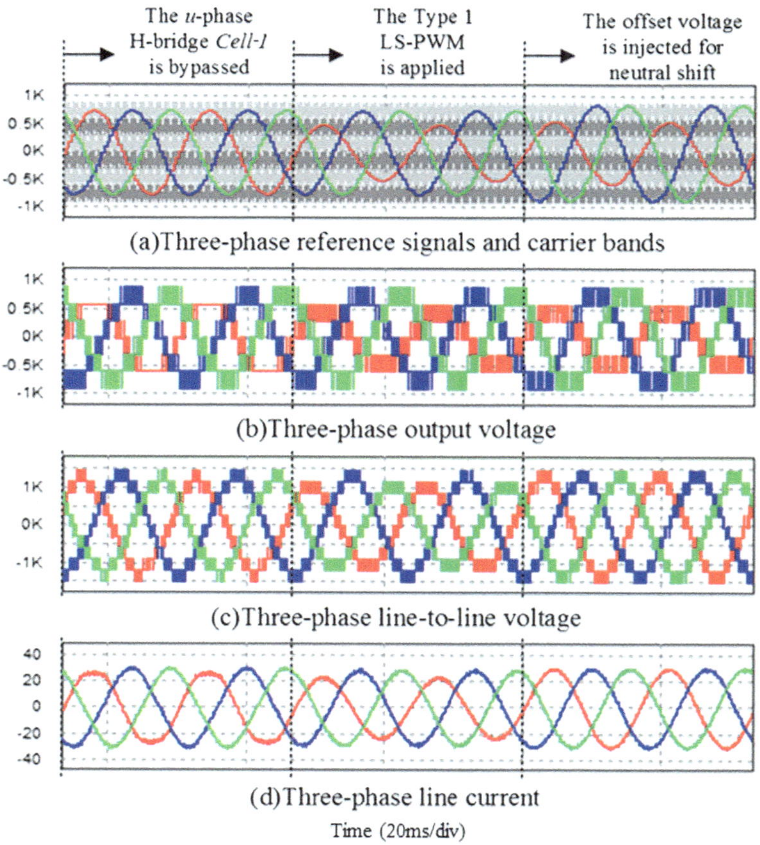

Fig. 6.36 Simulated output waveforms at the AC side of a CHMI, when the faulty *Cell-1* is bypassed and the type 1 LS-PWM is implemented (reprinted from [8], Fig. 9)

Figure 6.37 shows simulation waveforms under the same faulty switching device S_{11} condition as the simulation shown in Fig. 6.36. However, in this case, *Cell-1*, which has the faulty S_{11}, is not bypassed, and the healthy switching devices in *Cell-1* are operated normally. Therefore, a six-level output voltage waveform is generated in the u-phase.

The type 2 LS-PWM scheme is applied at the middle of the time scale. The u-phase reference signal is only modified during the modulation period of the faulty S_{11}. When the reference signal is higher than $2V_{cap}$ (600 V), the reference signal is decreased by a factor of 2/3 by (6.22). However, during this period, the reference signal is not modified.

Zero-sequence voltage is injected only for this period, as shown in Fig. 6.37. Therefore, the three-phase reference signals are increased in this period by an injection of offset voltage. With the Type 2 LS-PWM and neutral shift, balanced three-phase line-to-line voltages and currents are generated.

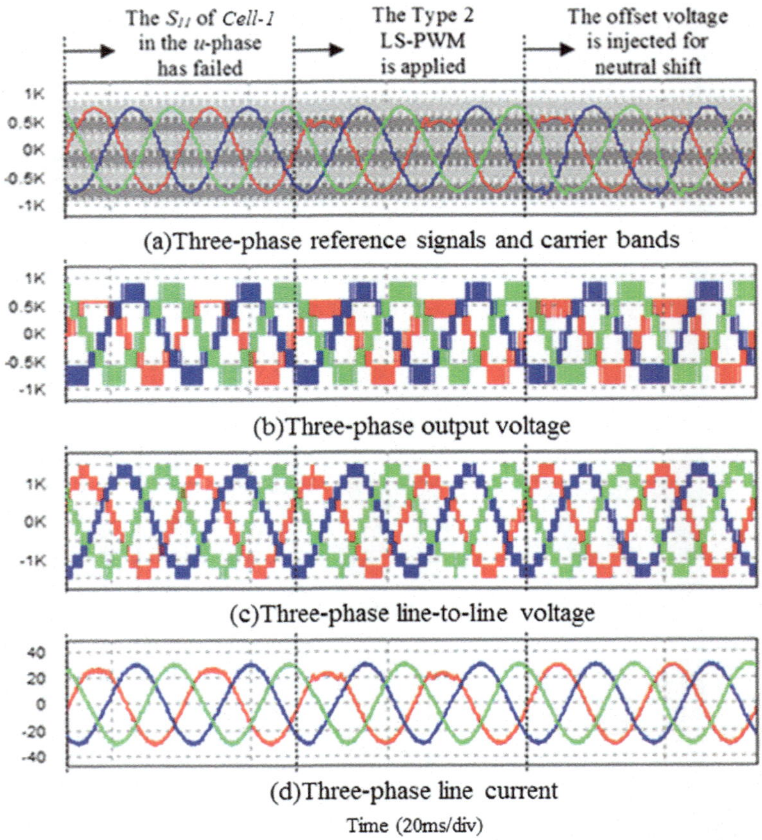

Fig. 6.37 Simulated output waveforms at the ac side of a CHMI, when the faulty *Cell-1* is not bypassed and the type 2 LS-PWM is implemented with the neutral shift (reprinted from [8], Fig. 10)

Simulations were performed to verify the performance of the fault-tolerant strategy under two faulty phase conditions, and the results are shown in Fig. 6.38. In this simulation, the type 2 LS-PWM is applied at the middle of the time scale. In the same way, the type 2 LS-PWM and neutral shift are only applied for the period where the reference signals of both the u-phase and v-phase are higher than $2V_{cap}$. Even though the faulty switching device condition occurred in w-phases, three-phase balanced line-to-line voltages and line currents are generated by the fault-tolerant strategy.

Figure 6.39a shows simulation results when the conventional rotation scheme is applied. It is assumed that the switching device S_{11} of *Cell-1* in the u-phase has failed. The faulty *Cell-1* conducts power for a very brief time when compared to the healthy *Cell-2* and *Cell-3*. Only 587.27 W of AVG power is transmitted through the faulty *Cell-1* when *Cell-2* and *Cell-3* conduct 4799.97 W each.

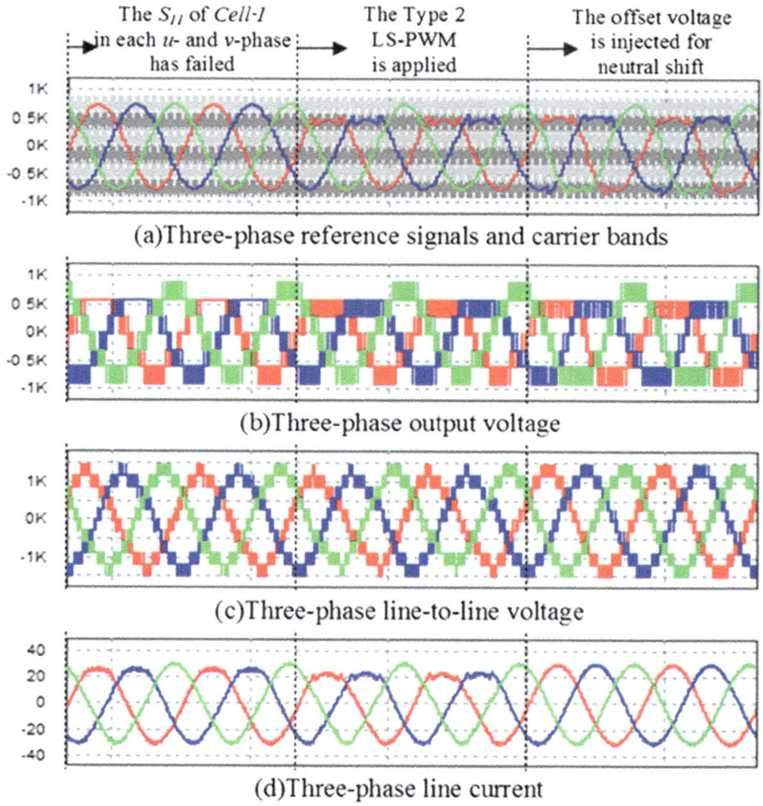

Fig. 6.38 Simulated output waveforms at the ac side of a CHMI, when one of the switching devices failed in each u-phase and v-phase and the type 2 LS-PWM is implemented (reprinted from [8], Fig. 11)

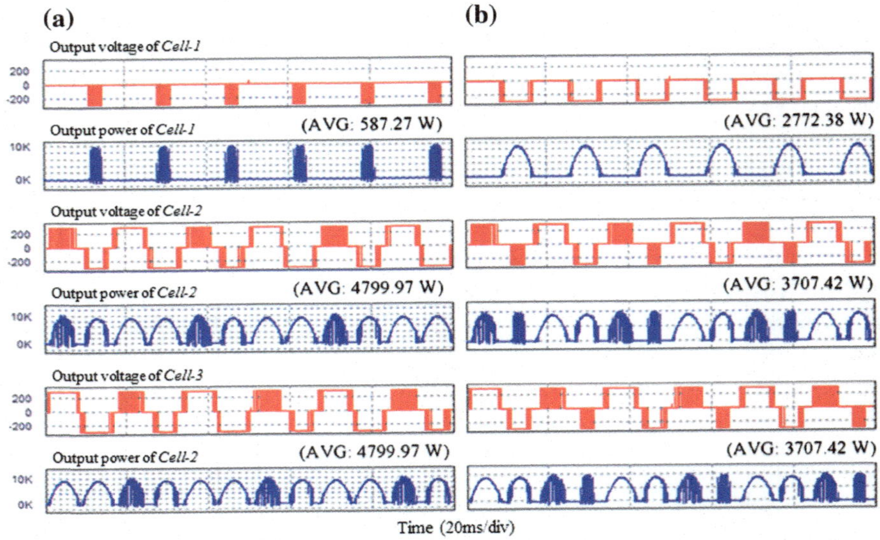

Fig. 6.39 Simulated output waveforms at the AC side of each H-bridge cell in the u-phase under the faulty *Cell-1* condition with the type 2 LS-PWM: **a** conventional rotation scheme, **b** proposed rotation scheme (reprinted from [8], Fig. 12)

Table 6.5 Comparison of power distributions depending on the rotation scheme

Rotation scheme	Conventional		Proposed	
Cell-1	587.27 W	0.122	2772.38 W	0.748
Cell-2	4799.97 W	1	3707.42 W	1
Cell-3	4799.97 W	1	3707.42 W	1

Figure 6.39b shows the resulting waveforms under the same simulation conditions as Fig. 6.39a, but with the rotation scheme is applied. *Cell-1* can take on a greater burden of power conduction with this rotation scheme. *Cell-2* and *Cell-3* can reduce their conduction burden by 22.77% due to the increase in the *Cell-1* conduction time. Table 6.5 shows the power-conduction ratio of the faulty cell under the same faulty conditions for the two different rotation schemes.

F. **Experimental results**

To test the feasibility of the tolerant control strategy for fault tolerance, a three-phase 6 kW seven-level CHMI system was constructed as shown in Fig. 6.22.

The experimental conditions are shown in Table 6.6. Each H-bridge inverter cell has an independent DC source of 30 V. This DC voltage is too small for the original purpose of a CHMI topology, but it is enough to show the switching results of the tolerant control.

Table 6.6 Experiment parameters

Parameter	Value
Individual DC source voltage	30 V
Total DC bus voltage	90 V
Amplitude modulation index	0.8333
Reference signal frequency	60 Hz
Switching frequency	2.5 kHz
Load	25 Ω resistors and 4 mH inductors

Figure 6.40 shows experimental waveforms that verify the effectiveness of the type 1 modulation method. At T_0, one of the switching devices in *Cell-1* failed and the faulty cell, *Cell-1*, is bypassed. The u-phase reference signal cannot be modulated at the outermost region. As a result, only a five-level output voltage is generated. The type 1 LS-PWM scheme is applied at T_1. The reference signal of the u-phase can be modulated from its peak-to-peak value even though the output voltage waveform of the u-phase still has five levels. After the offset voltage is injected for a neutral shift at T_2, balanced three-phase line-to-line voltages and line currents are generated.

Figure 6.41 shows experimental waveforms when the switching device, S_{11} of *Cell-1* in the u-phase, has failed. The faulty inverter cell, *Cell-1*, is not bypassed and the healthy switching devices in the cell are used normally. Due to the fault in S_{11}, the inverter cannot perform the switching operation when the reference signal exceeds $2V_{cap}$ in the u-phase. Therefore, the uppermost voltage level ($3V_{cap}$) of the u-phase is not generated, and the total six-level output voltage waveform is generated in the u-phase. The type 2 LS-PWM is applied at T_1. After the type 2 LS-PWM is implemented, the inverter carries out the switching operation continuously for every fundamental period.

The neutral shift is applied from T_2. The zero-sequence voltage is injected when the reference signal of the u-phase exceeds $2V_{cap}$ as explained earlier. Figure 6.41b–c show that three-phase line-to-line voltage and current waveforms are balanced after all of the steps of the fault-tolerant strategy are completed. Although the fundamental components of the line-to-neutral voltages are not balanced, those of the line-to-line voltages and currents are balanced well. It may be a little difficult to evaluate the performance of the fault-tolerant control from waveforms of the line-to-line voltages. Since *Cell-1* takes charge of the outermost voltage level, the failure effects of *Cell-1* are the smallest among the inverter cells. However, the three-phase line current waveforms show that the control scheme performs well for fault-tolerant operation. As previously mentioned, the tolerant control and neutral shift should be implemented at the same time once the fault is detected.

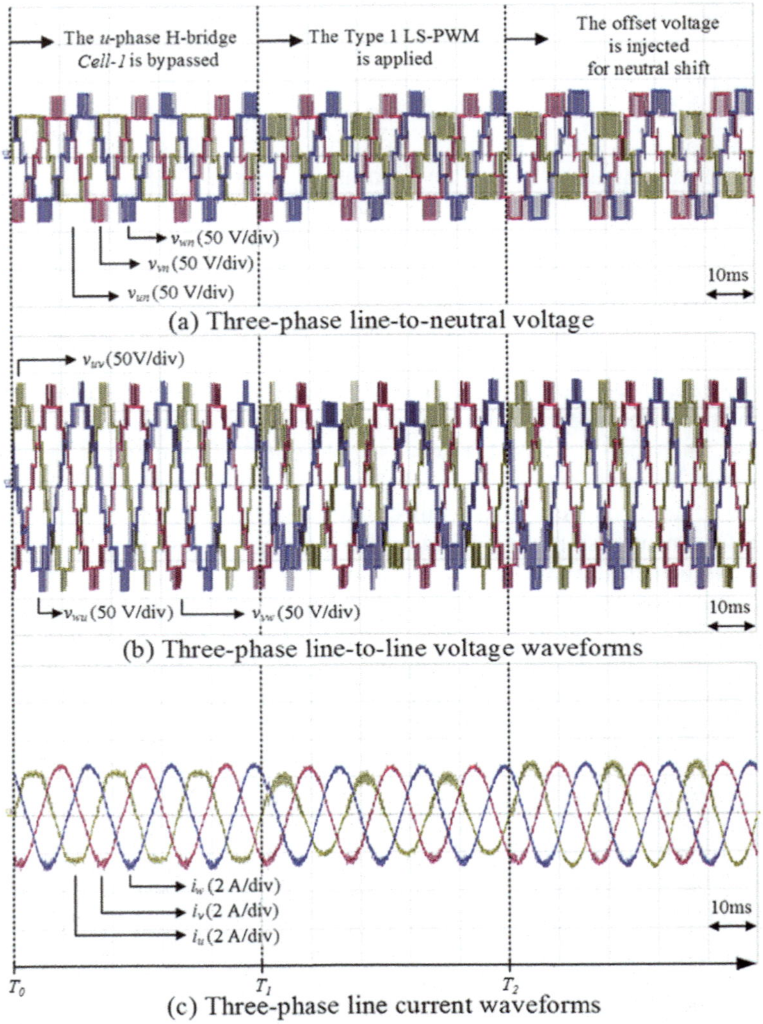

Fig. 6.40 Experimental output waveforms at the ac side of a CHMI, when the faulty *Cell-1* is bypassed and the type 1 LS-PWM is implemented with the neutral shift (reprinted from [8], Fig. 14)

Figure 6.42 shows output voltage waveforms for each of the H-bridge inverter cells in the u-phase when the proposed rotation scheme is applied. These are the same waveforms as the simulation results in Fig. 6.12b, which shows that the proposed rotation scheme can be experimentally realized. Since the switching

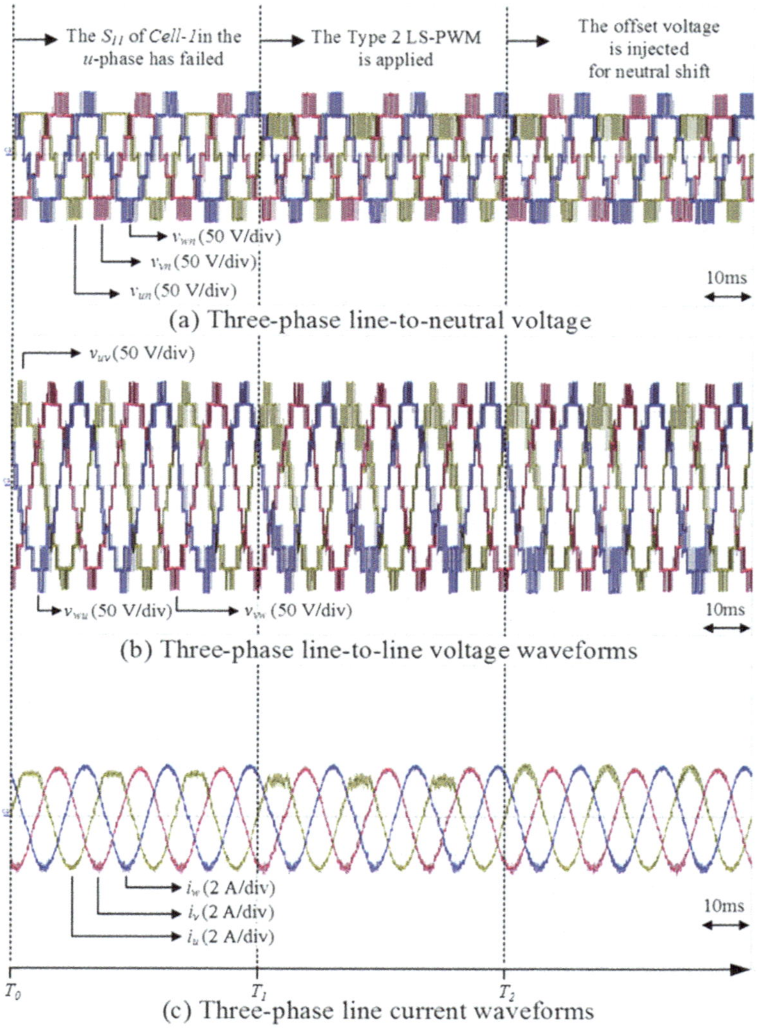

Fig. 6.41 Experimental output waveforms at the ac side of CHMI, when the faulty *Cell-1* is not bypassed and the type 2 LS-PWM is implemented with the neutral shift (reprinted from [8], Fig. 15)

device S_{11} failed, the faulty *Cell-1* takes charge of the innermost modulation level in the negative domain. If S_{12} or S_{13} of *Cell-1* fails, the faulty *Cell-1* generates the innermost voltage level in the positive domain.

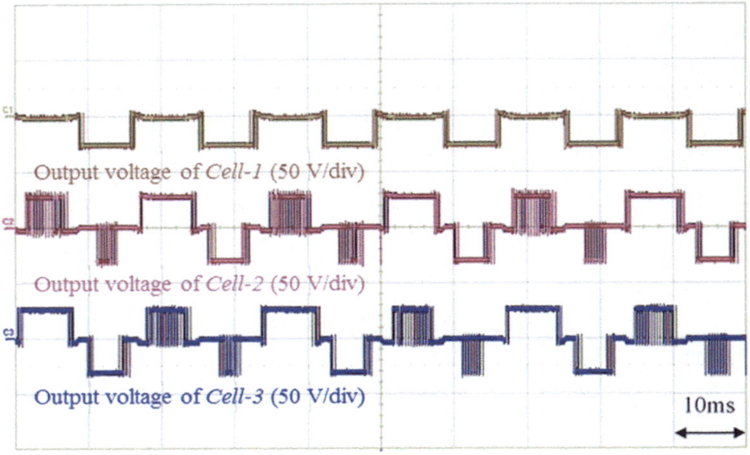

Fig. 6.42 Experimental output voltage waveforms of each H-bridge cells in the u-phase when the proposed rotation scheme is applied under the faulty *Cell-1* condition (reprinted from [8], Fig. 16)

References

1. Wu, B. 2006. High-Power Converters and AC Drives. Hoboken, NJ, USA: Wiley.
2. Rodriguez, J., S. Bernet, B. Wu, J.O. Pontt, and S. Kouro. 2007. Multilevel voltage-source-converter topologies for industrial medium-voltage drives. *IEEE Transaction Industrial Electronics* 54 (6): 2930–2945.
3. Lee, J.S., and K. B. Lee. 2015. Open-switch fault tolerance control for a three level NPC/T-type rectifier in wind turbine systems. *IEEE Transaction Industrial Electronics* 62 (2): 1012–1021.
4. Youssef, M.Z., K. Woronowicz, K. Aditya, N. A. Azeez, and S. S. Williamson. 2016. Design and development of an efficient multilevel DC/AC traction inverter for railway transportation electrification. *IEEE Transaction on Power Electronics* 31 (4): 3036–3042.
5. Malinowski, M., K. Gopakumar, J. Rodriguez, and M.A. Perez. 2010. A survey on cascaded multilevel inverters. *IEEE Transactions on Industrial Electronics* 57 (7): 2197–2206.
6. Lee, J.S., H.W. Sim, and K.B. Lee. 2014. Cascaded H-bridge multilevel inverter for increasing output voltage levels. In *Proceedings of IEEE Energy Conversion Congressand Exposition Conference* (pp. 365–370).
7. Sim, H.W., J.S. Lee, and K.B. Lee. 2016. Detecting open-switch faults: Using asymmetric zero-voltage switching states. *IEEE Industry Applications Magazine* 22 (2): 27–37.
8. Kim, S.M., J.S. Lee, and K.B. Lee. 2016. A modified level-shifted PWM strategy for fault tolerant cascaded multilevel inverters with improved power distribution. *IEEE Transactions on Industrial Electronics* 63 (11): 7264–7274.
9. Yazdani, A., H. Sepahvand, M.L. Crow, and M. Ferdowsi. 2011. Fault detection and mitigation in multilevel converter STATCOMs. *IEEE Transactions on Industrial Electronics* 58 (4): 1307–1315.
10. Lezana, P., R. Aguilera, and J. Rodríguez. 2009. Fault detection on multicell converter based on output voltage frequency analysis. *IEEE Transactions on Industrial Electronics* 56 (6): 2275–2283.

11. Khomfoi, S., and L.M. Tolbert. 2007. Fault diagnosis and reconfiguration for multilevel inverter drive using AI-based techniques. *IEEE Transactions on Industrial Electronics* 54 (6): 2954–2968.
12. Maharjan, L., T. Yamagishi, H. Akagi, and J. Asakura. 2010. Fault-tolerant operation of a battery-energy-storage system based on a multilevel cascade PWM converter with star configuration. *IEEE Transactions on Power Electronics* 25 (9): 2386–2396.